The American Class Structure

OTHER WADSWORTH TITLES OF RELATED INTEREST
IN SOCIOLOGY

Margaret L. Andersen and Patricia Hill Collins, *Race, Class, and Gender: An Anthology*

Susan A. Basow, *Gender Stereotypes: Traditions and Alternatives,* 2nd ed.

Gary Kessler, *Voice of Wisdom: A Multicultural Reader*

Bernice Lott, *Women's Lives: Themes and Variations in Gender Learning*

Martin N. Marger, *Race and Ethnic Relations: American & Global Perspectives,* 2nd ed.

Martin N. Marger, *Elites and Masses: An Introduction to Political Sociology,* 2nd ed.

Charlotte O'Kelly and Larry S. Carney, *Women and Men in Society: Cross-Cultural Perspectives on Gender Stratification,* 2nd ed.

Barbara J. Risman and Pepper Schwartz, *Gender in Intimate Relationships: A Microstructural Approach*

Robert Staples, *The Black Family: Essays and Studies,* 4th ed.

The American Class Structure

A New Synthesis

FOURTH EDITION

Dennis Gilbert
Hamilton College

Joseph A. Kahl
Cornell University

Wadsworth Publishing Company
Belmont, California
A Division of Wadsworth, Inc.

Editor: *Serina Beauparlant*
Consulting Editor: *Charles M. Bonjean*
Senior Editorial Assistant: *Marla Nowick*
Production: *Hoyt Publishing Services*
Print Buyer: *Karen Hunt*
Copy Editor: *David Hoyt*
Cover: *Chris Ingersoll*
Compositor: *Graphic Composition, Inc.*
Printer: *R. R. Donnelley & Sons Co.*

For additional credits, see page 355.

 This book is printed on acid-free recycled paper.

Japanese translation of first edition by Kunio Inamoto
(Tokyo: Jiji-Tsushinsha, 1958)

© 1957, Joseph A. Kahl (New York: Rinehart)
© 1993, 1987, 1982, Dennis Gilbert and Joseph A. Kahl

4 5 6 7 8 9 10—96 95 94

Library of Congress Cataloging-in-Publication Data

Gilbert, Dennis.
   The American class structure: a new synthesis / Dennis Gilbert,
Joseph A. Kahl.—4th ed.
      p.   cm.
   Includes bibliographical references and indexes.
   ISBN 0–534–16374–2
   1. Social classes—United States. 2. United States—Social
conditions.   I. Kahl, Joseph Alan.   II. Title.
HN90.S6G54   1992
305.5'0973—dc20   92–14141
                                                   CIP

Contents

Chapter 4 Wealth and Income 85

Chapter 5 Socialization, Association, Lifestyles, and Values 111

Chapter 6 Succession and Mobility: Structural Opportunities 141

Chapter 10   The Poor, the Underclass, and Public Policy   267

Chapter 11   The American Class Structure: A Synthesis   305

Bibliography

Name Index

Subject Index

Credits

Preface

Our goal in this book is to present readers with a synthesis of the most pertinent social science research on the class structure of the United States. The book is not an encyclopedic survey of all stratification research, nor is it an exercise in stratification theory. Instead, it draws a simple organizing framework from classical theory and singles out the most essential empirical studies of social stratification in the contemporary United States for detailed critical examination. The book focuses on two questions about those studies: What were the main conclusions? By what procedures were those conclusions reached? This approach conveys important factual information, a perspective for interpreting that information, and an understanding of the research process that underlies the conclusions of social scientists. By concentrating on a few studies of exemplary value, we hope to avoid intellectual indigestion, and we invite students to share a sense of participation in the search for knowledge. New questions and new answers are constantly arising, so the process of research is as important as are momentary conclusions.

Coherence is given to our interpretation of the American class system through a conceptual scheme that was chosen on utilitarian grounds: It starts with the main theorists of the field, Marx and Weber, and isolates nine "variables" that can be studied one by one and then combined to show major patterns. The "structure" in the title of the book refers to those patterns of convergence among the variables that create social classes. The patterns are tendencies, "ideal types," never fully realized in any one situation but discernible when one steps back from detail to think about underlying forces. We constantly emphasize the influence of economic forces on social and political consequences but do not neglect important points of feedback where social facts shape economic trends.

We hope that we will serve two audiences: undergraduate students studying the subject for the first time and more advanced students in research-oriented seminars who seek a quick review of the major literature on the United States as preparation for more careful study of selected themes, perhaps in a comparative perspective. For the benefit of both groups, we have kept the book as concise as possible so that other materials can be added by instructors, either from one of the convenient readers that collect research articles or from monographic studies that pursue one or another aspect in depth.

Our approach to bibliography follows from our purposes: Instead of being exhaustive, we are highly selective. We refer in the body of the text to the classic studies that have had major influence on the research tradition. We often add a recent article or two, which in turn contains references to a wide range of contemporary investigation. The seminar student can use the latter as convenient guides to current work. At the end of each chapter, we suggest helpful readings about the United States and occasionally some that allow pointed comparisons with other countries.

This is the fourth edition of *The American Class Structure* to appear. The initial edition, written by Kahl, was published in 1957 and remained in print for nearly 25 years. It synthesized the findings of the first generation of systematic stratification studies conducted by American social scientists. The second edition, by Gilbert and Kahl, came out in 1982. It preserved the approach of the first but took a completely new look at a body of research that had grown enormously in sophistication and volume in the intervening years. For this second revision of the 1982 text (a third edition was published in 1987), we have incorporated fresh material into every chapter and updated the suggested readings. We have also expanded our treatment of women and minorities, and we have examined change in the class system.

Looking back over the last decade, it does not seem to us that there have been any critical breakthroughs in the study of social stratification. Although some very fine studies have appeared, we would be hard pressed to name any that have altered the way sociologists approach stratification. But during these years, the American class structure itself has been in the midst of a transformation unprecedented at least since the 1940s. In revising *The American Class Structure* for this new edition, measuring and interpreting change has been our first concern. We examine recent trends in income, wealth, poverty, social mobility, and other areas. Some of the data we have used are published here for the first time.

In the prefaces to previous editions, we thanked many friends, colleagues, and students whose advice, criticism and stimulation made *The American Class Structure* a better book. In preparing this edition, we received able research assistance from Eddie Gale and Ad Hardin. We are grateful to Robert Hauser, whose unpublished analysis of recent mobility data is the basis of much of the discussion in Chapter 6, and to Keiko Nakao, who provided the new occupational prestige scores presented in Chapter 2. We are also indebted to Charles Bonjean, who read the original version of this edition with great care and made many valuable suggestions. We gratefully acknowledge the insightful comments of the following reviewers: John W. Bedell, California State University, Fullerton; Davita Silfen Glasberg, University of Connecticut, Storrs; Helen A. Moore, University of Nebraska, Lincoln; Timothy Wickham-Crowley, Georgetown University; and Kenneth L. Wilson, University of Alabama, Birmingham.

Finally, a minor technical note: We have simplified many of our tables by rounding numbers, because this makes them easier to read and is more consistent with the limited precision of the original data. Readers should ignore the so-called rounding errors (e.g., columns that do not really total 100 percent) that this procedure sometimes introduces.

Joseph A. Kahl   Dennis Gilbert

List of Figures

List of Tables

1 The Dimensions of Class

All communities divide themselves into the few and the many. The first are the rich and well-borne, the other the mass of the people. . . . The people are turbulent and changing; they seldom judge or determine right. . . . Give, therefore, to the first class a distinct, permanent share in the government. They will check the unsteadiness of the second, and as they cannot receive any advantage by a change, they therefore will ever maintain good government.

Alexander Hamilton (1780)

On the night in 1912 when the *Titanic* sank on her maiden voyage across the Atlantic, social class proved to be a key determinant of who survived and who perished. Among the women (who were given priority over men for places in the lifeboats), 3 percent of the first-class passengers drowned, compared to 16 percent of the second-class and 45 percent of the third-class passengers. Of the victims in first class, all but one had refused to abandon ship when given the opportunity. On the other hand, third-class passengers had been ordered to stay below deck, some of them at the point of a gun (Lord 1955: 107, cited in Hollingshead and Redlich 1958: 6).

The divergent fates of the *Titanic*'s passengers present a dramatic illustration of the connection between social class and what pioneer sociologist Max Weber called *life chances*. Weber invented the term to emphasize the extent to which our chances for the good things in life are shaped by social class.

Contemporary sociology has followed Weber's lead and found that the influence of social class on our lives is indeed pervasive. Table 1-1 gives a few examples. These data compare people at the bottom, middle, and top of the class structure. They show, among other things, that people at the bottom are more frequently the victims of violent crime, less likely to be in good health, and more likely to feel lonely. Those at the top are healthier, safer, and more likely to send their children to college; it is no wonder that they are, on average, happier with their lives and more conservative in their political outlooks.

Thoughtful observers have recognized the importance of social classes since the beginnings of Western philosophy. They knew that some individuals and families had more money or more influence or more prestige than their neighbors. The philosophers also realized that the differences were more than personal or even familial, for the pattern of inequalities tended

TABLE 1–1   Life Chances by Social Class[a]

	Lower Class	Middle Class	Upper-Middle and Upper Class
In excellent health[b]	28%	37%	53%
Victims of violent crime per 1000 population[c]	50	29	21
Psychologically "impaired" per 100 "well"[d]	470	135	46
Feel lonely frequently or sometimes[e]	46%	35%	27%
Obesity in native-born women[f]	52%	43%	9%
Children, 18–24, in college[g]	15%	38%	54%
Dissatisfied with personal life[h]	22%	15%	5%
Favor liberal economic policies[i]	48%	38%	28%

[a]Classes defined by income.
[b]Self-assessment. U.S. Public Health Service 1990: 164.
[c]U.S. Justice 1990: 239.
[d]Moderate to serious symptoms. Strole et al. 1978: 308.
[e]DeStefano 1990: 33.
[f]Burnight and Marden 1967: 81.
[g]U.S. Census 1986a: 141.
[h]Gallup 1988: 234.
[i]In 1984 survey, favored increased federal spending on at least three of five social programs (food stamps, social security, medicare, public schools, job creation). Center for Political Studies, "American National Election Study, 1984" (computed from tape).

to congeal into strata of families who shared similar positions. These social strata or classes divided society into a hierarchy; each stratum had interests or goals in common with equals but different from, and often conflicting with, those of groups above or below them. Finally, it was noted that many of the political activities of people in society flowed from their class interests. As Hamilton said, the rich sought social stability to preserve their advantages, but the poor worked for social change that would bring them a larger share of the world's rewards.

This book is an analysis of the class structure of the United States today. It examines the distribution of income, prestige, power, and other stratification variables among the different classes in the country. It will point out the ways in which the variables react upon one another; for instance, how a person's income affects beliefs about social policy or how one's job affects the choice of friends or spouse. And it will explore the question of movement from one class to another; for a society can have classes and still permit individuals to rise or fall among them.

In order to talk about a complex system and show how one part of it influences another, we must have a separate concept or word for each main aspect and a special method for estimating or measuring it. It is helpful to start the analysis by examining significant theories of stratification, to identify the major facets of the subject as a guide to concept formation. We will look at the theories propounded by Karl Marx and Max Weber, whose work established the intellectual framework that has been used by most subsequent scholars. (See the Suggested Readings at the end of the chapter for other key theorists.)

KARL MARX

Although the discussion of stratification goes back to ancient philosophy, modern attempts to formulate a systematic theory of class differences began with the work of Karl Marx in the nineteenth century; most subsequent theorizing has represented an attempt either to reformulate or to refute his ideas. Marx, who was born in the wake of the French Revolution and lived in the midst of the Industrial Revolution, emphasized the study of social class as the key to an understanding of the turbulent events of his own day. His studies of economics, history, and philosophy convinced him that societies are mainly shaped by their economic organization and that social classes form the link between economic facts and social facts. He also concluded that fundamental social change is the product of conflict between classes. Thus, in Marx's view, an understanding of classes is basic to comprehension of how societies function and also how they are transformed.[1]

In Marx's work, social classes are defined by their distinctive relationships to the means of production. Taking this approach, capitalists, or the *bourgeoisie,* are a class consisting of the owners of the means of production, such as mines or factories. Likewise, workers, or the *proletariat,* are a class consisting of those who must sell their labor power to the owners of the means of production in order to earn a wage and stay alive. Marx maintained that in modern, capitalist society, each of these two basic classes tends toward an internal homogeneity that obliterates differences within them. Minor businessmen lose out in competition with big ones, creating a smaller bourgeoisie of monopoly capitalists. In a parallel fashion, machines get more sophisticated and do the work that used to be done by skilled workers, so gradations within the proletariat fade in significance. But notice that these are statements about trends, about long-run tendencies. At any given moment, distinctions within each class that stem from historical residues—even from markedly different earlier epochs—may influence the situation in important ways that shape behavior. Sometimes Marx called these subdivisions *fractions* of a class, and sometimes he seemed to consider them as momentarily separate classes. Generally his descriptions of particular situations in his writings as a journalist and pamphleteer show more complexity in economic and political groupings than do his writings as a theorist of history analyzing long-term trends.

Why did Marx look to production for the basis of social classes? In the

---

[1] This perspective stressed the notion that circumstances create ideas more than the other way around and was labeled "the materialist conception of history"; it was the opposite of the prevailing view of his epoch (especially in Germany), which assumed that the evolution of political, religious, and philosophical ideas was the force behind historical change.

most general sense, because he regarded production as the center of social life. He reasoned that people must produce in order to survive, and they must cooperate in order to produce. The individual's place in society, relationships to others, and outlook on life are shaped by his or her work experience. More specifically, those who occupy a similar role in production are likely to share economic and political interests that bring them into conflict with other participants in production. Capitalists, for instance, reap profit (in Marx's terms, *expropriate surplus*) by paying their workers less than the resale value of what they produce. Therefore, capitalists share an interest in holding wages down and resisting legislation that would reinforce the power of unions to press their demands on employers.

From a Marxist perspective, the manner in which production takes place (that is, the application of technology to nature) and the class and property relationships that develop in the course of production are the most fundamental aspects of any society. Together they constitute what Marx called the *mode of production.* Societies with similar modes of production ought to be similar in other significant respects and should therefore be studied together. Marx's analysis of European history after the fall of Rome distinguished three modes of production, which he saw as successive stages of societal development: *feudalism,* the locally based agrarian society of the Middle Ages, in which a small landowning aristocracy in each district exploited the labor of a peasant majority; *capitalism,* the emerging commercial and industrial order of Marx's own lifetime, already international in scope and characterized by the dominance of the owners of industry over the mass of industrial workers; and *socialism,* the technologically advanced, classless society of the future, in which all productive property would be held in common.

Unlike many later writers who believed that the level of technology by itself was the crucial determinant of social organization, Marx emphasized that modes of production entail patterns of both technology and social relations and that each can vary independently of the other. An agrarian society in which each producer cultivated land that he owned himself would not represent a feudal mode of production. Likewise, Marx viewed socialism as a new mode of production, which could be built on the industrial technology already developed under the capitalist mode of production.

Marx regarded the mode of production as the main determinant of a society's *superstructure* of social and political institutions and ideas. He used the concept of superstructure to answer an old question: How do privileged minorities maintain their positions and contain the potential resistance of exploited majorities? His reply was that the class that controls the means of production typically controls the means of compulsion and persuasion—the superstructure. He observed that in feudal times, military and political power was monopolized by the landowners; with the rise of modern capitalism, political power was captured by the bourgeoisie when they gained control of the national government. In each case, the privileged class could use the power of the state to protect its own interests. For instance, in Marx's own time the judicial, legislative, and police authority of European governments dominated by the bourgeoisie was employed to crush the early labor movement, a pattern that was repeated a little later in the United States. As Marx expressed it in the *Communist Manifesto* in 1848: "The executive of the modern State is but a committee for managing the common affairs of the whole bourgeoisie" (Marx 1979: 475).

But Marx did not believe that class systems rested on pure compulsion. He allowed for the persuasive influence of ideas. It was in this regard that Marx made one of his most significant contributions to social science, the concept of *ideology*. Marx argued that human consciousness is a social product. It develops out of our experience of cooperating with others to produce and to sustain social life. However, social experience is not homogeneous, especially in a society that is divided into classes. The peasant does not have the same experience as the landlord and therefore develops a distinct outlook. One important feature of this differentiation of class outlooks is the tendency for members of each group to regard their own particular class interests as the true interests of the whole society. What makes this significant is that one class has superior capacity to impose its self-serving ideas on other classes. The class that dominates production, Marx argued, also controls the institutions that produce and disseminate ideas, such as schools, mass media, churches, and courts. As a result, the viewpoint of the dominant class pervades thinking in areas as diverse as the laws of family life and property, theories of political democracy, notions of economic rationality, and even conceptions of the afterlife. In Marx's words, "the ideas of the ruling class are in every epoch the ruling ideas" (Marx 1979: 172). In extreme situations, ideology can convince slaves that they ought to be obedient to their masters, or poor workers that their true reward will eventually come to them in heaven.

Marx maintained, then, that the ruling class had powerful political and ideological means to support the established order. Nonetheless, he regarded class societies as intrinsically unstable. In a famous passage from the *Communist Manifesto*, he observed in 1848 (Marx 1979: 473–474):

> The history of all hitherto existing society is the history of class struggles.
> Freeman and slave, patrician and plebian, lord and serf, guildmaster and journeyman, in a word, oppressor and oppressed stood in constant opposition to one another, carried on an uninterrupted, now hidden, now open fight, a fight that each time ended either in a revolutionary reconstitution of society at large, or in the common ruin of the contending classes.
> In the earlier epochs of history, we find almost everywhere a complicated arrangement of society into various orders, a manifold gradation of social rank. In ancient Rome, we have patricians, knights, plebeians, slaves; in the Middle Ages, feudal lords, vassals, guild-masters, journeymen, apprentices, serfs; in almost all of these classes, again, subordinate gradations. . . .
> Our epoch, the epoch of the bourgeoisie, possesses, however, this distinctive feature: it has simplified the class antagonisms. Society as a whole is more and more splitting up into two great hostile camps, into two great classes directly facing each other: Bourgeoisie and Proletariat.

As these lines suggest, Marx saw class struggle as the basic source of social change. He coupled class conflict to economic change, arguing that the development of new means of production implied the emergence of new classes and class relationships. The most serious political conflicts develop when the interests of a rising class clash with those of an established ruling class. Class struggles of this sort can produce a "revolutionary reconstitution of society." Notice that each epoch creates within itself the growth of a new class that eventually seizes power and creates a new epoch; thus, change is explained by an internal dynamic that Marx called *the dialectic*.

Two eras of transformation through class conflict held particular fascination for Marx. One was the transition from feudalism to modern capitalism in Europe, a process in which he assigned the bourgeoisie (the urban capitalist class) "a most revolutionary part" (Marx 1979: 475). Into a previously stable agrarian society, the bourgeoisie introduced a stream of technological innovations, an accelerating expansion of production and trade, and radically new forms of labor relations. These changes were resisted by the feudal landlords, who felt their own interests threatened by those of the bourgeoisie. The result was a series of political conflicts (the French Revolution was the most dramatic instance), through which the European bourgeoisie wrested political power from the landed aristocracy.

Marx believed that a second, analogous era of transformation was beginning during his own lifetime. The capitalist mode of production had created a new social class, the urban working class, or proletariat, with interests directly opposed to those of the bourgeoisie. This conflict of interests arose, not simply from the struggle over wages between capital and labor, but from the essential character of capitalist production and society. The capitalist economy was inherently unstable and subject to periodic depressions with massive unemployment. These economic crises heightened awareness of long-term trends separating rich from poor. Furthermore, capitalism's blind dependence on market mechanisms built on individual greed created an alienated existence for most members of society. Marx was convinced that only under socialism, with the means of production communally controlled, could these conditions be overcome.

The situation of the proletarian majority made it capitalism's most deprived and alienated victim and therefore the potential spearhead of a socialist revolution. However, in Marx's view, the objective situation of a class does not automatically cause its members to recognize their shared class interests and the need for militant class action—in short, to develop *class consciousness* leading toward political revolt. Some of Marx's most fruitful sociological work, to which we will return in Chapter 9, is devoted to precisely this problem. He asked: What intrinsic tendencies of capitalist society are most likely to produce a class-conscious proletariat? Among the factors he isolated were the stark simplification of the class order in the course of capitalist development; the concentration of large masses of workers in the new industrial towns; the deprivations of working-class people, exacerbated by the inherent instability of the capitalist economy; and the sophistication gained by the proletariat through participation in working-class organizations such as labor unions and mass parties.

What, in sum, can be said of Marx's contribution to stratification theory? His recognition of the economic basis of class systems was a crucial insight. His theory of ideology and his conception of the connection between social classes and political processes, although oversimple as stated, proved a fruitful starting point for modern research. As for his conception of change, a series of twentieth-century revolutions—including those in Mexico (1910), Russia (1917), and China (1949)—have established the significance of class conflict for radical social transformation. However, revolutions have typically occurred in peasant societies during early stages of industrialization under foreign influence rather than in the advanced industrial countries where Marx anticipated them. In the industrial countries, the proletariat used labor unions and mass political parties to defend its interests, thus rechanneling the forces of class conflict into the legal procedures of parlimentary politics; it successfully resisted impoverishment. A century after his death, it is apparent that Marx was a better sociologist than he was a prophet. He identified many of the central processes of capitalist society, but he was unable to foresee all the consequences of their unfolding, and his own vision of a humane socialist future was certainly not actualized in any socialist countries.

MAX WEBER

The great German sociologist Max Weber, who wrote in the early years of the twentieth century, was interested in many of the same problems that had fascinated Marx—among them, the origins of capitalism, the role of ideology, and the relationship between social structure and economic processes. Weber frequently benefited from Marx's work, even while reaching rather different conclusions. In the field of stratification, his special contributions were (1) to introduce a conceptual clarity that was often lacking in Marx's references to social classes and (2) to highlight those situations in which ideology, particularly religion, made an independent contribution to historical change.

Weber made a crucial distinction between two orders of stratification: *class* and *status*. The first term had roughly the same meaning for both Weber and Marx. It refers to groupings of people according to their economic position. Class situation or membership is defined by the economic opportunities an individual has in the labor, commodity, and credit markets. They determine *life chances*, a term Weber used to emphasize those fundamental aspects of an individual's future possibilities that are shaped by class membership, from the infant's chances for decent nutrition to the adult's opportunities for worldly success.

Following Marx, Weber observed that the most important class distinction is between those who own property and those who do not. However, he noted that many significant distinctions can be made within these two categories. Among the propertied elite, for example, those who support themselves with stocks, bonds, and other securities (*rentiers*) are in a different class situation from those who live by owning and operating a business (*entrepreneurs*). The propertyless can be differentiated on the basis of the occupational skills that they bring to the marketplace: The life chances of an unskilled worker are vastly different from those of a trained engineer.

A class, then, becomes a group of people who share the same economically shaped life chances. Notice that this way of defining a class does not imply that the individuals in it are necessarily aware of their common situation. It simply defines a statistical category of people who are, from the point of view of the market, similar to each other. Only under certain circumstances do they become aware of their common fate, begin to think of each other as equals, and develop institutions of joint action to further their interests—in Weber's words, become a *community*.

By contrast, wrote Weber (1946: 186–193):

> Status groups are normally communities. They are, however, often of an amorphous kind. In contrast to the purely economically determined "class situation," we wish to designate as "status situation" every typical component of the life fate of men that is determined by a specific, positive or negative, social estimation of honor. . . .
>
> In content, status honor is normally expressed by the fact that above all else a specific style of life can be expected from all those who wish to belong to the circle. Linked with this expectation are restrictions on "social" intercourse (that is, intercourse which is not subservient to economic or any other of business's "functional" purposes). These restrictions may confine normal marriages to within the status circle and may lead to complete endogamous closure. . . .
>
> Of course, material monopolies provide the most effective motives for the exclusiveness of a status group. . . . With an increased inclosure of the status group, the conventional preferential opportunities for special employment grow into a legal monopoly of special offices for the members. . . .
>
> With some over-simplification, one might thus say that "classes" are stratified according to their relations to the production and acquisition of goods; whereas "status groups" are stratified according to the principles of their consumption of goods as represented by special "styles of life."

In those passages, Weber specified many of the interrelations between class and status, between economy and society. Because of class position, a person earns a certain income. That income permits a certain lifestyle, and people soon make friends with others who live the same way. As they interact with one another, they begin to conceive of themselves as a special type of people. They restrict interaction with outsiders who seem too different (they may be too poor, too uneducated, too clumsy to live graciously enough for acceptance as worthy companions). Marriage partners are chosen from similar groups, for once people follow a certain style of life, they find it difficult to be comfortable with people who live differently. Thus, the status group becomes an ingrown circle. It earns a position in the local community that entitles its members to social honor or prestige from inferiors.

Status groups develop the conventions or customs of the community. Through time they evolve appropriate ways of dressing, of eating, of living that are somewhat different from the ways of other groups. These ways get expressed as moral judgments reflecting abstract principles of value that separate "good" from "bad." The application of these principles to individuals establishes rankings of social honor or prestige. These distinctions often react back on the marketplace; in order to preserve their advantages, high-status groups attempt to monopolize those goods that symbolize their style of life—they pass consumption laws prohibiting the lower orders from wearing lace, or they band together to keep Jews or blacks out of prestigious country clubs or universities. (Weber regarded invidious distinctions among ethnic groups as a type of status stratification.)

A status order tends to restrict the freedom of the market, not only by its monopolization of certain types of consumption goods, but also by its monopolization of the opportunities to earn money. If they can get the power, status groups often restrict entry into the more lucrative professions or trades, for example, giving the son of a bricklayer more chance of gaining a union card than the son of a farmer. Even without such formal restriction, birth into a high-status family gives children advantages of education and of personal contacts that eventually help careers. Weber indicates that in its pure form, the class or economic order is universalistic and impersonal; it recognizes no status distinctions and operates solely on the basis of competitive skill. But the status order is exactly the opposite; it is based on particularistic distinctions among types of persons that make some "better" than others, and thus it tends to restrict freedom of competition in both production and consumption.

Weber did not view the status order as an automatic reflection of the class order, a superstructure. In fact, there are important ways in which class phenomena are subject to the influence of status phenomena. However, he said that in the long run, the status order is created by the class order; consumption, after all, is based on production. For example, although elite society reacts against the status claims of newly rich social climbers, it readily accepts their descendants if they have properly cultivated the conventions of the higher status group. On another level, the appearance of classes based on new sources of wealth—for instance, the emergence of an industrial bourgeoisie in Europe and America in the nineteenth century—signals a future restructuring of the status order as a whole.

Weber, like Marx, was interested in the relationship between class and status structures and political power. In fact, it would be accurate to say that for both men, stratification was essentially a political topic. But Weber was highly skeptical of the implication in Marx's work that all political phenomena could be traced back directly to class. For instance, he suggested that the institutions of the modern bureaucratic state exercise an influence on society that is not reducible to the control exercised by a single class. (In *The Eighteenth Brumaire*, Marx [1979: 594–617] himself reluctantly adopted this view for a special circumstance, but not Weber's corollary that a socialist state might be grimly similar to a capitalist state, reflecting bureaucratic domination of society.)

Weber opposed the "pseudo-scientific operation" of Marxist writers of his day, who assumed an automatic link between class position and class consciousness (Weber 1946: 184). A common economic situation can and sometimes does lead to an awareness of shared class interests and a willingness to engage in militant class action, but it need not. Indeed, the very notion of class interest was highly ambiguous for Weber. Given his flexible conception of class, based on economic life chances, he realized that modern society has several classes rather than just two and that they are continually changing. Furthermore, individuals can perceive their own situation and its convergence with that of other persons in a variety of ways that cannot be neatly divided into an accurate or "true" versus an inaccurate or "false" consciousness of interest. He noted that the development of class consciousness is a complex process contingent on numerous factors (some of them already noted by Marx), including the "transparency" of the social arrangements that form the basis for divergent interests, the rate of social change, the dominant value system, the spread of radical interpretations of social reality that put dramatic labels on social experiences, and finally, the presence of leaders and associations, such as trade unions and political parties, capable of organizing class action into long-term conflict. Indeed, he investigated the different styles of party organization with such thoroughness that he almost seems to have considered them a semiautonomous influence in the total process.

Implicit in Weber's approach to stratification is the idea that status considerations may undermine the development of class consciousness and class struggle. For example, the politics of the American South has long been shaped by the tendency of poor whites to identify with richer whites rather than with poor blacks who may share their economic position. Weber noted that political parties may be based on class, status, or other reasons for conflict over power. The major American political parties are amorphous coalitions that have never been as clearly oriented toward the pursuit of class interests as, say, the working-class parties of Western Europe.

In sum, Weber accepted Marx's idea of the underlying economic basis of stratification in the long run. However, he identified another order of stratification by differentiating between class and status. He argued that the two interact with each other and with the political process in complex ways not fully recognized by Marx.

NINE VARIABLES

The scientific study of social stratification in the United States began in the 1920s and 1930s with the pioneering investigations of small-town class systems (which we review in Chapter 2). Since that time, sociologists have collected a mass of empirical evidence about the character of the American stratification system, both local and national. Unfortunately, the facts are described from numerous points of view and frequently conveyed in a highly technical manner. Our purpose in this book is to present the best examples of the research in a coherent form; to do so, we need a standardized language but a minimum of technical jargon.

It is possible to combine the knowledge from the available empirical research by organizing it around a simple conceptual framework derived from our theorists, especially Weber, whose work suggests the difference between the economic, social, and political aspects of a stratification system. Under each of these rubrics, we can delineate a few basic variables that are significant for broad theories of stratification but still specific enough to be operationally defined and measured by the researcher. In the next few paragraphs, we will define these basic variables. In subsequent chapters, we will examine each one in detail, both as a concept and as a body of verified information about the United States.

The economic variables we will consider are *occupation, income,* and *wealth.* The first may be defined as a social role that describes the major work that a person does to earn a living. Modern societies are characterized by an elaborate and stratified system of occupations, most of them related in some way to production or trade. *Income* refers to monetary gain received in the course of a given period of time (e.g., $10,000 per year). Income is closely related to, but must be carefully distinguished from, *wealth,* which refers to assets held at a given point in time. The principal categories of wealth are monetary savings; personal property, such as homes or automobiles; and business assets, including real estate, equipment, inventory, and securities. Wealth can be viewed as an accumulation of past income. In certain forms, such as ownership of a business or of stocks and bonds, wealth becomes capital and is a source of new income. It is this possibility that makes wealth especially important for students of stratification.

Our consideration of the status aspect of the American class system will emphasize *personal prestige, association,* and *socialization.* The first is the most obvious of our stratification variables; if we study a local community, we notice immediately that some people have higher *personal prestige* than others.[2] People have high prestige when neighbors in general have an attitude of respect toward them. Another word that is used for this attitude is *deference,* or the granting of social honor. Prestige is a sentiment in the

---

[2]The word *status* is often used as a synonym for *prestige,* but this usage is ambiguous and should be avoided, since *status* has other uses in sociology.

minds of men and women, although they do not always know that it is there. The shrewd observer can often notice deference behavior that is not recognized by the participants, such as imitation of ideas or lifestyles. Consequently, it is necessary to study prestige in two ways: by asking people about their attitudes of respect toward others and by watching their behavior.

People who share a given position in the class and status structures tend to have more personal contact or *association* with each other than with those in higher or lower positions. Such patterns of differential contact are significant because they promote similarities in behavior and opinion and a sense of community among the members of a stratum. Association (or interaction) is a variable that directs our attention to everyday social processes that can be studied scientifically. By counting its frequency, measuring its duration, classifying its quality, and watching who initiates and who follows, we can draw systematic conclusions about association and can judge its consequences.

Socialization is the process through which an individual learns the skills, attitudes, and customs needed to participate in the life of the community. Although socialization takes place throughout the life cycle of the individual, we will be primarily interested in the socialization of children and teenagers. The early socialization experience of most members of a society is broadly similar; were it not, differences in expectations and behavior would be so great that social life would become impossible. However, social scientists have found significant variations in socialization among subgroupings within societies, especially complex societies. In the United States, research has uncovered class differences in socialization patterns, which may reinforce differences in values, channeling the young toward interactions with others of similar background and eventually toward assuming the class positions of their parents.[3]

Two variables related to the political aspect of stratification systems will be of interest to us: *power* and *class consciousness*. Weber defined *power* as the potential of individuals or groups to carry out their will even over the opposition of others (Weber 1946: 180). This classic definition implies that power is a significant dimension in quite varied social settings, from the power of parents over children to the power of defense contractors over national security policy. We will restrict our use of the concept to broad political and economic contexts, dealing, for instance, with the power of corporations over national economic priorities or the power of the federal government to reduce inequalities in the distribution of income.

The degree to which people at a given level in the stratification system are aware of themselves as a distinctive group with shared political and economic interests is the measure of their *class consciousness*. In some

---

[3]Formal education can be treated as part of the broad socialization of individuals or as part of their direct preparation for occupational roles. We will deal with it both ways, so we do not list it as a separate variable.

circumstances, similar people do not have much contact with one another and think primarily in individualistic rather than in group terms. In other circumstances, they become highly group conscious, and then they are likely to organize political parties, trade unions, and other sorts of associations to advance their group interests. In general, Americans are thought to be less class conscious than Europeans; our traditions of equality lead many of us to deny that classes exist. But the traditions concern more what ought to be than what is.

In addition to considering the eight variables just outlined, we will examine a final variable, the dual concept of *succession and mobility*. When children inherit the class position of their parents, we speak of succession; when they move up or down relative to their parents, we refer to mobility. This variable stands apart from the others because it has a clearly implied time dimension, measured in generations, and because it cannot be assigned to the economic, social, or political aspect of stratification systems. In studying succession and mobility, we will be interested in measuring the degree of inheritance that prevails in American society today. We will also seek to answer two questions: What factors shape the structure of opportunities that make mobility possible? What characteristics of individuals account for differences in their ability to take advantage of mobility opportunities?

THE VARIABLES AS A SYSTEM

We have chosen these nine variables—occupation, income, wealth, personal prestige, association, socialization, power, class consciousness, and mobility—on pragmatic grounds. They constitute the set of variables that most efficiently organizes the existing empirical data on the American class system, and they are congruent with the thinking of the major theorists. Each variable, with the possible exception of power, can be measured by distinct and separate operations, and each can be used to stratify a given population. Thereafter, it is possible to study ways in which position on one dimension influences that on the others.

The nine variables do not form a closed system, all on the same level of abstraction, in the strict sense of scientific theory. However, they do constitute a useful conceptual scheme. By gathering data on each of these variables—as we will do for the United States—we can develop a thorough description of our class system. Moreover, since all the variables are mutually dependent, they provide a framework for thinking about the dynamics of class systems. For instance, numerous studies have demonstrated a connection between the stratification of occupations and each of the remaining eight variables. There is also evidence that the occupational structure (that is, the relative proportions of individuals in different occupations) of the advanced industrial nations is gradually changing. An important question for students of stratification concerns the effect these changes are having on the other variables. Or, to pose a different sort of problem, we might look at our variables to help us understand how class systems—based, after all, on the unequal distribution of privilege—manage to persist. A few clues: The socialization of the less privileged appears to inculcate values supportive of the existing class order; mobility offers opportunities for dissatisfied individuals to change their class positions without changing the system; the privileged have disproportionate power over political and economic institutions.

There is much to be gained, then, by viewing these nine variables together. However, to say so is not to imply that they are of equal importance. They are not. We are convinced that societies that are similar in their technology and economic institutions have broadly similar class systems, that classes develop out of relationships to the means of production, and that status communities emerge from stable distributions of economic privilege. In short, stratification systems are based on economic differences within societies. Therefore, among the variables defined here, those related to the economic aspect of stratification systems (occupation, income, and wealth) are of fundamental significance. Nonetheless, the effects of these variables on the behavior of men and women are mediated through the other variables, which often exercise an independent or autonomous influence, especially in the short run. Thus, while recognizing the primacy of the economic variables, we believe that they can be studied most fruitfully as part of a more complex system.

When we are thinking about the way the nine variables interrelate to form a complex system of stratification, it is crucial to make a distinction between the system as it is and the system as it changes. Much of the research in the United States tends to take the system or structure as it is—as given—and then seeks to understand the way individuals are placed in and influenced by that system. Thus, we concern ourselves with the influence of occupation on voting behavior, with the relation between income and health and longevity, or the prediction of a son's career chances from his father's social position. The main methodological procedure is to collect data on a sample of individuals and assign each person a numerical score or rank on each of two or more variables, then calculate the correlation coefficients among them.

The alternative approach is to concentrate on historical changes in the structure itself, paying more attention to its overall shape and dynamics than to the differences among individuals. This approach is exemplified in Chapter 3 by our discussion of the transformation of the American class structure brought about by industrialization at the end of the nineteenth century. Here, various historical data are used to gain an understanding of the broad sweep of change in society as a whole.

Obviously, the two approaches are related, since the individuals we study are part of a structure that was determined by history and continues to change. Unfortunately, the languages used by the two approaches are less congruent than they appear to be, and much misunderstanding arises when people shift from one perspective to the other. Marx and Weber were more concerned with history than with description, yet they sometimes concerned themselves with the differential fate of individuals within a given system. American researchers are usually concerned with the differential fate of individuals, yet they cannot completely ignore historical trends. *Most theorists think historically; most empiricists think descriptively.* Let the reader be aware of this fundamental problem and notice which perspective dominates the discussion as we move from one researcher to another.

WHAT ARE SOCIAL CLASSES?

A stratified society is one marked by inequality, by differences among people that are evaluated by them as being higher or lower. The simplest form of inequality is based on the division of labor that always appears according to age and sex. Young children are everywhere subordinate to their elders; old people may have a high or a low position, depending upon cultural values; women are often ranked below men.

There is another form of inequality that appears in every society (other than the very smallest and most primitive ones), which ranks families rather than individuals. A family shares many characteristics among its members that greatly affect their relationships with outsiders: the same house, the same income, the same values. If a large group of families are approximately equal in rank to each other and clearly differentiated from other families, we call them a *social class*.[4]

Logically, it is possible for a society to be stratified without having distinct classes, for there could be a continuous gradation between high and low without any sharp lines of division, but in reality this is most unlikely. The sources of a family's position are shared by many other similar families; there is only a limited number of types of occupations or of possible positions in the property system. One either works the soil, or uses one's hands to manufacture things, or trades, or performs some function of intellectual, military, or political leadership that allows one to live from the productive labor of others. There is a tendency for the persons of each type to become similar to their fellows and distinct from the members of other types; the similarities are shared within families and often inherited by children. In other words, the various stratification variables tend to converge and jell; they form a pattern, and it is this pattern that creates social classes.

The pattern formed by the objective connections among the variables is heightened by the way people think about social matters, for popular thought creates stereotypes. Bankers are conceived of as a homogeneous group, and distinctions between big bankers and little ones are ignored. Similarly, poor people tend to lump together all bosses, and rich people

---

[4]Here we depart from the strict terminology used above that differentiates *class* from *status* in favor of ordinary English, which combines the two and uses *class level* or *social stratum* interchangeably.

overlook the many distinctions that exist among those who labor for an hourly wage.

It would not be true to go so far as to say that every family has an equal score on all nine variables. At least in our society, the stratification system is too vague and too fluid for that to occur with high regularity. But in the long run, social life is such that a family *tends* to equalize its position on all of the variables. The forces toward convergence, toward the crystallization of the pattern, will be emphasized in this book, despite the fact that many disturbing influences, mostly the results of rapid social change, keep the patterns from becoming as clear-cut in reality as in theory.

The classes that we talk about are ideal-type constructs. They are intellectual inventions, based on observation of reality, that describe how the classes would look if the system were freed of extraneous influences. Ideal types are complex models of the interrelations among variables; they are guides to research and frameworks for synthesizing the results of research. They are seldom phrased as specific propositions that can be completely proved or disproved. Consequently, it is not surprising that different authors use different sets of class divisions.

Comparing one society to another, we see that stratification systems differ in two ways: the distinctness of their strata and the amount of mobility that occurs between strata. These two characteristics are closely related but not identical. The type of stratification system that is most rigid on both characteristics is called a *caste system*. In caste systems, each stratum is markedly different from the others; its members have a special occupation that is hereditary, they are endogamous (marry within the caste), and they have many special cultural characteristics (often including religious rites) that are unique to themselves. In law as well as fact, caste membership determines life. A caste system is composed of a number of quite separate social communities that live side by side in economic interdependence. The pure form of caste system was approached only in India, but many other societies have had castelike characteristics. Slaves are often a separate caste, but a caste system need not be an outgrowth of slavery.

At the other extreme is a modern, "open" class system. Here there are no legal recognitions of group inequality, and there are minimal differences between the cultural ways of life of the classes. Furthermore, there is much movement from one class to another, both during the lifetime of a person and from one generation to another. There are many societies with such open class systems, and they may vary in terms of the degree of difference between classes or in the amount of mobility that occurs among them, or both.

AN AMERICAN CLASS STRUCTURE

Some readers will have concluded by now that there is as much art as science in the study of social stratification—and they are probably right. We can make factual statements about, say, the distribution of income or patterns of association. But efforts to combine such information into broader statements about the class system are inevitably influenced by the viewpoint of the author.

We will, for example, be examining several general models of the class structure. Each tells us how many classes there are, how they can be distinguished from one another, and who belongs in each class. Some class models are more convincing than others, perhaps because they make better use of the facts, but there is really no way to establish that a particular model is "true" or another "false." However, it is easier to demonstrate that a given model is useful.

Our own model of the American class structure is presented in the final chapter. It represents a synthesis of what we have learned writing this book. So that the reader will know where we are headed, we will summarize it here.

In brief, we conceive of the class structure in terms of six economically defined classes:

> *Capitalist class.* Investors, heirs, and top executives. Very high incomes, largely derived from return on assets.
>
> *Upper-middle class.* University-trained professionals and managers with comfortable incomes.
>
> *Middle class.* White-collar workers, including lower-level professionals and managers, and skilled blue-collar workers. Mainstream living standards.
>
> *Working class.* Semiskilled blue-collar workers and some low-level white-collar workers. Employed at routinized, closely supervised jobs. Incomes just below the mainstream.
>
> *Working poor.* Laborers, service workers, and other low-wage workers. Unstable employment. Modest incomes.
>
> *Underclass.* People with limited connection to the labor force. May work erratically or part-time. Often depend on government programs. Poverty-level incomes.

IS THE AMERICAN CLASS STRUCTURE CHANGING?

We will repeatedly come back to this question in the book. In particular, we will want to find out how the transformation of the U.S. economy in the last two decades has affected the class structure. In recent years, some observers have argued that the United States is becoming a less egalitarian, more rigidly stratified society. They say that poverty is growing, the middle class is shrinking, and wealth is becoming more concentrated. One of the country's keenest political analysts predicts that social inequality may become the dominant political issue of the 1990s (Phillips 1990). We will examine data on wealth, income, mobility, poverty rates, political attitudes, and other factors to see whether the American class system is changing, and if so, how.

SUGGESTED READINGS

Bendix, Reinhard, and Seymour Martin Lipset, eds. 1966. *Class, Status and Power: Social Stratification in Comparative Perspective.* 2nd ed. New York: Free Press.
> *A large collection of articles on important aspects of stratification in various countries. Part I covers basic theory, including Marx, Weber, and Davis and Moore's functionalist explanation of stratification.*

Coser, Lewis. 1978. *Masters of Sociological Thought.* 2nd ed. New York: Harcourt Brace Jovanovich.
> *Contains short personal and intellectual biographies of Marx and Weber, which put them into the context of their times.*

Giddens, Anthony. 1973. *The Class Structure of the Advanced Societies.* New York: Harper & Row.
> *Attempt to rethink the main ideas of Marx and his critics to arrive at an acceptable modern interpretation of class structure.*

Giddens, Anthony, and David Held, eds. 1982. *Classes, Power and Conflict: Classical and Contemporary Debates.* Berkeley: University of California Press.
> *Marxist-oriented collection covering many key issues.*

Lenski, Gerhard. 1966. *Power and Privilege: A Theory of Social Stratification.* New York: McGraw-Hill.
> *Ambitious attempt to explain the development of stratification in each of several evolutionary stages, based on technology.*

Marx, Karl. 1979. *The Marx-Engels Reader.* Edited by Robert C. Tucker. 2nd ed. New York: Norton.
> *Convenient collection of the writings of Marx and his partner, Friedrich Engels. Particularly relevant are: The German Ideology, Part I; Wage Labour and Capital; Manifesto of the Communist Party; and Engels' Socialism: Utopian and Scientific.*

Weber, Max. 1946. *From Max Weber: Essays in Sociology.* Edited by H. H. Gerth and C. Wright Mills. New York: Oxford University Press.
> *A selection of Weber's most important sociological writings (except for his book* The Protestant Ethic and the Spirit of Capitalism). *Especially relevant are: Class, Status, Party; Bureaucracy; The Protestant Sects and the Spirit of Capitalism.*

2 Position and Prestige

Fame did not bring the social advancement which the Babbitts deserved. They were not asked to join the Tonawanda Country Club nor invited to the dances at the Union. Himself, Babbitt fretted, he "didn't care a fat hoot for all these highrollers, but the wife would kind of like to be among those present." He nervously awaited his university class-dinner and an evening of furious intimacy with such social leaders as Charles McKelvey the millionaire contractor, Max Kruger the banker, Irving Tate the tool-manufacturer, and Adelbert Dobson the fashionable interior decorator. Theoretically he was their friend, as he had been in college, and when he encountered them they still called him "Georgie," but he didn't seem to encounter them often. . . .

Sinclair Lewis (1922: 190)

In the small group, in the local community, in the society as a whole, we notice that some people are looked up to, respected, considered people of consequence, and others are thought of as ordinary, unimportant, even lowly. Everywhere we see notables and nobodies.

Prestige is a sentiment in the minds of people that is expressed in interpersonal interaction: Deference behavior is expected by one party and granted by another. Obviously, it can occur only when there are values shared by both parties that define the criteria of superiority; deference at pistol point is not the result of prestige. But this does not necessarily mean that both parties agree about all aspects of the situation. For instance, the subordinate person may feel that the superordinate person *should not* have the right to deference and may try to foment a revolution to take it away. Nevertheless, so long as the subordinate recognizes that the superordinate *does* have the right to claim deference and feels constrained by group norms to grant it, a prestige difference exists. The degree of consensus can range from a situation in which deference is given joyously as a recognition of moral worthiness that reflects the will of God, to one in which deference is given grudgingly, against part of one's own will that cries out against these claims to special privilege. However, stable situations are characterized by the development of legitimating ideologies, which assure approval as well as acceptance for prestige inequalities.

The facts of institutionalized social life that are recognized in deference behavior usually include inequality in wealth and authority. Ideologies are reinforced by sanctions, and people of low prestige grant deference partly because they know that in the long run it is to their advantage to do so. Through symbolic submission to superiors, they gain favor in the eyes of those who are in a position to grant rewards—goods and services, promotion, protection, salvation, or perhaps just good will and a reassuring smile. The distinction between deference at pistol point and deference because of prestige is the distinction between usurpation of power by naked force and authority gained through control of legitimate institutions. In neither case is the will completely free. Of course, those who manage the institutions that create and spread cultural values—newspapers, television, schools, and churches—are aware of the power of prestige symbols to reinforce legitimacy. Consensus can be manufactured as a way of promoting stability and reinforcing privilege.

The most direct way to study prestige is to go into a local community where everyone knows everyone else and observe how they treat one another. That observation would be aided by some questions around the theme "What do you think of Jackson as compared to Albers?" As the observer moves from one sphere of local life to another, he or she will soon begin to see a pattern that groups individuals and families into clusters of people who are relatively homogeneous; we call them *social strata* or *prestige classes*. Within each, people feel similar and find it easy to communicate and share activities. Between strata, however, there may be some sense of distance and even tension: "They are not *our* sort of people."

Despite the diversity of roles people play, and the differences in evaluation thereof made by different observers in the community, there is usually a trend toward consistency, for individuals carry with them some of their habits and reputations as they move from one role to another. Insofar as these roles are public and watched by the community at large, they tend to blend into an overall prestige reputation. For example, a man who is a successful banker gains the general respect of the community, but to keep it he should be a good family man, maintain the appearance of his house, and not curse too loudly when he muffs a shot on the golf course. Since his skill as a typist is not considered important, it neither adds to nor detracts from his general prestige reputation.

The prestige hierarchy of a local community represents the synthesis of all the other stratification variables. It is the result of the evaluation by the people of the totality of the class structure. Such exhaustive mutual evaluations are only possible in small communities whose members have detailed knowledge of one another's lives and backgrounds. In other social contexts, people must depend on more limited and superficial clues to assign prestige to others, and the sociologists who study them must invent indirect methods of ranking. Nevertheless, an understanding of such local prestige structures is an excellent beginning for the study of the dynamics of stratification, for if we can picture in detail the hierarchy of prestige, we can then look behind it for the factors that created it, in terms of both local circumstances and national conditions.

In this chapter, we will examine some classic investigations of prestige structures in small towns, a related study conducted more recently in two major cities, and a series of national surveys of the prestige associated with occupations. In each case, we will want to answer two questions: *What* were the results? *How* were they obtained? The answer to the second question allows us to assess the validity of the reply to the first.

W. LLOYD WARNER: PRESTIGE CLASSES IN YANKEE CITY

The concern of American sociologists with the prestige aspect of stratification can be traced back to a series of community studies conducted by W. Lloyd Warner and his students and colleagues. Their research began in the early 1930s. Warner had already completed a three-year study of Australian aborigines and wanted to employ the techniques of social anthropology— the study of the "whole man" in the complete context of his sociocultural life—in the United States. Although Warner and his associates realized that in many ways, the metropolis was the typical social environment of modern society, they felt that the study of the great city as a whole from the viewpoint of social anthropology was too vast an undertaking for their techniques and resources, so they decided to start with a small community. They chose one "with a social organization which had developed over a long period of time under the domination of a single group with a coherent tradition" (Warner and Lunt 1941:5). Their first study, of a New England town of 17,000 called "Yankee City," was followed by studies of even smaller communities in the South and Middle West (Warner et al. 1949b contains references to all of them).

Yankee City was once a famous seaport. It had a long history in New England commerce, having been a center of trade, fishing, and, more recently, manufacturing, especially of shoes and silverware. In many ways, its glory was in the past. In recent years, it had become merely a small city not too far from Boston, and many of its young people left for the more exciting life to be found in Boston and New York. Ethnically the town was relatively homogeneous but not perfectly so. Some families had been there for 300 years. Half its inhabitants had been born in the community, and another quarter came from other parts of New England and the United States. But the remaining quarter were from French Canada, Ireland, Italy, and Eastern Europe.

When Warner began the research, he made explicit to his staff the hypothesis that

> the fundamental structure of our society, that which ultimately controls and
> dominates the thinking and actions of our people, is economic, and that the
> most vital and far-reaching value systems which motivate Americans are to
> be ultimately traced to an economic order. Our first interviews tended to
> sustain this hypothesis. They were filled with references to "the big people
> with money" and to "the little people who are poor." They assigned people
> high status by referring to them as bankers, large property owners,
> people of high salary, and professional men, or they placed people in a low
> status by calling them laborers, ditchdiggers, and low-wage earners. Other

similar economic terms were used, all designating superior and inferior positions (Warner and Lunt 1941: 81).

However, after the research team had been in Yankee City for a while, they began to doubt that social standing could so easily be equated with economic position, for they found that some people were placed higher or lower in the prestige scale than their incomes would warrant. Furthermore,

> Other evidences began to accumulate which made it difficult to accept a simple economic hypothesis. Several men were doctors; and while some of them enjoyed the highest social status in the community and were so evaluated in the interviews, others were ranked beneath them although some of the latter were often admitted to be better physicians. Such ranking was frequently unconsciously done and for this reason was often more reliable than a conscious estimate of a man's status. . . .
>
> We finally developed a class hypothesis which withstood the later test of a vast collection of data and of subsequent rigorous analysis. By class is meant two or more orders of people who are believed to be, and are accordingly ranked by the members of the community, in socially superior and inferior position. Members of a class tend to marry within their own order, but the values of the society permit marriage up and down. A class system also provides that children are born into the same status as their parents. A class society distributes rights and privileges, duties and obligations, unequally among its inferior and superior grades. A system of classes, unlike a system of castes, provides by its own values for movement up and down the social ladder. In common parlance, this is social climbing, or in technical terms, social mobility. The social system of Yankee City, we found, was dominated by a class order (Warner and Lunt 1941: 82–84).

Warner concluded from his interviews and observations, in other words, that prestige rank was the result of a combination of variables that included wealth, income, and occupation but also differential patterns of interaction and associated distinctions in social behavior and lifestyle.

One of the most important of the interaction patterns was determined by kinship. Not only did children get assigned the status of their parents, but certain families had a prestige position that was not entirely explainable by their current wealth or income and seemed to flow from their ancestry.

When an individual had an equivalent rank on all the variables, his townsmen had no difficulty in deciding how much prestige to give him.[1] But when he had somewhat different scores on the several variables, there was difficulty in knowing exactly how to place him. This usually meant that the person was mobile and was changing his position on one variable at a time; eventually he would be likely to get them all into line. Consequently, time was an important factor in stratification placement. For example, if a

---

[1] We use the masculine pronoun, since at the time of the study most women did not have jobs, and they were usually assigned the ranking of their husbands.

man who started as the son of a laborer became successful in business, he would be likely to move to a "better" neighborhood, join clubs of other business and professional men, and send his children to college. However, if he himself did not have a college education and polished manners, he would never be fully accepted as a social equal by the businessmen who had Harvard degrees. His son, however, might well gain the full acceptance denied the father.

After several years of study by more than a dozen researchers, during which time 99 percent of the families in town were classified, Warner declared that there were six groupings sharp enough to be called classes (Warner and Lunt 1941: 88):

1. Upper-upper, 1.4 percent of the total population. This group was the old-family elite, based on sufficient wealth to maintain a large house in the best neighborhood, but the wealth had to have been in the family for more than one generation. This generational continuity permitted proper training in basic values and established people as belonging to a lineage.

2. Lower-upper, 1.6 percent. This group was, on the average, slightly richer than the upper-uppers, but their money was newer, their manners were therefore not quite so polished, and their sense of lineage and security was less pronounced.

3. Upper-middle, 10.2 percent. Business and professional men and their families who were moderately successful but less affluent than the lower-uppers. Some education and polish were necessary for membership, but lineage was unimportant.

4. Lower-middle, 28.1 percent. The petty businessmen, the school teachers, and the foremen in industry. This group tended to have morals that were close to those of Puritan Fundamentalism; they were churchgoers, lodge joiners, and flag wavers.

5. Upper-lower, 32.6 percent. The solid, respectable laboring people, who kept their houses clean and stayed out of trouble.

6. Lower-lower, 25.2 percent. The "lulus" or disrespectable and often slovenly people who dug clams and waited for public relief.

This proportionate distribution among the classes represents not only a long New England history, but also the special conditions of the Great Depression of the 1930s, which, for example, probably inflated the size of the lower-lower class.

Once the general system became clear to him, Warner said, he used a man's clique and association memberships as a shorthand index of prestige position. Thus there were certain small social clubs that were open only to upper-uppers, while the Rotary was primarily upper-middle in membership, the fraternal lodges were lower-middle, and the craft unions were upper-lower. It seems that in cases of doubt, intimate clique interactions were the crucial test: A repeated invitation home to dinner appeared to be, for Warner, the best sign of prestige equality between persons who were not relatives.

PRESTIGE CLASS AS A CONCEPT

Warner maintained that the breaks between all these prestige classes were quite clear-cut, except for that between the lower-middle and the upper-lower. At that level, there was a blurring of distinctions that made placement of borderline families quite difficult. Of course, the placement of mobile families at all levels was difficult.

When he said that the distinctions between the classes were clear-cut, he meant in the minds of the people of Yankee City. He saw his job of scientific observer as mainly one of staying around long enough to find out what the people really thought. In this, he ran into difficulties. Since there are certain value traditions which maintain that social inequality is un-American, some people deny that it exists while at the same time behaving as though it does.

> In the bright glow and warm presence of the American Dream all men are born free and equal. Everyone in the American Dream has the right, and often the duty, to try to succeed and to do his best to reach the top. Its two fundamental themes and propositions, that all of us are equal and that each of us has the right to the chance of reaching the top, are mutually contradictory, for if all men are equal there can be no top level to aim for, no bottom one to get away from (Warner et al. 1949b: 3).

It is because he recognized this value conflict that Warner said in the passage quoted above that "ranking was frequently unconsciously done and . . . was often more reliable than a conscious estimate of a man's status. . . ." If rankings are unconscious, how do we know about them? The answer is: We observe words and behavior and then note the recurring patterns. Thus, sometimes class distinctions are explicitly described in the words of the people in the community, and sometimes they are abstractions created by the researcher to make sense out of the accumulated data.

The analyst has to create scientifically neat concepts or variables that approximate, as closely as possible, aspects of the real behavior of the subjects when they evaluate each other. Then the researcher has to formulate standard rules that show how to add up scores on each separate variable to arrive at the totality that will best predict the overall prestige judgment concerning a person that is made by others in the community. To be systematic and approach the goal of variables that make good statistics, the scientist is going to perform mental operations that are not identical to the automatic, half-conscious, and often contradictory thinking of the subjects. The aim of research is to predict, with reasonable accuracy, what most subjects will do in most circumstances. Furthermore, analysts must use variables that have theoretical meaning in other contexts, for only thus can they explain as well as describe.

In fact, concepts are always categories invented by social scientists to help explain what they have to say, but they are not invented out of pure imagination. They are based on careful observation, plus scientific reasons for abstracting from those observations a few simple factors that are worth studying in detail. In his more cautious moods, Warner admitted the degree of scientific abstraction that lay behind his six classes: "Structural and status analysts construct scientific representations (or 'maps') which represent their knowledge of the structure and status interrelations which compose the community's social system" (Warner and Lunt 1941: 34). A bit farther on, he admitted that some residents were more class-conscious than others and paid more attention to status distinctions (p. 69).

We can conclude that Warner's classes are ideal-type constructs that help him organize a vast amount of data on attitudes and behavior; they are not mere descriptions of the mental categories used by the inhabitants of Yankee City and other small towns. As such, they stand as hypotheses, and other observers have a right to test them against the data. The only sense in which these classes can be called real is to claim that they organize the data more usefully than any alternative set of hypotheses.

ARE THERE SIX CLASSES?

How can Warner's scheme be tested? How can we find out whether his six classes make a better fit to reality than five, or seven, or six that are cut differently from his? Much of the rest of the book will deal with aspects of this question, for the full answer involves behavior in terms of values, interactions, class consciousness, and so on. Here we can give only a quick preview, emphasizing direct perception of prestige by local informants who rank people they know and live with.

In *Deep South*, by Warner's colleagues Allison Davis, Burleigh R. Gardner, and Mary R. Gardner, there appears an interesting chart showing the "social perspectives of the social classes," or the way the people at each level perceive the people at other levels. It is reproduced here as Figure 2–1. The chart suggests that the people at each level did not recognize six classes in Old City, a southern town of 10,000 inhabitants, but either four or five, though the scientists said that six classes existed in the white population. The people in the town made finer distinctions between persons close to themselves than between those who were far away. However, all the distinctions that were made coincided, in spite of the fact that they might have had different names. Thus, the lower-lowers made a distinction between "society" and the "way-high-ups but not society," but they did not subdivide "society." According to the authors, the line between society and the way-high-ups but not society was precisely the same as the line drawn

FIGURE 2–1   The Social Perspectives of the Social Classes

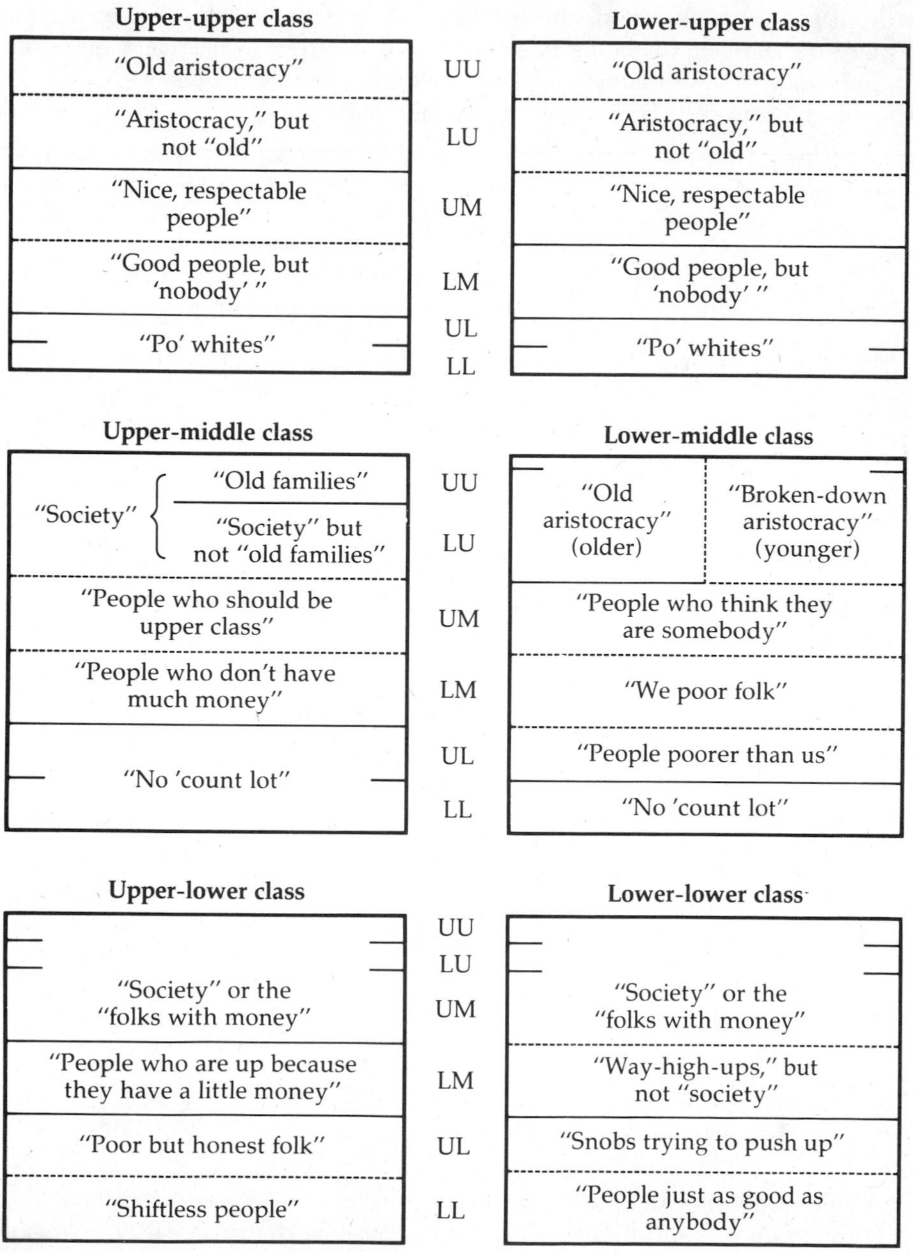

Upper-upper class		Lower-upper class
"Old aristocracy"	UU	"Old aristocracy"
"Aristocracy," but not "old"	LU	"Aristocracy," but not "old"
"Nice, respectable people"	UM	"Nice, respectable people"
"Good people, but 'nobody' "	LM	"Good people, but 'nobody' "
"Po' whites"	UL / LL	"Po' whites"

Upper-middle class		Lower-middle class
"Society" { "Old families" / "Society" but not "old families"	UU / LU	"Old aristocracy" (older) / "Broken-down aristocracy" (younger)
"People who should be upper class"	UM	"People who think they are somebody"
"People who don't have much money"	LM	"We poor folk"
"No 'count lot"	UL	"People poorer than us"
	LL	"No 'count lot"

Upper-lower class		Lower-lower class
	UU	
	LU	
"Society" or the "folks with money"	UM	"Society" or the "folks with money"
"People who are up because they have a little money"	LM	"Way-high-ups," but not "society"
"Poor but honest folk"	UL	"Snobs trying to push up"
"Shiftless people"	LL	"People just as good as anybody"

SOURCE: Reprinted from page 65 of *Deep South: A Social-Anthropological Study of Caste and Class*, by Allison Davis, Burleigh B. Gardner, and Mary R. Gardner. This chart and accompanying text from pages 71–73 of the same work are reprinted by permission of The University of Chicago Press. Copyright 1941 by The University of Chicago.

by upper-uppers between "nice, respectable people" and "good people but nobody" (upper-middle and lower-middle). That is, when names of specific families were mentioned, both lower-lowers and upper-uppers placed them in the same one of these two groups. However, the upper-uppers made further subdivisions not recognized by the lower-lowers into "old aristocracy," "new aristocracy," and "nice, respectable people."

Davis and the Gardners wrote in *Deep South* (1941: 71–73):

> While members of all class groups recognize classes above and below them, or both, the greater the social distance from the other classes the less clearly are fine distinctions made. Although an individual recognizes most clearly the existence of groups immediately above and below his own, he is usually not aware of the social distance actually maintained between his own and these adjacent groups. Thus, in all cases except that of members of the upper-lower class the individual sees only a minimum of social distance between his class and the adjacent classes. This is illustrated by the dotted line in Figure [2–1]. Almost all other class divisions, however, are visualized as definite lines of cleavage in society with a large amount of social distance between them.
>
> In general, too, individuals visualize class groups above them less clearly than those below them; they tend to minimize the social differentiations between themselves and those above. . . . In view of this situation it is not surprising that individuals in the two upper strata make the finest gradations in the stratification of the whole society and that class distinctions are made with decreasing precision as social position becomes lower.
>
> Not only does the perspective on social stratification vary for different class levels, but the bases of class distinction in the society are variously interpreted by the different groups. People tend to agree as to where people are but not upon why they are there. Upper-class individuals, especially upper-uppers, think of class divisions largely in terms of time—one has a particular social position because his family has "always had" that position. Members of the middle class interpret their position in terms of wealth and time and tend to make moral evaluations of what "should be. . . ." Lower-class people, on the other hand, view the whole stratification of the society as a hierarchy of wealth. . . .
>
> The identity of a social class does not depend on uniformity in any one or two, or a dozen, specific kinds of behavior but in a complex pattern or network of interrelated characteristics and attitudes. Among the members of any one class there is no strict uniformity in any specific type of behavior but rather a range and a "modal average." One finds a range in income, occupation, educational level, and types of social participation. The "ideal type" may be defined, however, for any given class—the class configuration—from which any given individual may vary in one or more particulars.

How do investigators discover these "modal averages" of behavior: How do they find the classes? They first observe differing patterns of general behavior or style of life—the high-society crowd who hang around the country club versus the little shopkeepers who belong to the Elks. Then they pay particular attention to the names of specific families who belong to each group. If different informants use labels with varying shades of moral evaluation to describe the different groups, the analyst can equate the labels by interpretation: the "high-society crowd" of one informant and the "old aristocracy" of another turn out to be identical, because *both* the behavior described and the individuals who are said to belong are the same.

The fact that all people do not recognize six levels is not so important, so long as the breaks that they do make all fit together in a consistent way. Naturally, people make the finest distinctions regarding those whom they know best and tend to merge others into broader categories. The analysts can take this into consideration and, if they wish, can subdivide a group according to the views of those in and immediately adjacent to it. The problem is like that of accommodating for perspective when making a map from aerial photographs. (Incidentally, these differences in perspective are not mere disturbances to be ironed out by appropriate techniques—they are social facts well worth studying in themselves. If we knew more about them, we would know more about the dynamics of social perception that underlie all prestige judgments.) The practical problems in social mapping reduce to two: Do all observers put Albers above Jackson in the hierarchy, and if they distinguish between their ranks at all, do they all divide Jackson's group from Albers' at the same place? These are the questions of *ranking consistency* and of *cutting consistency*.

THREE METHODS OF PRESTIGE PLACEMENT

The Warner group evolved its techniques for stratifying local communities out of its practical experience in the field and thus did not follow sharply explicit procedures in a consistent way throughout all the studies. Toward the end of the series of researches, the group published a field manual to guide other researchers (Warner et al. 1949b). Their key method was called *Evaluated Participation*, which included various devices for matching the reputation of one family against another on the basis of what others said about them, along with their participation in the community through informal networks of association and formal memberships in organizations. These procedures produced the rank order of families. The final decisions about cutting the ranking into class levels were made by the investigators, using various symbols that were common in the language of the residents to divide their town into strata, as well as the clustering of network ties into groupings that tended to be mutually exclusive. To follow the method well, a team of researchers had to live for a while in the community and accumulate voluminous records on many (if not all) residents.

Simultaneously with the laborious procedures of Evaluated Participation, the Warner team developed a shortcut device for ranking local citizens, called the *Index of Status Characteristics*. Once the evaluated rankings were established, they discovered that they could accurately predict them by a numerical index composed of ratings on occupation, source of income (inherited income, wages and salaries, public relief, etc.), house type, and residential area. When the ratings on each of the four factors were pooled, weighted, and averaged, the resultant score came quite close to predicting the relative position of families on Evaluated Participation (for details, see Warner et al. 1949b). Not only is the index a useful device, but it suggests that despite Warner's emphasis on social participation, the main bases of prestige placement are economic.

A simpler procedure for studying a community prestige hierarchy was developed by August B. Hollingshead, who did research in the same midwestern town the Warner group called "Jonesville"; Hollingshead called it "Elmtown." The Hollingshead method developed a control list of class-ranked families, which could be employed as a measuring rod to place other families. The control list was constructed in the following way (Hollingshead 1949: 25–41).

After three months of field work, Hollingshead sifted through his interview notes and selected the names of thirty families who were frequently mentioned. These families were spread throughout the prestige range, and there seemed to be considerable agreement about their relative positions. Hollingshead placed the names of the husbands and wives on small cards and went back to twenty-five of the people who had been interviewed previously. They were asked to place the thirty families in different stations or classes by dividing the cards into stacks of equivalent rank. The informants could use as many stacks as they wished. Nineteen informants used five groups; 77 percent of the placements of specific families were in agreement.

In a second round of testing with a fresh group of subjects, Hollingshead improved these results by eliminating from the control list ten families who had frequently received inconsistent ratings. He concluded that the level of agreement achieved was sufficient to sustain the judgment that the community had five classes and that residents could place their fellows fairly well into one or another of them.

Using the revised control list as a measuring rod, Hollingshead was able to place a sample of 535 Elmtown families in the class structure by asking a new group of informants to indicate which of the control-list families most approximated the social position of each additional family to be rated. When informants disagreed about the placement of a particular family, their judgments were averaged to produce a class rank. (Hollingshead noted that most of the disagreement centered on seventy-four families who were experiencing social mobility. This fact suggests more than the origin of some methodological headaches; it points to a significant source of ambiguity in perceptions of the class structure of modern societies.) Using his technique, Hollingshead proceeded to classify the families of all the students in the high school, where he was conducting his research.

Hollingshead's approach of asking informants to sort cards with the names of local families into stacks of equivalent prestige is a neat technique that establishes both ranking *and* cutting in one quick operation. The

Warner Index of Status Characteristics is also easy to use, but its development assumes a previous and difficult operation of Evaluated Participation, unless one is willing to accept the guess that the index can be used in new communities without reevaluating it. In Jonesville-Elmtown, the methods produced roughly similar results: Over 80 percent of the families were placed in equivalent categories by all three techniques. Both investigators ended up with five classes, since neither found the distinction between established wealthy families and those with new fortunes, which subdivided the upper classes in Yankee City and Old City in the South. Other than that, the class structures in all three towns were strikingly similar. (A general review of measurement techniques is Jackson and Curtis 1968; see also Rainwater 1974 and Nock and Rossi 1978.)

COLEMAN AND RAINWATER: CLASS STRUCTURE OF THE METROPOLIS

Although a few studies of small towns have been conducted in more recent years, they did not attempt to parallel the work of Warner and Hollingshead, and we do not know how much communities of the type they studied have changed since World War II. However, two former students of Warner have completed an urban study that clearly shows the mark of their mentor, to whose memory their book is dedicated.

The Warner group had earlier published a study of the black community in Chicago (Drake and Cayton 1945), but Coleman and Rainwater in *Social Standing in America* (1978) were more ambitious; they described two complete metropolitan areas. In the early 1970s, they collected qualitative and quantitative data on public conceptions of social class through 600 interviews conducted in Boston and 300 in Kansas City. The procedures employed were designed to provide representative samples of adults in Greater Boston and Greater Kansas City—that is, of the respective Standard Metropolitan Statistical Areas (SMSAs) designated by the Bureau of the Census. The interviews were standardized and followed a fixed schedule of questions in the style of a social survey, but many were open-ended and allowed the respondents to choose their own words.

The hierarchy of prestige classes that Coleman and Rainwater stitched together from their analysis of the interviews is rather complex, so we offer a simplified, schematic version in Table 2–1. (Note that the annual incomes were recorded in 1971 dollars. Tripling these amounts will give roughly equivalent values in inflated 1990 dollars.) Inspection of the table shows that the basic structure of the hierarchy is parallel to the one found by Warner in Yankee City. For instance, in both studies, the upper-upper and lower-upper classes correspond to a distinction between established families and "new money," although the distinction may be noticed only by those who are themselves close to the top. In Boston, the upper-class respondents spoke of the former as "the tip-top—as close to an aristocracy

TABLE 2-1 Coleman and Rainwater's Metropolitan Class Structure

Class	Typical Occupations or Source of Income	Typical Education	Annual Income 1971	% of Sample
I. Upper Americans				
Upper-upper Old rich, aristocratic family name	Inherited wealth	Ivy League college degree; often postgraduate	Over $60,000	2%
Lower-upper Success elite	Top professionals; senior corporate executives	Good colleges; often postgraduate	Over $60,000	
Upper-middle Professional and managerial	Middle professionals and managers	College degree; often post-graduate	$20,000 to $60,000	19
II. Middle Americans				
Middle class	Lower-level managers; small-business owners; lower-status professionals (pharmacists, teachers); sales and clerical	High school plus some college	$10,000 to $20,000	31
Working class	Higher blue-collar (craftsmen, truck drivers) lowest-paid sales and clerical	High school diploma for younger persons	$7,500 to $15,000	35
III. Lower Americans				
Semipoor	Unskilled labor and service	Part high school	4,500 to $6,000	13
The bottom	Often unemployed; welfare	Primary school	Less than $4,500	

NOTE: Percentages adjusted for undersampling of "Lower Americans," acknowledged by authors.
SOURCE: Coleman and Rainwater 1978.

as you'll find in America . . . Yankee families that go way back; the WASPs who were here first . . . the bluebloods with inherited income—they live on stocks and bonds" (p. 150). The same respondents described the lower-uppers as "a mix of highly successful executives, doctors, and lawyers with incomes of $60,000 at least, many $100,000 and more. . . . They have help in the house, fancy cars, frequent and expensive vacations, and at least two houses. . . . They're not considered top society because they don't have the right background—they're newer money, with less tradition in their lifestyle" (p. 151).

By contrast to those at the top, Coleman and Rainwater delineated a bottom class characterized by dependence on irregular, marginal employment or public relief, often shifting from one to the other. As in Yankee City, Boston and Kansas City families in this class were regarded as less than respectable and described in terms suggesting that they were physically and morally "unclean." However, many of Coleman and Rainwater's informants made a distinction between families on the very bottom and a class of semipoor families who worked more regularly and were slightly more orderly in their lifestyles.

As portrayed by Coleman and Rainwater, then, "upper America" and "lower America" neatly parallel corresponding prestige groupings in Yankee City. The same would appear to be true of "middle America," where Coleman and Rainwater's middle class and working class are equivalent to Warner's lower-middle and upper-lower classes. However, it was in the middle range of the class structure that they had the hardest time organizing the views of Kansas City and Boston respondents into prestige categories. In judging prestige, city respondents at this level gave almost exclusive emphasis to income and standard of living and paid relatively little attention to other stratification variables, such as occupation and association, that had seemed important to Warner.

Coleman and Rainwater reported that their middle American respondents recognized three levels among themselves, often called "people at the comfortable standard of living," "people just getting along," and "people who aren't lower class but are having a real hard time" (pp. 158–159). However, the two sociologists found these categories inadequate and insisted on the more traditional distinction between middle class and working class, which, they suggest, can each be subdivided along income lines. Their conclusion that the middle class/working class distinction is more fundamental than income differences is based on evidence that the survey collected on differences in lifestyles and patterns of association. For instance, among families at the same "comfortable" income level (approximately $11,000 to $20,000), they noted important differences in consumption patterns. The middle-class families were likely to spend more on living room and dining room furniture and less on television sets, stereos, and refrigerators and other appliances that were attractive to working-class families at these income levels. Working-class families own larger and more expensive automobiles and more trucks, campers, and vans. Moreover, income equality in middle America does not appear to produce social equality; patterns of friendship, organizational membership, and neighborhood location parallel lifestyle distinctions (pp. 182–183). Some respondents implicitly recognized these differences. A "comfortable" working-class man observed (p. 184):

> I'm working class because that's my business; I work with my hands. I make good money, so I am higher in the laboring force than many people I know. But birds of a feather flock together. My friends are all hard-working people. . . . We would feel out of place with higher-ups.

The wife of a white-collar man was more explicit (p. 184):

> I consider myself middle class. My husband works for a construction company in the office. Many of the construction workers make a lot more than he does. But when we have parties at my husband's company, the ones with less education feel out of place and not at ease with the ones with more education. I think of them as working class.

The difficulties Coleman and Rainwater encountered in delineating the prestige hierarchy of middle America raise general, and by now familiar, questions about the methods by which the sociologist defines the structure of prestige classes in a community. They described their approach to prestige measurement as close to Warner's Evaluated Participation. However, the metropolitan context of their research imposed an important limit on their ability to replicate in a precise way Warner's methodology. Since the respondents in Boston and Kansas City did not know the same people in their communities, the technique of matched agreements comparing specific families (which Hollingshead had systematized with his control families) could not be employed. This made synthesizing individual judgments of the class system more difficult; the data consisted of verbal statements about general symbols rather than details about particular others in the community.

Coleman and Rainwater were clear that their version of the prestige hierarchy was not a mirror image of the class structure as understood in the community but rather an ideal type defined by the researchers. "Ultimately, to number and name the American social classes is a task for the social scientists. The status structure is too complex to be comprehended fully by average persons from their inevitably narrow vantage point" (p. 120). They approached this "task" in two stages. First they attempted "to piece together hundreds of 'narrow' views on the subject and construct from them a single picture that is a summation of public impressions about the social class hierarchy" (p. 120). Subsequently this "image-based picture" was elaborated by combining it with data on behavior to produce "a picture that, in our view, is more nearly the truth about classes in America" (p. 120). As in Warner's work, the outcome is significantly dependent on the "clinical" judgment of the investigators.

The basic vehicle for the initial step was an open-ended question in the survey instrument: "How many classes would you say there are [in America], and what names would you give to them (or what names have you heard other people use)? What else could you say to describe each of the classes you've mentioned?" (p. 120). The most frequent answer to the first part of the question was that there were three classes. However, this response was given by only one third of the informants. Other answers included two, four, five, six, seven, and nine classes or a flat refusal to say. One Bostonian replied, "You have too many classes for me to count and name. . . . Hell! There may be fifteen or thirty. Anyway, it doesn't matter a damn to me" (p. 121). Even respondents who suggested the same number of classes differed in the way they described and named the classes, especially when the number was greater than three.

The disparate character of responses to both parts of the question suggests, as Coleman and Rainwater admit, that there is no "clear public consensus as to the name and number of social classes in the United States" (p. 20), at least in large cities. However, the variations among images of the class system that emerged in Boston and Kansas City were in some ways systematic. As in the earlier studies by the Warner group, the researchers found that respondents tended to construct more class boundaries close to their own position in the prestige hierarchy and fewer at a distance from themselves. Furthermore, when they asked people for reactions to a series of hypothetical class names, they found that the names "were 'cheapened' in their assumed reference by respondents of successively lower-status" (p. 127). Thus, upper Americans understood middle-middle class to refer to those earning $15,000 or more a year, typically college graduates. But middle Americans associated the label with incomes of $12,000 to $13,000 and lower Americans assumed incomes of $9,000 to $10,000 (p. 127).

Coleman and Rainwater noted that the simplest division, that into a two-class model, emphasized the distinction between "us" and "them" and was usually offered by people slightly below average in their own position who expressed both "awe and animus" toward those above them in the hierarchy with more money (p. 121); this perception is linked to a conflict model of society. Citizens who thought in terms of a three-class model also generally used income as the basic criterion for dividing up the social system, but they added another class below themselves at the bottom—people who were poor, who lived in the slums on welfare—and gained some sense of pride by having someone to be above. Models that contained four or more classes usually used criteria other than money to make additional distinctions and often came from respondents in higher positions. They talked of education, the way people lived, family background, and other factors that modified the influence of money by itself. These respondents implicitly used an individualistic and competitive instead of a conflict model of society. (For a theoretical elaboration of this difference, see Ossowski 1963; for a revealing study in England, see Bott 1954 and 1964.)

Coleman and Rainwater imposed order on the many conceptions of the class system proposed by their respondents. As they viewed the problem, "All these seeming inconsistencies and variations should be thought of not as confusion but as a rich lode of imagery about the status system; represented therein are the best efforts of usually inarticulate people to express categorically and concretely the underlying continuum of status" (p. 123). The trick was to find the criteria for cutting points in the continuum, explicit or implicit in the comments of respondents, which indicate the class groupings that were most salient for them. This could be done without assuming that Bostonians and Kansas Citians agreed completely among themselves on the number of classes or the proper names for them. Approaching the matter in this way, Coleman and Rainwater located thirteen logical dichotomies, of which only six "emerged with sufficient frequency and strength of supporting argument to be considered centrally significant" (p. 124). They were the following:

1. A source-of-income cut, placing those with inherited wealth socially above those with self-earned wealth.
2. A level-of-income cut, establishing "people who really have it made" financially above those who are "doing well but aren't really well-to-do."

3. A cut based on educational credentials and associated occupational accomplishments—"the degrees versus the non-
   degrees"—ranking the former above the latter.
4. A cut based on publicly perceived income level and associated
   standard of living—"people who have a comfortable salary and
   all the necessities plus a few luxuries" versus families who are
   "just getting along."
5. Another cut based on income and standard of living—"people
   getting by who have a decent house and the husband has
   a decent-paying job" versus "people who have to work at jobs
   that don't pay well. . . . Their housing is not slum, but it's
   undesirable."
6. A cut based on source of income, this time private earnings versus public charity—"the poor who are working and largely self-
   supporting" versus "the welfare class."

These six cuts combine to suggest a seven-class model close to the one
outlined in Table 2–1. It is worth noting that, for the most part, the classes
delineated by these cuts are not simply *quantitatively* distinguished from
each other, as they might be on a single scale, such as income. The cuts
imply *qualitative* differences among classes. For example, the first cut separates new money from old money; the third cut divides those with from
those without college degrees. A full application of all the cuts allowed
Coleman and Rainwater to further subdivide the seven classes shown in
Table 2-1, but we need not follow them that far.

PRESTIGE OF OCCUPATIONS

Warner, Coleman and Rainwater, and many other investigators have
stressed the importance of occupation for the prestige evaluations Americans make of one another. Especially in metropolitan settings, where
people do not have a detailed knowledge of one another's income, family
background, lifestyle, associations, and so forth, they are forced to fall back
on a few shorthand indicators of personal prestige, such as occupation.
They know, of course, that occupation is a fair indicator of two other
sources of prestige: income and education. Physicians are typically affluent; not many janitors hold college degrees. They may also associate particular lifestyles and patterns of interaction with occupations, such as the
general distinction between blue-collar and white-collar workers. These expectations account for the emphasis they place on occupation in making
prestige assessments. For the sociologist engaged in a large-scale research
operation, occupation is especially useful: It is more visible than income,
and it can be studied with census data as well as direct social surveys and
qualitative field studies. Furthermore, since census data are available for
earlier periods, we can use occupation as an indicator in historical research.

There have been numerous studies of occupational prestige, going back at least fifty years, but the best known are national polls conducted under the auspices of the National Opinion Research Center (NORC) at the University of Chicago, beginning with a 1947 survey (NORC 1953) supervised by Cecil North and Paul Hatt. In this section, we will examine preliminary results from a 1989 NORC survey of 1,500 people and review some of the published analyses of earlier NORC studies of occupational prestige (Reiss 1961; Hodge et al. 1966; Nakao and Treas 1990).

Keiko Nakao and Judith Treas, who conducted the 1989 survey, describe their procedure as follows:

> Each respondent was asked to evaluate 110 occupations according to their *"social standing"* and to sort out small cards bearing the occupational titles onto a nine-rung ladder of social standing (from "1" for the lowest to "9" for the highest) (pp. 1–2; emphasis added).

The individual ratings from this procedure were averaged and converted into occupational prestige scores, ranging, theoretically, from 0 to 100. In practice, however, there was enough skepticism about the high-prestige jobs and sufficient respect for the low-prestige jobs that the scores rarely went above 80 or below 20.

Table 2–2 presents a sample of the occupational prestige scores from the 1989 study.[2] The results at the top and bottom of the scale are consistent with what we have seen in the Warner group's community and metropolitan studies (which, of course, are based on much more than occupation). The highest-ranking occupations are professional and managerial (physician, professor, plant manager, hospital administrator), ordered by the level of expertise or administrative responsibility they entail. Virtually all assume a college education or better.

The lowest are unskilled, manual jobs, such as garbage collector, janitor, and filling station attendant. Between these extremes are the less demanding office or sales positions (bookkeeper, secretary, cashier in a supermarket) and the skilled manual jobs (electrician, plumber, welder). But note that there is no clear distinction between white-collar and blue-collar jobs (in Warner's terms, lower-middle and upper-lower). The electrician ranks above the bookkeeper, the carpenter above the traveling salesman, and the truck driver above the store clerk. Obviously, when faced with this sort of task, respondents are interested in something more than just where someone works or the color of a shirt collar (Glenn and Alston 1968).

When interviewees in the first NORC survey were asked the main factor they had weighed in making their ratings, the most frequent replies

---

[2]Most of the occupations in the table are from a list of forty representative jobs that all respondents were asked to rate. A few were chosen from 700 additional occupations that were divided into groups and rated by subsets of respondents. The authors found sufficient consistency in ratings of the forty to conclude that this procedure, designed to maximize the number of occupations rated, was statistically reliable.

TABLE 2–2   National Opinion on Prestige of Occupations

Occupation	NORC Score
Physician	86
Department head in a state government	76
Lawyer	75
College professor	74
Chemist	73
Hospital administrator	69
Registered nurse	66
Accountant	65
Public grade school teacher	64
General manager of a manufacturing plant	62
Policeman	59
Superintendent of a construction job	57
Airplane mechanic	53
Farm owner-operator	53
Electrician	51
Manager of a supermarket	48
Secretary	46
Bookkeeper	46
Insurance agent	46
Plumber	45
Bank teller	43
Welder	42
Post office clerk	42
Travel agent	41
Barber	36
File clerk	36
Assembly-line worker	35
Housekeeper in a private home	34
House painter	34
Cook in a restaurant	34
Cashier in a supermarket	33
Bus driver	32
Logger	31
Salesperson in furniture store	31
Carpenter's helper	30
Salesperson in shoe store	28
Garbage collector	28
Bartender	25
Cleaning woman in private home	23
Farm worker	23
Janitor	22
Telephone solicitor	22
Filling station attendant	21
Table clearer in restaurant	21

SOURCE: Nakao and Treas 1990.

were pay, service to humanity, education, and social prestige, but none of these criteria was volunteered by more than 18 percent of the sample (NORC 1953: 418). Whatever the bases of their judgments, the surveys demonstrate that respondents did have a scale in mind on which they could place occupations with a rough consensus. Although there were significant differences among individuals in their relative ratings of occupations, sociologists were more impressed with the great consistency of the average ratings that were given to occupations by relevant subgroups of the population. The correlations between the ratings made by random pairs of individual raters ran about 0.6, but the correlations between the average ratings made by the prosperous and the poor, people in high- and low-prestige occupations, blacks and whites, men and women, residents of the Northeast and the South, and city and country dwellers were all 0.95 or above. Even those who proposed different criteria for judging occupations did not differ in the way they ranked occupations (Reiss 1961: 189, 193; Treiman 1977: 60–74; Nakao and Treas 1990: 12).

What systematic differences there were can be summed up in two principles: (1) People tended to raise in rank their own and closely related occupations, and (2) people agreed with each other more concerning occupations that were well-known. For example, in 1947 many respondents did not know how to place "nuclear physicist" (answers in response to a question about this included "assistant to a physic" and "he's a spy"); yet they had no trouble deciding about a "scientist."

However, even in less esoteric fields brief occupational titles of the sort employed in the NORC surveys are somewhat ambiguous. Confronted by "lawyer," the respondent does not know if the reference is to a small-town general-practice lawyer or a partner in a major Wall Street firm. Of course, we can never capture with a survey the richness of detail that Warner reports from a community study, since surveys force us to depend on a few simple categories. On the other hand, there is no substitute for the systematic knowledge a survey can provide. It is especially useful for making broad comparisons between different communities or different historical periods, but we must always remember that we are using social symbols about general types of jobs, leaving out a lot of concrete detail.

The consistency of prestige ratings across social subgroups is matched by their stability over time. The correlation between scores in the 1947 and 1963 NORC surveys was virtually perfect (0.99). The correlation for occupations rated in both the 1963 and the 1989 surveys was reportedly at a similarly high level.[3] Moreover, retrospective comparisons between the NORC studies and several earlier studies suggested that there had been virtually no changes in the rankings of occupational prestige since 1925 (Hodge et al. 1966).

INTERNATIONAL COMPARISONS

Occupational prestige is a convenient index for stratification studies because it transcends local communities. The consensus throughout the country found by the NORC confirms that statement. But the intriguing

---

[3]Phone conversation with Keiko Nakao. At this writing, the analysis of the 1989 survey has yet to be published.

possibility arises that perhaps the meaning of occupation transcends national boundaries as well. A massive study by Donald Treiman dwarfed earlier attempts to deal with this question (Inkeles and Rossi 1956; Hodge et al. 1966; Treiman 1977). Treiman collected data from occupational prestige studies in sixty separate societies. From country to country, he correlated the relative prestige ratings given the same occupations. The average correlation between all pairs of countries were 0.8, slightly lower than found in earlier studies but remarkably high given the diversity of societies involved. Further analysis produced suggestive but by no means conclusive evidence that countries at the same level of economic development have similar patterns of occupational prestige.

The consistency of occupational prestige judgments across societies and between subgroups of the same society suggests that the ratings correspond to some fundamental aspect of social structure that is quite widely perceived, even though, as we have seen, divergent explanations may be offered for it. Treiman argues that the tendency toward a common international system of ranking occupations is rooted in international similarities in the division of labor. A complex division of labor attaches quite specific social functions to each occupation. This occupational differentiation brings varying control over scarce resources (specifically skills, authority, and property). For example, a physician must have special skills and authority to carry out that job. But differential control of resources means occupational differences in power. Those with greater power can obtain greater privileges for themselves; for example, the doctor claims high income. Finally, differential power and privilege are the sources of differential prestige. The physician's skill, authority, and high income become the bases of high prestige. (For critiques of the occupational prestige literature, see Goldthorpe and Hope 1972, and Coxon and Jones 1978.)

DUNCAN'S SOCIOECONOMIC INDEX FOR OCCUPATIONS

The merits of applicability in different places, stability over time, and intergroup consensus have made occupational prestige rankings an attractive stratification measure for all sorts of sociological studies. But investigators who used the NORC ratings ran into one very serious difficulty: Only ninety occupations were rated in the original NORC surveys. When researchers encountered occupations not listed, they were forced to assign approximate prestige scores on the basis of "similarity" to an NORC occupation.

Otis Dudley Duncan (1961) provided one solution to this problem that has been widely used by others. Duncan noted that both the average education and the average income of those who worked in various occupations were highly correlated with the prestige scores of the occupations. Taking advantage of this fact, he combined income and education in a simple regression formula designed to estimate how much education and income were required to produce the prestige of each occupation on the NORC list.

His income-plus-education formula turned out to be a very efficient predictor of occupational prestige: It correlated 0.91 with the ratings received by the NORC occupations for which income and educa- tion data could be found. Duncan therefore felt confident in employ- ing the same formula to estimate prestige scores for occupations *not* covered by the surveys, using census data to get typical incomes and ed- ucations for the people who worked in hundreds of different occupa- tions.

Duncan was, in effect, using education and income as stand-ins for prestige (the index is called SEI—Socioeconomic Index). His justification for doing so—beyond the obvious pragmatic argument that it seemed to work—was that education and income are both functionally and tempo- rally related to occupation. Education is the necessary preparation for en- tering an occupation; income is the reward that flows from an occupation. We have, of course, already examined studies that point to close ties be- tween prestige and income and education. If the NORC ratings do in fact provide an accurate prestige measure, Duncan's estimation formula con- firms that income and education are strongly (and approximately equally) associated with the prestige of occupational titles.

OCCUPATIONS AND SOCIAL CLASSES

The NORC surveys reveal how the public places occupations on a contin- uum, but the surveys leave us in the dark as to how people might cut that continuum into occupational classes. The problem is obviously similar to the one we raised earlier concerning the grouping of families in Yankee City or Boston into prestige classes. Some evidence on this question was gathered in a national survey of approximately 2,000 adults conducted by the University of Michigan's Survey Research Center in 1975 (Jackman 1979). The SRC questionnaire asked respondents to place a series of occu- pations into one of five class categories: poor, working class, middle class, upper-middle class, and upper class. This question and several others in the SRC study replicate items in a 1946 survey conducted by Richard Cen- ters, whose influential work on class consciousness and political attitudes we will examine in Chapter 9 (Centers 1949). The results from the 1975 survey are presented in Table 2–3.

The data indicate that people do not have a difficult time associating occupations with social classes (there are few "don't knows") and that there is considerable popular agreement about where occupations should be placed in the five-class system that was suggested by the interviewer. In virtually all cases, the most frequent class assignment (indicated in italics) accounted for over 50 percent of the responses.

As we can anticipate by now, class placement of occupations is easiest at the top and bottom of the scale. There is, for example, strong agreement that high-ranking corporate officers are upper class and janitors and assembly-line workers are working class. The greatest ambiguity centers

TABLE 2–3   Social Class Assignment of Occupations

Occupations	Poor	Working	Middle	Upper Middle	Upper	Don't Know	Total
			Social Class Assigned (percent)				
Corporation directors and presidents	—	1	3	22	72	2	100
Doctors and lawyers	—*	2	4	36	57	2	100
Executives and managers	—*	3	14	56	26	2	100
Supervisors in offices and stores	—*	17	56	23	3	1	100
Small businessmen	2	18	61	18	1	1	100
Schoolteachers and social workers	—*	19	60	17	3	1	100
Foremen in factories	1	40	47	10	1	2	100
Plumbers and carpenters	1	44	40	12	2	1	100
Workers in offices and stores	2	60	33	3	1	1	100
Assembly-line factory workers	5	75	16	2	1	2	100
Janitors	25	67	6	1	—*	1	100
Migrant farm workers	73	22	2	1	1	2	100

*Less than 1. Totals may not add to 100 due to rounding. Numbers in italics represent modal category. Sample size = 1,850.

SOURCE: Jackman 1979: 449.

on the middle-ranking occupations, especially skilled manual workers and foremen. Note also that the distinction between working class and middle class does not neatly correspond to the difference between blue-collar and white-collar jobs. Here, as when ranking occupations in the NORC surveys, people are more discriminating than that. The majority place the lower white-collar positions in offices and stores in the working class.

The 1975 data differ little from the original results that Centers obtained in the 1940s, except for minor variations due to the exact wording of the question. Both studies support the idea that the public perceives more than a rank order of occupational titles: It groups those titles into class categories, that have widespread meaning, even though the agreement about the placement of particular occupations is less than perfect.

THE PERCEPTION OF RANK AND STRATA

Over the years, investigators have tried various devices to systematize the creation of groups or strata or classes out of an undifferentiated rank order. They generally find that the rank order is seen by the public with more clarity and consistency than are subdivisions into groups. Some people divide the world into conflicting groups of "us" versus "them," and they are likely to use labels such as working class versus middle class. Others stress the openness of our society and the chance for people to keep climbing a notch or two, and they are likely either to deny sharp class divisions or to use a lot of them with minor distinctions separating one from another. In other words, perception of classes is intimately linked with degrees and styles of class consciousness, which we will explore further in Chapter 9.

When a sociologist presents a subject with either a list of the neighboring families in a small town or a list of occupations, and requests that the subject arrange them in categories of equivalent prestige, the stimulus presented is partly structured and partly ambiguous. During a total life experience, a subject will have run into a number of situations that indicate prestige differences, but these situations have not been so clear and so consistent that they completely determine the subject's perceptions. The very nature of our society is such that the worker in overalls can put on a business suit for special occasions. As the person changes from one social role to another, his or her behavior changes somewhat—but not completely. Meeting the boss on Sunday, the worker will show deference, even if they both are wearing the same kind of suit. Consequently, the respondent in the survey will have a general idea of a prestige hierarchy, but not a perfectly sharp one with fixed labels for each level plus standard status symbols to indicate who belongs where. The degree of existing ambiguity allows the subject to project some personal views into the data and shape them slightly to suit individual desires and values. Therefore, people at different levels give somewhat different answers for two reasons: Their outer experience has been different, and their inner frames of reference are different.[4] Both stimuli and perceiver vary from one respondent to another.

There is no such thing as an "objective" status structure that can be viewed by the completely neutral observer. Prestige is embedded in attitudes about the relationships between persons, positions, and symbols, and it varies according to the perspective from which they are viewed. The neutral observer has to find a way of summarizing the agreements and disagreements but should not foolishly conclude that the result is a reality that is more "real" than the subjective perceptions of the respondents (and the behavior they determine).

From the many studies we have reviewed, three conclusions stand out and are confirmed by every one of them. (1) In American society, there is a prestige hierarchy of both persons and occupations. (2) This hierarchy or rank order is divided by most citizens into a few categories or classes, but there exists considerable ambiguity about just how to define and differentiate them. (3) There is more agreement about rank order than about the criteria used in making ranking decisions, and more agreement about

---

[4]There is, in fact, another cause of varied and inconsistent answers: People do not pay full attention to a task of this type, and a lot of random error creeps in.

ranks than about strata. There is enough consensus to allow us to arrange a rank order that permits further operations to seek the variables that correlate with it and to begin to study the causes of that covariation. We then find that the covariation between occupational prestige and personal prestige is higher than for most pairs of variables that sociologists are interested in. Either occupational prestige largely causes personal prestige, or else they both flow from the same underlying causes.

There are some tentative principles that seem to explain the differences in perceptions of both ranking and grouping, and these appear in a number of the different studies, though all the data are not completely consistent. The principles can be summarized as in the list that follows; they interact with and sometimes offset the effects of one another.

1. People perceive a rank order.
2. People tend to enhance their own position:
    a. By raising their own position relative to others.
    b. By varying the size of their own group. Here the evidence is not consistent. Apparently there is a tendency to narrow the group when thinking of individual persons about whom invidious distinctions can be made (especially those lower on the scale), and enlarge it when thinking of general categories of persons who are closely similar to one another.
    c. By perceiving separate but equal groups, thus accepting difference but denying hierarchy.
3. People agree more about the extremes than about the middle of the prestige range. This is probably a result both of clarity of stimulus and of aspects of perceptual organization. There are more people in the middle, and they are less publicized. Also, errors at the extremes can be made in only one direction.
4. People agree more about the top of the range than the bottom and make more distinctions about the top than about the bottom. This probably reflects stimulus reality, for those at the top are more conspicuous and also more differentiated.
5. People lump together into large groups those who are furthest from them.
6. The better persons and occupations are known, the more agreement there is concerning them.
7. People have more consensus about the relative rank of persons or occupations that are closely related to each other in some functional way, such as the hierarchy of doctors, nurses, and orderlies.
8. People at the top are more consistent with one another, and make more divisions into groups, than those on the bottom.
9. People in the middle or at the bottom are more likely to conceive of class differences in financial terms, while those at the

top are more conscious of prestige distinctions from lineage
and style of life.

10. Mobility is a source of ambiguity in perception of the prestige
order. People find it difficult to "place" mobile individuals. Perception of high rates of mobility leads to the conclusion that
class boundaries are amorphous or nonexistent.

These ten principles connecting social facts with the way people perceive those facts are sufficient to explain why there is no straightforward
answer to the question that is asked so often: How many social classes are
there in America? The moment we try to answer the question with data
that come from the views of ordinary citizens, we are bound to come up
with ambiguities that lead us to reply, "It all depends on how you view the
situation." Only those theorists who have a scheme that allows them to
stand above the system and decide for themselves what is important can
arrive at a clear-cut answer to the question. Since the theorists disagree
with one another, once more we end up saying, "It all depends on how you
view the situation."

CONCLUSIONS

The studies of the Warner group found that personal prestige in small
towns seems to reflect more directly than anything else the styles of life
and clique memberships involved in consumption behavior. But in a modern society with a complex division of labor, consumption behavior reflects
more than anything else the income derived from occupational roles.
These roles are sufficiently standardized to permit a direct prestige ranking
of occupational titles. That ranking has remained relatively constant in the
United States for the last fifty years, and it bears striking similarity to the
ranking in other industrial countries.

However, occupation implies more than just income. The full significance of work is best determined by those who understand it—a person's
colleagues on the job. They use such criteria as the skill the job demands,
the talent and training necessary to produce the skill, the responsibility
and authority over others that the job entails, the pay it brings, and even
the nature of the product. The pay has a double function: It is granted as a
reward for skill, responsibility, and authority, but it soon becomes a symbol
of them and a direct stimulus for effort—we say, "He's a $50,000-a-year-
man." The product of work is judged by general social values; we believe
that it is more important to save life than to beautify it (thus a doctor
is ranked higher than a hairdresser), and we grant honor to those whose
decisions have wide social impact (which puts Supreme Court Justices at
the top).

In a small community, people can be ranked on the totality of their
roles—occupational, familial, and civic. In the larger cities, neighbors do

not know so much about one another and tend to use shorthand symbols. Sometimes they judge in terms of income; sometimes they talk of styles of life that reflect both income and the education and values that shape the way income is spent; and sometimes they recognize that occupation may be the best index of all, since it reflects education, skill, authority, and income. (Indeed, technical studies show that occupation is the best single indicator of general social position; see Kahl and Davis 1955.)

The vitality of our materialistic and competitive culture requires a reasonably close relationship between these causes and consequences of occupational roles. People must believe that, on the average, fair rules apply, so that those who have high skill and use discipline to train their skill will be rewarded with important jobs that pay well, and that money can buy a style of life that is admired in the community. Without such a chain of connections, motivation would suffer and legitimacy would weaken. We note tensions, for example, when those who spend lots of money seem to have gained it in ways that do not reflect proper occupational achievements— such as racketeers whose methods are suspect, or children of rich families who spend without earning. Those who turn their backs on conspicuous consumption may even withdraw from the rigors of occupational competition and go live in communes.

Many of the tensions derive from contradictions between principles of legitimacy and the actual operations of power. Salaries, for instance, are determined by people who have authority in organizations, and profits reflect decisions by owners of businesses, so it should not surprise us that people at the top receive lots of money and then assert that they deserve it and general prosperity requires it. Sometimes these determinations are considered fair and just by subordinates, but sometimes they are considered to be outrageous exploitation. Furthermore, the family system passes on property through the generations, which is usually considered to be more natural and appropriate by the rich than by the poor. Marx stressed that the rules of property are set by those with power; Weber emphasized the connection between occupational rewards and general values of legitimacy, rationality, and efficiency. When the ordinary facts of daily life are perceived by growing segments of the population to depart too far from the rules of fair play, the tensions produce conflict and change, as we shall see in subsequent chapters.

SUGGESTED READINGS

Aldrich, Nelson, Jr. 1988. *Old Money: The Mythology of America's Upper Class*. New York: Vintage.
   Old money and not-so-old money. The ethos of the national upper class explained from within.
Curtis, Richard, and Elton Jackson. 1977. *Inequality in American Communities*. New York: Academic Press.

Compares several communities on various indicators of stratification. Authors conceive stratification as a continuum rather than as a structure of classes.

Ossowski, Stanislaw. 1963. *Class Structure in the Social Consciousness.* New York: Free Press.

A systematic account of competing basic conceptions of class structure in the history of Western social thought.

Veblen, Thorstein. 1934. *The Theory of the Leisure Class.* New York: Modern Library. (First published in 1899.)

This book introduced the idea that "conspicuous consumption" was the way to buy prestige in competitive America.

Warner, Lloyd, et al. 1973. *Yankee City.* New Haven: Yale University Press.

Handy abridgement of the entire series of volumes on Yankee City.

3 Social Class, Occupation, and Social Change

The prudent, penniless beginner in the world, labors for wages awhile, saves a surplus with which to buy tools or land, for himself; then labors on his own account another while, and at length hires another new beginner to help him. This, say its advocates, is free labor— the just and generous, and prosperous system, which opens the way for all—gives hope to all, and energy, and progress, and improvement of condition to all.

Abraham Lincoln (address in Milwaukee, 1859)

A post-industrial society, being primarily a technical society, awards place less on the basis of inheritance or property . . . than on education and skills.

Daniel Bell (1976: xviii)

We may have created too many dumb jobs for the number of dumb people to fill them.

Quoted in Braverman (1974: 35)

Occupation is a crucial and convenient stratification variable. We have already found that it is closely tied to personal prestige, education, and income. In later chapters, we will see that occupation shapes social processes as disparate as the exercise of political power and the way parents bring up their children. For the researcher, as we have noted, occupation is a convenient variable, because it is not bound to the particular circumstances of a local community but has a national definition that is relatively stable over time. Stability of definition makes occupation a useful measuring rod for historical change, especially since we have occupational data (on both local and national levels) that stretch back many decades. These data allow us to analyze shifts in the occupational structure (the proportional distribution of occupations in the labor force), which have important implications for the evolution of the class system in general. Before turning to the national studies, it will be helpful to look at local research that shows how occupational changes are linked to many aspects of community life.

THE LYNDS: MIDDLETOWN

One sociological community study particularly emphasizes the occupational changes that have occurred in the past: the research on "Middletown." (For a reassessment of the study, see Caplow 1980.) It began when Robert S. Lynd and Helen Merrell Lynd went to Indiana to describe a "typical" American community as it existed in 1924. At that time, Middletown (a pseudonym) had 35,000 inhabitants. In order to have a baseline for contrast, the Lynds reconstructed life in 1890, when the town had only 11,000 people and was going through the first stages of industrialization. They eventually returned to the city and wrote a second book about it, telling of its growth to 47,000 by 1935 and its reactions to the days of boom and bust that had followed the first research. Thus we have three points of time to contrast: 1890, 1924, and 1935 (Lynd and Lynd, 1929 and 1937).

The Lynds lived in the town for over a year, meeting and talking to as many people as possible. They interviewed all the important people and many of the unimportant ones; they read newspapers, diaries, local histories; they went to various ritual gatherings of church and civic groups and luncheon clubs. Occasionally they passed out a questionnaire to get standardized information about such matters as budget behavior or attitudes of students in the high school, but the emphasis was more on qualitative than quantitative data.

In their first book, *Middletown*, the Lynds said this about life in 1924 (1929: 21–24):

> As the study progressed, it became more and more apparent that the money medium of exchange and the cluster of activities associated with its acquisition drastically condition the other activities of the people. Rivers begins his study of the Todas with an account of the ritual of the buffalo dairy, because "the ideas borrowed from the ritual of the dairy so pervade the whole of Toda ceremonial." A similar situation leads to the treatment of the activities of Middletown concerned with getting a living first among the six groups of activities to be described. . . . [1]
>
> At first glance it is difficult to see any semblance of pattern in the workaday life of a community exhibiting a crazy-quilt array of nearly four hundred ways of getting its living. . . . On closer scrutiny, however, this welter may be resolved into two kinds of activities. The people who engage in them will be referred to throughout this report as the working class and the business class. Members of the first group, by and large, address their activities in getting their living primarily to *things,* utilizing material tools in the making of things and the performance of services, while the members of the second group address their activities predominantly to *people* in the selling or promotion of things, services, and ideas. . . . There are two and one-half times as many in the working class as in the business class—71 in each 100 as against 29. . . .
>
> While an effort will be made to make clear at certain points variant behavior within these two groups, it is after all this division into working class and business class that constitutes the outstanding cleavage in Middletown. The mere fact of being born upon one or the other side of the watershed roughly formed by these two groups is the most significant single cultural factor tending to influence what one does all day long throughout one's life; whom one marries; when one gets up in the morning; whether one belongs to the Holy Roller or Presbyterian church; or drives a Ford or a Buick; whether or not one's daughter makes the desirable high school Violet Club; or one's wife meets with the Sew We Do Club or with the Art Students' League; whether one belongs to the Odd Fellows or to the Masonic Shrine; whether

---

[1]The others are: making a home, training the young, using leisure, engaging in religious practices, and engaging in community activities.

one sits about evenings with one's necktie off; and so on indefinitely throughout the daily comings and goings of a Middletown man, woman, or child.

One of the central themes of the first Middletown volume was that, from 1890 to 1924, there were basic changes in the work pattern of both the business and working classes—changes, incidentally, that resulted in a wider gap between them. These changes flowed from three causes: larger population, more machinery, and increasing emphasis on money. The Lynds described in vivid detail a case study of the great transformation of modern life—industrialization.

Middletown in 1890 was a market town that was just beginning to turn to manufacturing. The work habits and values of its people were extensions of the traditions of their farmer parents. Those farmers were people who had conquered a wilderness: There had been land for all who would work it, and from such plenty, there emerged a society that lacked gradations of rank and privilege, a society that stressed individual initiative and progress, family solidarity, simplicity of manners and style of life, equality among neighbors. As trading and handicraft manufacturing succeeded farming as the base of livelihood, the old traditions could easily continue. A man earned whatever his own efforts deserved. True, there developed a gradation in income that extended from unskilled through skilled laborers to bosses (who were often ex-craftsmen) and a few professional people. This gradation, however, was not sharply divided into levels; a person often moved through several steps in a few years, and it was understood that the system was open to everybody in fair and equal competition. People started at the bottom of some line of endeavor and worked their way up. Income and prestige were direct outcomes of competence at work, and everybody could understand and agree that as competence increased with age and experience, it brought the right and necessity of teaching and directing the work of people who were less skillful, and thereby it earned a higher income. The advent of machines began to change all this (1929: 31–32, 73–74):

> "When tradition is a matter of the spoken word, the advantage is all on the side of age. The elder is in the saddle" (Goldenweiser). Much the same condition holds when tradition is a matter of learned skills of hand and eye. But machine production is shifting traditional skills from the spoken word and the fingers of the master craftsman of the Middletown of the 90s to the cams and levers of the increasingly versatile machine. And in modern machine production it is speed and endurance that are at a premium. A boy of nineteen may, after a few weeks of experience on a machine, turn out an amount of work greater than that of his father of forty-five. . . .
>
> The demands of the iron man for swiftness and endurance rather than training and skill have led to the gradual abandonment of the apprentice-master craftsman system; one of the chief characteristics of Middletown life in the 90s, this system is now virtually a thing of the past. . . . With the passing of apprenticeship the line between skilled and unskilled workers has become so blurred as to be in some shops almost nonexistent.

There were basic changes among the business class as well. The old businessman was a small merchant or manufacturer, whose capital consisted mostly of his personal savings. He had started as a worker and had become a businessman. His relations with both employees and customers were personal, even intimate. The new businessman operated in terms of bank credit, had too many employees and customers to know them personally, and had ties with other businessmen all over the country. True, the petty grocer and his kind still existed in 1924, but the major part of production and exchange in Middletown was passing into the hands of the new businessman (who was sometimes a branch manager of a national corporation).

Money had become the significant link between people. In the old days, each family was more self-sufficient; they processed most of their own food and clothing (from purchased raw materials, of course); they entertained themselves at home and with the neighbors; when they did buy things, they paid cash. By 1924, "credit was coming rapidly to pervade and underlie more and more of the whole institutional structure within which Middletown earned its living" (1929: 45). Businessmen were much more dependent upon the banks and individuals upon the credit agencies and merchants who allowed them to buy on credit. More articles of use were bought and fewer were homemade; more activities had changed from family and neighborhood affairs to commercial propositions. All of this was well symbolized by the automobile, and the Lynds devote some brilliant pages to the changes in life that centered around the family car. They sum up the transition to a money-centered life in these words (1929: 80–81):

> For both working and business class no other accompaniment of getting a living approaches in importance the money received for their work. It is more this future, instrumental aspect of work, rather than the intrinsic satisfactions involved, that keeps Middletown working so hard as more and more of the activities of living are coming to be strained through the bars of the dollar sign. Among the business group, such things as one's circle of friends, the kind of car one drives, playing golf, joining Rotary, the church to which one belongs, one's political principles, the social position of one's wife, apparently tend to be scrutinized somewhat more than formerly in Middletown for their instrumental bearing upon the main business of getting a living while conversely, one's status in these various other activities tends to be much influenced by one's financial position. As vicinage has decreased in its influence upon the ordinary social contacts of this group, there appears to be a constantly closer relation between the solitary factor of financial status and one's social status. A leading citizen presented this matter in a nutshell to a member of the research staff in discussing the almost universal local custom of "placing" newcomers in terms of where they live, how they live, the kind of car they drive, and similar externals: "It's perfectly natural. You see, they know money, and they don't know you."

By 1924, Middletown was becoming too large and its productive system too complex and mechanized for community prestige to flow automatically from skill at work. People did not understand just what the activities of others were; a grocer knew little about glass blowing or automobile-parts manufacturing, and a glass blower knew little about financing a grocery store on credit. Concurrently, the money nexus was becoming increasingly important as more spheres of life became parts of the commercial market. The result was that people began to use money as a sign of accomplishment, a common denominator for prestige. The question "How much does he earn?" was heard more frequently than "How much skill does he have?"

The Lynds did not find it necessary to divide the population any further than business class and working class when they wrote about Middletown in 1924. True, they recognized some gradation within each group, but they said that the working class was becoming more homogeneous through time as machines degraded skill, and the business class remained a small and basically undifferentiated group. They felt that no other distinction in Middletown was as important as this one.

The system could not work unless people believed that in a general way, authority and income on the job and style of life and prestige off the job were automatic results of relatively free individualistic competition. This belief was brought to Middletown from the farms and villages; it was the American heritage of the frontier. Although some people were beginning to question this belief in 1924, the vast majority of both workers and businessmen clung to it. Free enterprise was not a mere theoretical discussion by economists of the workings of the market; it was a quasi-religious belief of the people.

MIDDLETOWN REVISITED

In 1935, the Lynds returned for a restudy. Both they and Middletown had changed. They had gone somewhat to the left in their political credo; as Robert Lynd wrote in the second book, *Middletown in Transition:* "Middletown believes that *laissez-faire* individualism is the best road to 'progress.' The present investigator holds the view, on the other hand, that our modern institutional world has become too big and too interdependent to rely indiscriminately upon the accidents of *laissez-faire* . . ." (1937: xvii). Middletown had grown larger, reaching 47,000 people; its industrial plant was more mechanized, more centralized into larger units, more subservient to national corporations; it had gone through the boom of the late 1920s and the devastating crash of the early 1930s. Yet the interplay between the Lynds' assumptions and the reality of Middletown produced a report that stresses continuity in spite of change. The reader is surprised at how Middletown was able to take so much in its stride, to face such deep threats to its social structure, and yet come out with so many of the same beliefs— somehow more tense and strained and defensive in their expression, but basically unaltered. "The city's prevailing mood of optimism makes it view prosperity as normal, while each recurrent setback tends to come as a surprise which local sentiment views as 'merely temporary'" (1937: 13). In terms of this philosophy, Middletown never really admitted that it had a depression until it was all over.

The trend toward mechanization that was dominant from 1890 to 1924 continued on through 1935. It could be easily measured through occupational statistics. These showed that the city's increase in population went mostly into the service industries, for the machines were getting so efficient that a productive labor force that was only slightly larger could produce vastly more goods.

The details were these: Using the census figures for 1920 and 1930, the Lynds reported that the proportional increase of workers in the production industries was 26 percent, whereas the increase in the service industries was 66 percent. These service workers were becoming more professionalized through time; there were more schoolteachers, more social workers, more dental hygienists. The new efficiency of the productive machinery could support more services of a "luxury" type.

Within the factories, there were changes in the composition of the work force. The trend toward semiskilled machine tenders continued—proportionally more workers were in that category, fewer either unskilled or highly skilled. But a lot of half-trained workers working on a complex production line needed a few very well-trained people to invent the machines and to direct and coordinate their work. These technicians, engineers, and managers were usually trained for their jobs in school and college, rather than being promoted from the ranks. For an increase of only 10 percent in production workers who labored for an hourly wage, there was an increase of 31 percent in managers and officials, 128 percent for technical engineers, and 900 percent for chemists, assayers, and metallurgists. At the same time, the number of independent owner-manufacturers dropped 11 percent (1937: 68). Thus, we see the march of technology: A larger mass of machine tenders gathered into fewer but bigger factories directed by a tiny group of specialized technicians and executives. The gap between worker and manager had widened. And the managers had changed. They were not so often self-made owner-businessmen; they were more likely to be members of "the new middle class" of technicians and administrators, trained in college, who worked for a salary. No longer could the complex system of Middletown be adequately described by the simple division into business class and working class. (For Middletown's occupational system in 1970, see Caplow and Chadwick, 1979.)

MIDDLETOWN'S NEW CLASSES

As a result of the changes they observed in Middletown, the Lynds described the class structure in 1935 as follows (1937: 458–460):

   1. A very small top group of the "old" middle class is becoming an upper class, consisting of wealthy local manufacturers, bankers, and local head managers of one or two of the national corporations with units in Middle-

town, and a few well-to-do dependents of all the above, including one or two outstanding lawyers. . . .

2. Below this first group is to be found a larger but still relatively small group, consisting of established smaller manufacturers, merchants, and professional folk . . . and also most of the better-paid salaried dependents of the city's big-business interests. . . . These two elements in Group 2 constitute socially a unity but, in their economic interests, often represent somewhat divergent elements. . . .

3. . . . Middletown's own middle class in purely locally relative terms: the minor employed professionals, the very small retailers and entrepreneurs, clerks, clerical workers, small salesmen, civil servants—the people who will never quite manage to be social peers of Group 2 and who lack the constant easy contact with Group 1 which characterizes Group 2.

4. Close to Group 3 might be discerned an aristocracy of local labor: trusted foremen, building trades craftsmen of long standing, and the pick of the city's experienced highly skilled machinists of the sort who send their children to the local college as a matter of course.

5. On a fifth level would stand the numerically overwhelmingly dominant group of the working class; these are the semiskilled or unskilled workers, including machine operatives, truckmen, laborers, the mass of wage earners.

6. Below group 5 one should indicate the ragged bottom margin, comprising some "poor whites" from the Kentucky, Tennessee, and West Virginia mountains, and in general the type of white worker who lives in the ramshackle, unpainted cottages on the outlying unpaved streets. These are the unskilled workers who cannot even boast of that last prop to the job status of the unskilled—regular employment when a given plant is operating. [Most of the city's few blacks would also fit here.]

The Lynds add some general comments to the schematic outline just given (1937: 460–461):

Psychologically, Groups 1, 2, and 3 cling together as businessfolk, over against groups 4, 5, and 6. . . . If the nascent "class" system of "Magic Middletown" appears to follow somewhat the above lines, Middletown itself will turn away from any such picture of the fissures and gullies across the surface of its social life. It is far more congenial to the mood of the city, proud of its traditions of democratic equality, to think of the lines of cleavage within its social system as based not upon class differences but rather upon the entirely spontaneous and completely individual and personal predilections of the 12,500 families who compose its population.

The Lynds offer us a scheme of six groups or classes emerging from "nascent" tendencies in the occupational system. Interestingly enough, this scheme based on occupational niches is almost the same as the Warner scheme based on prestige, consumption style, and interaction networks.

The processes by which occupation and income are transformed into style-of-life symbols, interaction networks, and personal prestige are sharply illustrated by the story of the emergence into prominence of the "X family." They were the leading group of businessmen in Middletown. They had founded their fortune before the turn of the century by starting a small glass-manufacturing plant on a capital of $7,000. There were five brothers, four who developed the business and one who practiced medicine. The legends of Middletown contain many tales of the simplicity and humbleness of these early business pioneers who took advantage of the natural gas that was the base for the city's early industrialization (1937: 75): "One of the city's veteran clothing dealers is fond of telling how one of the brothers borrowed a light-weight overcoat for a week-end party of young people in 1889. 'He didn't feel he could afford a new coat that year, as he was just starting in business.' " Then there is the newspaper editorial that appeared in 1925 when one of the brothers died: "He always worked on a level with his employees. He never asked a man to do something he would not do himself."

These men built up a great production machine that became world famous. They grew rich, they contributed vast sums to local charities and development projects; they became active community leaders. "In their conscientious and utterly unhypocritical combination of high profits, great philanthropy, and a low wage scale, they embody the hard-headed *ethos* of Protestant capitalism with its identification of Christianity with the doctrine of the goodness to all concerned of unrestricted business enterprise. In their modesty and personal rectitude, combined with their rise from comparative poverty to great wealth, they fit perfectly the American success dream" (1937: 75–76).

The brothers and their sons could not avoid becoming embroiled in all aspects of Middletown life. They became active in politics. They put up the money that kept the banks from collapsing in the depression, thereby gaining control. They were leaders in the fight against unionization. They took over the biggest department store in town when it failed in the early 1930s. They opened a new residential subdivision for themselves and soon were surrounded by all the ambitious families in town who could afford to move. They contributed money to the local college (part of the state system), and one of them was elected president of the state university board of trustees. They bought a substantial interest in the local newspaper. They did not completely run the town, because there were other industries present, including manufacturing plants that were part of the General Motors empire. And the mass of the people still voted. But "the business class in Middletown runs the city. The nucleus of business-class control is the X family" (1937: 77).

It was not until the second generation that the X family gained leadership in consumption affairs (1937: 96):

> Around the families of the four now grown-up sons and two sons-in-law of the X clan, with their model farms, fine houses, riding clubs, and airplanes, has developed a younger set that is somewhat more coherent, exclusive, and self-consciously upper-class. The physical aggregation of so many of these families in the new X subdivisions in one part of town has helped to pinch off psychologically this upper economic sliver of the population from the mass of business folk. And the pattern of their leisure, symbolized by their riding clubs and annual horse show, tends to augment their

> difference. . . . Particularly as regards the male members of the older generation, there has always been a continuous preoccupation with business; and they did few of the things associated with a wealthy leisured class. It is the new note of a more self-conscious leisure built upon endowed wealth, and obviously expensive, that the younger generation is bringing to Middletown. They, too, work hard, but they play expensively and at their own sports, with somewhat more definitely their own social set.

Great wealth has to be used. Whether it is ploughed back into business, exchanged for elegance in leisure, used to patronize health, the arts, and education, or devoted to the training of a new generation who will be taught how to spend it gracefully, it enhances the power and prestige of those who own it. The second generation of owners is bound to be different: They cannot have the motives of ambitious individuals born in poverty. They will have the capital wealth that brings power, the income that brings luxury, and the values of those who have been reared to expect both; these values have been polished by attendance at national universities of high prestige, rather than the local college. Such people will be slightly different from first-generation entrepreneurs of the same age—a difference that will be less important during the day at business than during the evening at play. They will be bound to seek the company of others like themselves. They will, in other words, create and become an upper class with links to the national elite.

INDUSTRIALIZATION AND THE TRANSFORMATION OF THE NATIONAL CLASS STRUCTURE

The changes the Lynds chronicled in Middletown reflect the sweeping transformation of American society brought on by our conversion into an industrial nation. Although the process of industrialization was under way in the 1840s and 1850s, it was in the period after the Civil War that it was consolidated. Between 1870 and 1929, the population tripled, but the output of the American economy grew tenfold, largely as the result of the colossal gains in productivity made possible by new industrial technologies (Ross 1968). By 1900, the United States was the world's leading industrial nation (Hacker 1970: 192).

Closely associated with this achievement were other developments of critical significance for the American class system, most of which we have seen played out on a small scale in Middletown. For instance, Americans changed from predominantly rural residents and farmers to city dwellers employed at urban occupations. In 1870, approximately three quarters of the population lived in rural places (under 2,500 inhabitants), and more than half the labor force was employed in farming. By 1930, most Americans lived in towns and cities, and nearly 80 percent of the labor force was engaged in nonfarm occupations (Ross 1968: 26). Over the same period, occupational tasks were becoming increasingly subdivided and specialized. Thus, the steelworker and the metallurgist were among a multitude of successors to the blacksmith.

As the industrial economy expanded, the country's labor requirements grew faster than the native population. This labor deficit was met by encouraging immigration from abroad, a policy so successful that by 1920, 22 percent of the industrial labor force was foreign born.

Finally, the period saw a leap in the scale of economic organization. Large corporations emerged as the dominant force in the American economy. Industrial technology requires large enterprises, but it was the drive to control national markets and generate monopoly profits that produced such giants as Standard Oil, United States Steel, E. I. du Pont de Nemours, and General Electric, through a wave of corporate mergers at the turn of the century (Ross 1968: 40–41).

In the next few pages, we will try to show how the class structure was affected by these changes. First we will look at three broad class groupings (upper class, working class, and middle class) that were decisively transformed by industrialization. Then we will examine how the occupational structure was altered. In effect, we will be looking at the same phenomena from two different perspectives.

THE NEW UPPER CLASS

Our account of the upper class draws on the work of the late C. Wright Mills, whose classic study, *The Power Elite* (1956), is examined in greater detail in Chapter 8. Writing of the history of the upper class, Mills observed that the United States had never had a national aristocracy of the type known in Europe. In the absence of a feudal past or a single national center of wealth, power, and prestige like Paris or London, an enduring "pedigreed" class never developed. But at the time industrialization got under way, there were a series of relatively stable and compact regional upper classes across America:

> Up the Hudson, there were patroons, proud of their origins, and in Virginia, the planters. In every New England town, there were Puritan shipowners and early industrialists, and in St. Louis, fashionable descendants of French Creoles living off real estate. In Denver, Colorado, there were wealthy gold and silver miners. And in New York City, as Dixon Wecter has put it, there was a class made up of coupon-clippers, sportsmen living off their fathers' accumulation, and a stratum [of families] . . . trying to renounce their commercial origins as quickly as possible (Mills 1956: 48).

The prestige of these regional upper classes rested on old family fortunes, some on the eastern seaboard reaching back to colonial times. By gradually assimilating new moneyed families, the established families were able to keep prestige and wealth in close correspondence. But this system was overwhelmed in the post–Civil War era.

Mills noted that prior to the Civil War, there were very few millionaires. George Washington, among the richest men of this time, left an estate of $530,000 in 1799. In 1840, there were only thirty-nine millionaires in New York City and the entire state of Massachusetts. However, the postwar expansion of the industrial economy created unprecedented opportunities for rapid accumulation of new wealth. By 1892, a survey revealed more than 4,000 millionaires in the country (Mills 1956: 101–102). The wealth of the Lynds' "X" family and the fortunes associated with names such as Rockefeller, Woolworth, Carnegie, Morgan, and Vanderbilt originated in this period.

The post–Civil War fortunes destroyed the neat coincidence of family lineage, wealth, and prestige. The established families initially resisted the social pretentions of the new rich, but this was by no means easy, since the size and national scope of the new wealth dwarfed the older family fortunes. Not by chance, this era produced one of the most famous attempts to define membership in upper-class society, Ward McAllister's list of the top 400. McAllister, who established himself as an arbiter of New York society on the basis of his close ties to the prestigious Mrs. John Jacob Astor, decided in the 1880s that there were " 'only about four hundred people in fashionable New York Society. If you go outside that number, you strike people who are either not at ease in a ballroom or else make other people not at ease" (Mills 1956: 54). The list was a defensive attempt to preserve the prestige of the old families against the pretensions of the social climbers. In 1892, McAllister finally published his list, which in fact contained only 300 names and was basically a directory of the best-known pre–Civil War families. Only nine of the richest men of the day (those with fortunes in excess of $30 million) were among the 300 (Mills 1956: 54).

McAllister's effort to specify the makeup of upper-class society was obsolete even before it was formulated. By 1883, Mrs. Astor has already overcome her disdain for Mrs. Vanderbilt's new railroad money and accepted an invitation from her to a fancy-dress ball. Gestures of this sort, Mills argued, have prevented the emergence of an American aristocracy. "Always in America, society based on descent has been either bypassed or bought-out by the new and vulgar rich" (Mills 1956: 52). In the late nineteenth and early twentieth centuries, the family fortunes built after the Civil War were merged with the old society to create a new prestige class, national in scope like the new economy and centered in major cities such as New York, Philadelphia, Chicago, and San Francisco. (This class comprises the "upper-uppers" and some of the "lower-uppers" that Coleman and Rainwater identified in Kansas City and Boston.) The progress of this merger can be seen in the *Social Register*, an elite directory that first appeared during this era and proved to be the most successful and enduring effort to specify the socially elect. The first *Social Register*, published in New York in 1887, contained 881 families in a careful mix of new and old. Within a few years, registers were issued in other major cities. During the period 1890–1920, the rate of admissions to the *Social Register* was high, reflecting rapid assimilation of the new rich; since then, annual admissions have declined to more modest levels. By 1940, over three quarters of the families covered by Lundberg's study of America's great fortunes (most accumulated between the Civil War and World War I) were listed in the *Social Register* (Baltzell 1958: 20).

THE NEW WORKING CLASS

The economic transformation that forced a restructuring of the upper class also recast the bottom of the American class structure. Limited before the Civil War to the mills of New England, factory production spread across the country and created a modern industrial working class. In 1860, there were 1.5 million industrial wage earners in the country; in 1900, there were 5.5 million (Brooks 1971: 39). By 1930, 17 million workers, constituting over a third of the work force, were employed (largely in manual occupations) in the sectors most directly affected by industrialization: manufacturing, mining, and utilities and transportation.

Many of the new industrial workers were immigrants, actively recruited by business and government and drawn by the economic opportunities offered by an expanding economy. In the decades immediately after the Civil War, the immigrants were typically English, Irish, German, or Scandinavian, people historically and culturally similar to the native-born population. After 1890, however, the character of immigration changed, and most of the immigrants of the late nineteenth and early twentieth centuries came from southern and eastern Europe. These "new immigrants," as they have been called, were Italians, Jews, and Slavs, and by the time they arrived, the best jobs in the industrial economy were controlled by others. The new immigrants were therefore relegated to the least attractive manual-labor jobs, while native-born Americans and the offspring of the old immigrants dominated skilled occupations and supervisory positions (Link and Cotton 1973: 13–15; Hacker 1970: 88–92). Thus, a close association grew up between stratification and ethnicity, with the highest positions being occupied by the members of the most culturally Americanized groups. This system was reinforced by active discrimination against many immigrants and their children in schools and politics (Brooks 1971: 76–77).

The working conditions and wages offered America's expanding working class were dismal indeed, especially in the jobs available to recent immigrants. A contemporary who visited Italian and Jewish sweatshops in New York City described people crowded into rooms

> small as boat's cabins; crouched over their work in a fetid air which an iron stove made still more stifling and in what dirt; hunger hollowed faces; shoulders narrowed with consumption, girls of fifteen as old as grandmothers, who had never eaten a bit of meat in their lives (Allen 1952: 52–53).

In Chicago, another observer described the early-morning procession of industrial workers:

> heavy brooding men, tired anxious women, thinly dressed, unkempt little girls, and frail, joyless lads passed along, half awake, not uttering a word as they hurried to the factories (Allen 1952: 52–53).

A Commission on Industrial Relations, appointed by President Wilson, concluded in 1915 that "a large part of our industrial population are . . . living in conditions of actual poverty. . . . It is certain that at least one third and possibly one half of the families of wage earners employed in manufacturing and mining earn in the course of the year less than enough to support them in anything like a comfortable and decent condition" (Link and Cotton 1973: 38). The commission discovered that the children of the poor were dying at three times the rate of middle-class children and that 12 to 20 percent of all children in six major cities were underfed and undernourished. In 1897, the average workweek in industry was nearly sixty hours. Long hours and the indifference of employers to safety conditions produced an appalling record of industrial accidents. An incomplete survey showed that at least half a million workers were killed, crippled, or seriously injured on the job in 1907 (Link and Cotton 1973: 38).

In the early years of the new century, some improvement began in these conditions. State legislatures, over the determined opposition of upper-class employers, started to pass legislation controlling safety conditions, providing compensation for injured workers, prohibiting child labor, and regulating the labor of women. As industrial productivity increased, hours of work declined and wages went up. However, the life of the average worker remained unenviable. A survey of industrial accidents in 1913 (more reliable than the 1907 study) found that 25,000 persons had been killed and 700,000 seriously injured at work (Link and Cotton 1973: 38). Moreover, the benefits of increased pay and reduced hours were very unevenly distributed among workers; for the majority of workers who were unskilled and not unionized, the gains were slight, especially relative to growing industrial productivity (Ross 1968: 37–40).

The contrast between the conditions of the working class and the opulent lifestyles of the new rich convinced many Americans that class divisions were becoming sharper in America. This impression was reinforced by the evidence of class conflict between the emerging industrial working class and its upper-class employers. The period from the Civil War to World War I saw the most violent labor confrontations in American history. For example, in 1877, when railroads cut the wages of their already overworked (fifteen to eighteen hours per day) and underpaid workers, a wave of strikes hit the nation's rail system. State militia and federal troops (called out by officials sympathetic to the railroads to "maintain order") provoked bloody confrontations in which scores of people were killed or injured. Millions of dollars of damage was done to company property. In 1892, at Homestead, Pennsylvania, a private army of 300 Pinkerton detectives hired by management fired on striking employees of the Carnegie Steel Corporation (later United States Steel). In the ensuing battle, three Pinkertons and seven workers were killed. A few days later, the company prevailed upon the governor to send state militia to take over Homestead, which was under the (peaceful) control of a strike committee. Multiple indictments against strike leaders, charging crimes from murder to "treason against the state of Pennsylvania" broke the strike, and the union itself, by depleting the union treasury with legal expenses. However, not all strikes ended in management victories. When textile workers in Lawrence, Massachusetts, struck in 1912, martial law was immediately declared, and three dozen strikers were arrested and summarily sentenced to prison. After several violent incidents, including an attack by police on a group of women and children from striking families, the strikers gained public sympathy and won concessions from employers (Brooks 1971: 50–55, 84–92, 118–121).

In the wake of the 1877 rail strikes, E. L. Godkin, the prestigious editor of *The Nation*, wrote: "We have had an uprising not against political oppression or unpopular government but against society itself." These events should serve to disabuse Americans of the illusion that "this was the one country in which there was no proletariat, no dangerous class." The problem, as Godkin saw it, was that the United States had been invaded by people of alien "blood" (immigrants, that is) who had no respect for "American ideals." The solution? Reinforcement of the army and militia and a reduction of loose talk "about the laborer and his rights" (Litwack 1962: 53–56).

If such violent incidents were exceptional, they were nonetheless symptomatic of the era. Workers who sought decent wages and working conditions had to contend with powerful companies, which could generally count on the support of state and federal governments, the courts, and, as Godkin's remarks suggest, the press. Only gradually did the federal government begin to assume a more balanced attitude toward labor conflicts. As late as the 1930s, the legal right of workers to form unions was still in question.

In the late nineteenth and early twentieth centuries, the United States came as close as it ever would to Marx's conception of a capitalist society. Readers of the *Communist Manifesto* would recognize the emergence of bourgeoisie and proletariat in an era of rapid industrial development, the creation of a political and ideological superstructure beholden to the bourgeoisie, and evidence of violent class conflict. However, events did not take the course an orthodox Marxist might have anticipated. Despite Godkin's observation about an "uprising against society," American workers proved to be more interested in earning reasonable wages than in creating a socialist society, and they used both their unions and their votes to help achieve their goals. In addition, two tendencies prevented stark polarization between two classes, a trend that Marx regarded as a prerequisite to proletarian revolution. One was differentiation *within* the working class among ethnic groups, among the races, and among workers at different skill and wage levels, which kept the entire working class from developing a homogeneous and combative class consciousness. The other was the expansion of the middle class (as we have seen in Middletown) in response to the special needs of a sophisticated industrial economy, which offered the chance of advancement to workers of their children. Added to this were long-term gains in the standard of living of most groups that appeared to absorb their attention more than did the differences between rich and poor. We will return to these themes in Chapter 9.

THE NEW MIDDLE CLASS

For an understanding of the transformation of the middle class in industrializing America, we turn again to the work of C. Wright Mills. In *White Collar* (1951), drawing on the earlier work of Lewis Corey (1935; 1953), Mills distinguished between two groupings he called "old middle class" and "new middle class." The former is composed of small entrepreneurs: farmers, shopkeepers, self-employed professionals, and the like. The hallmark of this class is its independence, based on the entrepreneur's ownership of the property with which he or she works. The new middle class is composed of salaried white-collar people, an exceedingly heterogeneous grouping of office workers, salespeople, and salaried professionals and managers. Early in the nineteenth century, the old middle class comprised as much as 80 percent of the total population, but as the country's industrial society matured, the working class became the majority. At the same time, there was a shift in the internal composition of the middle classes, which Mills summarized with the labor force statistics shown in Table 3–1.

The changing balance within the middle class, Mills observed, represented the decline of property and the emergence of occupation as the principal basis of stratification. In the countryside, many small farmers

TABLE 3–1 The Middle Classes

	1870 (percent)	1940 (percent)
Old middle class	85	44
Farmers	62	23
Businessmen	21	19
Free professionals	2	2
New middle class	15	56
Managers	2	6
Salaried professionals	4	14
Salespeople	7	14
Office workers	2	22
Total	100	100

SOURCE: Mills 1951:65.

were forced off the land or became tenants as agricultural prices fell relative to the prices of the urban products that farmers buy. Behind declining prices were the domination of national economic policy by the new industrialists and financiers of the Northeast, intensified competition from foreign agriculture, and, ironically enough, significant advances in agricultural technology, which enabled fewer farmers to feed an expanding urban population.

In the cities, many businesspeople shared the financial plight of the small farmers. However, in this case the modest relative decline in the proportion of business owners in the labor force masked what was actually happening: Each year, millions of small business enterprises collapsed, only to be replaced by millions of new enterprises, most of them doomed to the same fate. In the developing post–Civil War economy, the small-business owner was relegated to the highly competitive retail sector of the economy, while much of the rest of the economy came under the domination of large corporations.

While the decline of the old middle class was evident in the fate of the farmer and small-business owner, the rise of the new middle class was tied to the triumph of the corporation. As the operations of corporations grew in productive efficiency, financial magnitude, and geographic scope, a decreasing proportion of the work force was required in the actual production of things, but an increasing proportion was needed to manage, design, sell, and keep account of production. Moreover the growth of the corporate economy imposed new tasks on government. The expansion of corporate and public bureaucracies meant rapid growth of the new middle class.

However, as the proportion of white-collar employees grew, their relative standing in the class system declined. Mills observed that, in the past, white-collar people had had strong claims on superior prestige. In contrast to blue-collar workers, they wore street clothes at work, engaged in occupational tasks that were more mental than muscular, and were likely to have direct contact with the entrepreneur, from whom they could, in effect, "borrow" prestige. White-collar workers typically had a high school education, and only a small proportion of the population stayed in school that long. Nearly all were white and native-born Americans who spoke good English. White-collar workers were paid substantially more than manual workers.

Virtually all of these traditional bases of white-collar prestige were weakened over time. The very growth of white-collar employment made white-collar status less special. As we will show later in this chapter, the character of work in the expanding white-collar bureaucracies increasingly approximated the experience of the assembly line. Contacts between most white-collar employees and ranking executives became infrequent. As the American-born sons and daughters of immigrants entered the labor market, the significance of nativity was reduced. The spread of education broke the new middle class's exclusive claim on the high school diploma. At the same time, the wage gap between blue-collar and white-collar workers began to close. In 1890, the average annual income of white-collar workers was approximately twice that of urban manual workers. By the mid-1930s, it was 1.6 times, and by the late 1940s, 1.2 times (Mills 1951: 72–73). Of course, well-paid professionals and managers, the top strata of the new middle class, earned salaries well above blue-collar wages. But the situation of the great mass of white-collar employees was approaching that of the working class, and the line between the two strata steadily grew more indistinct.

NATIONAL OCCUPATIONAL SYSTEM

The Lynds' chronicle of Middletown and the parallel national accounts by Mills and others suggest that the post–Civil War transformation of the American class system can be viewed as a result of a series of shifts in the distribution of occupations. We know, for instance, that there was a proportional decline in the number of farmers and a proportional increase in the number of factory workers and salespeople in the work force. Approaching social change this way has the advantage of being precise—we can count the number of people employed in different occupations at various points in time. It is also directly relevant to the technological and organizational shifts that have reshaped the economy. Fortunately, there is an excellent source of data on occupational change: the national census conducted by the federal government every ten years. In 1820, the Bureau of the Census began to ask questions about occupations, although systematic and comparable series of statistics do not reach back that far. In the early 1940s, an official of the Census Bureau, Dr. Alba M. Edwards, delved into the bureau's dusty archives and reorganized and standardized the material on occupations, going back to 1870 (Edwards 1943).

In order to produce a series of comparable historical statistics, Edwards needed to find a way of classifying thousands of specific occupations into a few niches, each one with homogeneous "social-economic status." He decided against using a single indicator of social standing. Instead he lumped together factors such as the nature of the work, the skill and training involved in it, the income it brought, and common opinion about its prestige. He made his decisions on the basis of general knowledge at the time, without systematic tests.

The occupational hierarchy that Edwards proposed has been repeatedly modified in detail by the Census Bureau, but it is still recognizable in hundreds of government publications on occupations and on factors related to occupation, from fertility to income distribution. Moreover, versions of the Edwards schema have been adopted by many nongovernment researchers.

The major occupational categories in the schema, along with examples of specific occupations typical of each category, are given in Table 3–2. Since this system of classification is so widely used, it is worth studying the examples to get a sense of the range of each occupational grouping.

TABLE 3–2 Major Occupational Groups, with Examples

Professionals
  Public school teachers
  Doctors
  Computer systems analysts
  Registered nurses
  Librarians
  Engineers

Managers and administrators
  Sales managers
  Public administrators
  Bar managers
  Corporate executives
  Proprietors of small retail establishments
  Accountants

Technicians (often classified with professionals)
  Laboratory technicians
  Draftspersons
  Computer programmers

Sales workers
  Retail-store clerks
  Insurance agents
  Manufacturers' sales representatives

Clerical workers
  Bank tellers
  Bookkeepers
  Secretaries
  Cashiers
  Data entry operators

Craft, precision production, and repair workers
  Carpenters
  Masons
  Foremen in construction or manufacturing
  Machinists
  Telephone repair workers
  Auto mechanics

Operatives
  Assembly-line workers in manufacturing
  Butchers
  Gas station attendants
  Truck drivers

Service workers
  Waiters and waitresses
  Police and firefighters
  Child care workers
  Maids and janitors
  Domestic servants

Laborers (excludes farm)
  Unskilled construction workers
  Freight and stock handlers
  Garbage collectors

Farm workers
  Farmers (owner-operator)
  Farm managers
  Unpaid farm-family laborers
  Migrant farm laborers

Edwards maintained that his classification was the most practical means for making a *scale* of occupations that would increase in prestige, education, and income as one ascended step by step. We can test this claim using current government data on income and education, together with occupational prestige scores drawn from the NORC survey. These statistics are presented in Table 3–3 for the original Edwards categories, plus the new group of service workers.

The major occupational groupings do form a rough scale on the three dimensions. When the categories are arranged as they are in the table, the prestige and education rankings are fairly consistent, but the rankings on earnings do not always coincide with the other two dimensions. For instance, managers earn more than professionals; workers in the blue-collar categories of crafts and operatives outearn some white-collar workers.

Another problem is disguised by the median statistics employed in the table. In practice, some of the categories are quite broad and include specific occupations that are not very similar. This creates particular problems for stratification analysis, especially in regard to the upper strata. For example, the professional grouping includes doctors and engineers, who rank high on all three dimensions. However, the most numerous professionals by far are public school teachers, who are characterized by modest earnings and prestige. Another example is the managerial category, which is large enough to encompass both the owner of a small hamburger stand and the president of a multinational corporation.

The Edwards schema, then, is less than perfect as a stratification measure. However, it is doubtful that Edwards or his successors at the Census Bureau could have done much better, given the difficulty of reducing the multitude of occupations represented in the work force (the government's *Dictionary of Occupational Titles* lists over 20,000 of them) to a simple set of

TABLE 3–3 Occupational Groups: Earnings, Education, and Prestige

Occupational Group	Median Income of Employed Workers (in dollars, 1989)	Median Years of Education (25–64 years old, 1987)	Median Prestige Score
Managers and administrators	30,508	14.6	75
Professionals	28,557	16.2 ⎫	82
Technical workers	23,021	14.2 ⎭	
Sales workers	14,538	13.5	67
Clerical workers	14,863	13.0	65
Craftsmen	21,443	11.9	68
Operatives	16,708	11.3	58
Laborers, except farm	8,810	11.2	46
Service workers	7,471	11.8	47
Farm workers	7,089	11.0	61/50*
All workers	16,943	13.2	

*First figure refers to farmers, second to farm laborers.
SOURCES: U.S. Census 1990a: 41–42; U.S. Labor 1990b: 43; Reiss 1961: 68.

categories. The original purpose of Edwards' scheme was to organize occupational data in order to make broad historical comparisons, and it
serves well for that purpose.

Table 3–4, which is based on Edwards' compilation and more recent
census data, reveals massive changes in the American occupational structure since 1870. The last set of figures, for the year 2000, are projections
made by the Labor Department's "occupational outlook" experts. The largest change by far is also the most predictable: Farm occupations have declined from over half the labor force in 1870 to under 3 percent in recent
years. As the percentage of agricultural workers declined, the proportion
of operatives, the grouping that includes most factory workers, expanded
until it became the single largest occupational category by 1950. However,
there is a small but significant surprise here. After 1950, the proportion of
operatives began to decline. This happened in spite of the fact that manufacturing production was expanding steadily. The other two major blue-
collar categories, craftsmen and nonfarm laborers, have followed a similar
curved trajectory, although the proportional decrease in laborers began
much earlier.

While the agricultural categories have continuously contracted and the
blue-collar groupings first expanded and then contracted, white-collar employment has grown steadily, as shown in Figure 3–1. By 1980, white-collar
workers constituted over half the labor force. Two classifications within
that group have grown especially rapidly in recent decades, professional/
technical and clerical, both having increased their proportion by about
fivefold.

The relatively modest growth of a third white-collar grouping, the

TABLE 3–4 Occupational Structure of the United States: 1870 to 2000

	Percent of the Labor Force							
	1870	1900	1930	1950	1972	1972*	1990	2000
Occupation:								
Professionals and technicians	3	4	7	9	14	13	17	17
Managers, officials, and proprietors	6	6	7	9	10	9	13	11
Sales workers ⎱	4	5	6	7	7	10	12	11
Clerical workers ⎰		3	9	12	17	16	16	16
Craftsmen and foremen	9	11	13	14	13	13	12	11
Operatives	10	13	16	20	17	16	11	9
Laborers, except farm	9	13	11	7	5	6	4	4
Service workers	6	9	10	10	13	13	13	16
Farmers and farm laborers	53	38	21	12	3	5	3	2
Total (rounded)	100	100	100	100	100	100	100	100
Number in labor force (millions)	12.9	29.0	48.7	59.0	81.7	82.1	117.9	136.2
Percent of labor force female	15	18	22	28	38	38	45	47

*The last three columns, based on the 1980 Census classification of occupations, are not directly
comparable with the first five, which are consistent with the 1970 Census classification.

sources: 1870 from Edwards 1943; 1900–1950 from U.S. Census 1975; 1972 from U.S. Census 1973
and U.S. Labor 1984; 1990 from U.S. Labor 1991; and 2000 from U.S. Labor 1990a.

FIGURE 3–1 Occupational Distribution of the Labor Force: Percent of Workers

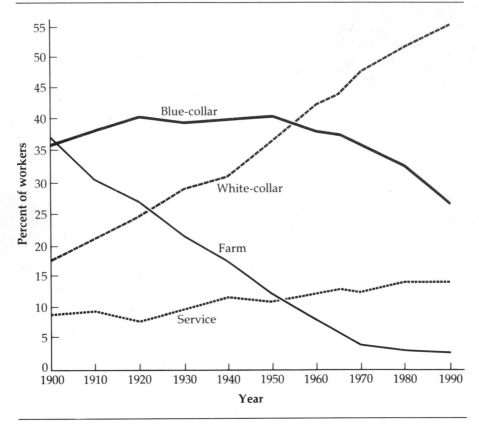

SOURCES: U.S. Census 1976: 387; U.S. Labor 1990b: 183–88.

managerial occupations, obscures a continuing evolution. The decline of the independent entrepreneur and complementary increase of salaried managers, noted by Mills in the period 1870–1940, has not abated. The proportion of managers who were self-employed dropped from 57 percent in 1957, to 27 percent in 1970, and finally to 9 percent in 1983 (Levitan and Taggart 1976: 67; U.S. Census 1984a: 395). Incidentally, that last figure of 9 percent approximates the proportion of the labor force as a whole that is currently self-employed (U.S. Census 1990c: 386). The United States has truly become a nation of employees.

The service worker category has also changed in composition as it has grown. At the beginning of the century, most of the workers falling under this rubric were domestic servants (or in the more neutral current language of the Census Bureau, "private household workers"). Today the largest and most rapidly expanding service occupations include hospital attendants, waiters and waitresses, janitors, and security guards. While making a hospital bed or washing windows in an office building may be no more satisfying than performing similar tasks in a private home, the change does imply a more democratic set of class relationships. Of course, there are still some privileged American households where domestic chores are done by hired servants. In 1988, the Census Bureau found nearly 1 million private household workers, enough people to fill a good-sized city (U.S. Census 1990c: 391).

There are several keys to these long-term patterns of occupational change. Earlier we noted the growth of large bureaucratic organizations associated with the modern corporation and the expansion of government. This tendency increases the demand for managers; certain professionals, such as accountants; and, above all, clerical workers. The phenomenal expansion of clerical employment is also linked to the increasing participation of women in the labor force, as noted above.

However, the fundamental cause of changes in the occupational structure since 1870 has been the transformation of the United States from an agricultural to an industrial and finally to a postindustrial society. In 1870, agriculture was still the mainstay of the economy, and most working Americans were farmers. By 1930, the United States was a mature industrial society, in which employment in manufacturing exceeded that in agriculture. As early as 1950, the United States could be described as a postindustrial society[2]—that is, one in which most workers are *not* employed in the goods-producing industries of agriculture, manufacturing, mining, and construction.

Both shifts were conditioned by technological change. The transition to industrial society was predicated on new technology, which substituted machine power for animal and human muscle. The emergence of a postindustrial society depended on continuing technological advances, which steadily increased productive efficiency and enabled a shrinking proportion of the labor force to meet the material needs of the rest of the population. This dependence on technology, of course, has increased the demand for scientific and technical personnel. Engineering, for instance, is now among the largest of professional occupations, exceeded in numbers only by teaching (U.S. Census 1990c: 389).

As the proportion of the labor force employed in producing goods has declined, employment has expanded in such areas as retailing, government, and the so-called service industries. By 1975, the last category, which includes the restaurant and hotel industries, education, health care, and legal services, accounted for more employment than agriculture and manufacturing combined (Levitan and Taggart 1976: 63). The net result of these changes has been to move workers out of the sectors that require large numbers of blue-collar production workers and into those that depend more on white-collar or service workers.

Technological change is not the only force driving occupational restruc-

---

[2]On the origins of this term, see Daniel Bell, *The Coming of the Post-Industrial Society* (New York: Basic Books, 1976), 13–40.

turing. Since the 1970s, American corporations have been struggling to adjust to sharply increased competition from producers in Europe and Asia. Their response has centered on bolstering sagging profits by reducing labor costs (Harrison and Bluestone 1988). Thus, many American companies have closed plants in the U.S. and shifted some of their operations to low-wage areas abroad. Alternatively, U.S. manufacturers buy components produced abroad or made by smaller, low-wage subcontractors in this country. The 1980s also saw a decline in the power of labor unions (see Chapter 9), making it easier for corporations to cut their labor force and compel wage concessions from their remaining workers.

These corporate strategies imply fewer domestic jobs in manufacturing and more low-wage positions throughout the economy. Harrison and Bluestone (1988: 122–123) tested this assumption by comparing the new, full-time jobs created in the 1960s, 1970s, and 1980s. They found that the proportion of new jobs paying under $11,000 (in 1986 dollars) was increasing from decade to decade.

WOMEN WORKERS AND PINK-COLLAR JOBS

A dramatic increase in the proportion of female workers has accompanied the emergence of the postindustrial economy. At the beginning of the century, women constituted only a tiny proportion of the labor force. As late as 1950, only 28 percent of all workers were women, but in recent decades women's labor force participation has accelerated, while the participation of men has begun to decline. By the end of the century, women will nearly match men's labor force participation and constitute close to half the labor force (Table 3–4; Blau 1984: 302).

Women were drawn into the labor market by the ballooning demand for clerical workers and later for service workers. In the 1970s and 1980s, many women joined the labor force in order to bolster family incomes that were being eroded by the declining real wages of male workers, combined with rising unemployment rates. The pull of economic forces was reinforced by the push of social forces: changing attitudes toward the employment of wives, a declining birth rate, and an increasing divorce rate (with concomitant growth in the proportion of female-headed families) have encouraged women to seek work outside the home.

In recent years, young women have increasingly educated themselves for attractive managerial and professional positions, at the same time that affirmative action programs have reduced the discriminatory barriers that created male occupational bastions. From 1970 to 1989, the proportion of females among people classified by the Census Bureau as managers grew from 18 to 40 percent; the proportion of females among physicians increased from 8 to 18 percent; and the proportion of females among lawyers climbed from 5 to 22 percent. Of course, we would expect some increase simply because of the expanding proportion of women in the labor force.

Women's overall situation in the workplace has changed less dramatically over the last two decades than might be inferred from the numbers of new female managers, doctors, and lawyers. In 1989, 59 percent of working women were employed as clerical, sales, or service workers—just 3 percent less than in 1970. The occupational segregation of women is even greater than might be assumed from their placement in the major occupational categories. For example, women appear to be amply represented among professionals (Table 3–5). In fact, nearly two thirds of female professionals are employed as nurses, teachers, social workers, or librarians. The high-prestige, high-income professions, including law, medicine, engineering, and architecture, are still dominated by men (U.S. Census 1984c: 166–175; U.S. Labor 1990b: 183–188).

In the labor force as a whole, women are crowded into a relatively small number of "pink-collar" occupations—fields that are almost entirely female. Secretaries, cashiers, hairdressers, and elementary school teachers are among pink-collar workers. A close look at the distribution of women in the approximately 300 occupations covered in the Labor Department's employment surveys reveals the following: 45 percent of women workers are employed in occupations that are over 80 percent female; over half work at occupations that are at least 70 percent female (U.S. Labor 1990b: 183–188).

Pink-collar jobs typically offer lower pay and lower prestige than other positions requiring similar levels of education and training. They seldom involve any exercise of authority and provide only limited opportunities for job advancement. A junior executive expects to move up a job ladder in the course of his or her career. A young secretary can have no such illusion (Howe 1977; Blau 1984).

The pink-collar phenomenon is of growing importance for the class system, because American families are becoming increasingly dependent

---

TABLE 3–5 Occupational Distribution by Sex, 1990

	Percent of Workers	
	Males	Females
Managers	13.8	11.1
Professionals	12.0	15.1
Technicians	3.0	3.5
Sales workers	11.2	13.1
Clerical workers	5.9	27.8
Craftsmen	19.4	2.2
Operatives	14.3	6.8
Service workers	9.8	17.7
Laborers, except farm	6.2	1.6
Farm occupations	4.4	1.0
Total	100.0	100.0
Number (millions)	64.3	53.0

SOURCE: U.S. Labor 1991.

on working women. In Chapters 4 and 10, we will see that pink-collar employment is one of the keys to understanding the skewed distribution of income and the persistently high poverty rates of recent years.

TRANSFORMATION OF THE BLACK OCCUPATIONAL STRUCTURE

Table 3–6 reveals a sweeping transformation of the black occupational structure over a relatively short period of time. In 1940, 80 percent of black workers were still concentrated in the four lowest occupational categories. By 1980, nearly 70 percent were in the upper six categories. The occupations of black workers had shifted even more rapidly than those of whites. As a result, the occupational distribution of blacks had moved much closer to that of whites.

The change for black women had been even more dramatic than this general picture for both sexes. As late as 1960, a third of employed black women cleaned white people's houses. Only a small percentage worked at the white-collar jobs that were typical of white women. By the 1980s, about half of black female workers held white-collar positions (Farley 1984: 48–49).

Black workers were affected by the same broad processes of socioeconomic change that were altering the world of white workers, but in ways that were peculiar to an oppressed minority. In the 1930s and 1940s, under the pressures of new industrial unions and the needs of economic mobilization for World War II, the discriminatory barriers that had kept blacks out of many industrial jobs began to fall. With the mechanization of southern agriculture in the 1940s and 1950s, large numbers of black sharecroppers and farm workers were forced off the land, typically into urban slums and low-wage urban employment. The civil rights movement of the 1960s pro-

---

TABLE 3–6   Occupational Structure of Black Workers, 1940–1980

	Percent of Workers		
	1940	1960	1980
Professionals and technicians	3	5	10.9
Managers, administrators, and proprietors	1	2	4.5
Sales workers ⎫	2	8	2.7
Clerical workers ⎭			18.5
Craftsmen and foremen	3	7	9.7
Operatives	10	21	20.1
Laborers, except farm	14	14	7.4
Service workers	34	34	⎫ 24.4
Farmers	15	3	⎬ 1.7
Farm laborers	17	5	⎭
Total	100	100	100

SOURCES: U.S. Census 1980: 75: U.S. Census 1983: 417.

duced antidiscrimination legislation and affirmative action programs in government and private industry, which opened many new jobs to blacks. At the same time, blacks were closing the educational distance between themselves and whites. In 1940, the average educational gap between blacks and whites in their late 20s was more than three years of schooling. By the early 1980s, the gap for this same age group had been reduced to half a year (Farley 1984: 17–18). As a result, many young blacks were able to qualify for attractive white-collar jobs.

Despite the progress of recent years, a considerable occupational gap remains between blacks and whites. Relative to whites, blacks are still underrepresented at the top of the occupational structure and overrepresented at the bottom (Table 3–7). There are, to take an extreme example, two white hospital orderlies for every white physician, but twenty-five black orderlies for every black physician (U.S. Labor 1990b: 183, 186). As this example suggests, the benefits of change have been unevenly distributed among black Americans. By 1990, the proportion of black professionals and managers was almost five times what it had been in 1940, but service jobs remained the largest single occupational category in 1990, as it had been 50 years earlier. Furthermore, in the early 1980s, many black workers had no job at all. Black unemployment had reached historically unprecedented rates, in excess of 15 percent, and was higher for young workers (U.S. Census 1990c: 396). Moreover, many black people of working age were not counted as unemployed by the government, because they had given up hope of finding a job and had stopped looking.

The net result of these diverse trends has been an increasing class differentiation within the black population. While large numbers of young, well-educated workers are moving into jobs that were open to only a few of their parents, the underclass of low-wage or unemployed workers may

TABLE 3–7 Occupational Distribution by Race, 1989

	Percent of Workers	
	Black	White
Managers	7.6	13.2
Professionals	8.6	13.8
Technicians	2.9	3.1
Sales workers	7.5	12.5
Clerical workers	17.5	15.4
Craftsmen	9.1	12.1
Operatives	16.0	10.6
Service workers	22.9	12.2
Laborers, except farm	6.7	3.9
Farm occupations	1.7	3.0
Total	100.0	100.0
Number (millions)	12.0	105.4

SOURCE: U.S. Labor 1990b: 183–188.

actually be growing. The implications of these trends are summarized by William J. Wilson:

> the impressive occupational gains made by Blacks during these [last] three decades have been partly offset by the effects of basic structural changes in our modern industrial economy, changes that are having differential impact on the different income groups in the Black community. Unlike more affluent Blacks, many of whom continued to experience improved economic opportunity even during the recession period of the 1970s, the Black underclass has evidenced higher unemployment rates, lower labor-force participation rates, higher welfare rates, and, more recently, a sharply declining movement out of poverty. The net effect has been a deepening economic schism in the Black Community that could very easily widen and solidify (Wilson 1980: 142).

Wilson makes a final point when he says (p. 182): "Many Blacks and white liberals have yet to recognize that the problem of economic dislocation is more central to the plight of the Black poor than is the problem of purely racial discrimination." We will return to that theme in Chapter 10 in the discussion of poverty in America.

OCCUPATIONAL CHANGE AND THE FUTURE OF THE CLASS STRUCTURE

How are the recent occupational changes affecting the American class structure? We have already shown how industrialization restructured the upper class and created an urban working class and a bureaucratic middle class of salaried white-collar workers. Many social scientists expect to find changes of similar magnitude associated with the emergence of a post-industrial society. The most optimistic of them see evidence of a more egalitarian class system taking form in recent occupational trends. Focusing on the growth of the professional, managerial, and clerical groupings and the shrinkage of the farm and nonfarm laborer categories, they anticipate the further decline of menial work and an overall upgrading of occupational skills and prestige. This line of thought was particularly common in the 1970s, when one writer celebrated the "white-collarization" of America, and another detected a "status upheaval" (Wattenberg 1974: 26; Bell 1976: 134). It might be inferred from such descriptions that the class structure is being reshaped to resemble an inverted pyramid, narrow at the bottom and broad at the top. Unfortunately this image finds little support in a closer examination of occupational trends.

The expansion of the professional-technical category, for example, does not indicate as large a proliferation of opportunities at the top of the class structure as some writers have assumed. The most significant growth in this occupational grouping has been among technicians and so-called semiprofessionals, such as teachers, librarians, social workers, accountants, and nurses, rather than doctors, lawyers, and other high-status professionals. From 1950 to 1980, technicians and semiprofessionals accounted for roughly two thirds of professional/technical employment growth (U.S. Labor 1981a). The Labor Department's projections suggest that this proportion will be maintained. The income and education associated with these positions are typically well below the averages for "professional and technical" workers. Moreover, the findings of the NORC occupational prestige surveys and Coleman and Rainwater's study of social standing in Boston and Kansas City rank semiprofessionals and technicians below people in the established professions and higher managerial posts (see Tables 2–1 and 2–2). In both the Coleman and Rainwater metropolitan prestige-class model and our own economic-class model (introduced in Chapter 1), these low-level professionals would be labeled middle class.

The less dramatic expansion of the managerial category may also leave a misleading impression of growth at the top. Available data on income, education, and prestige suggest that many of the new managerial opportunities are middle-ranking within the class structure (Tables 2–1 and 2–2). This conception would certainly fit the managers of small retail establishments or the relatively low-level government officials who are counted in this occupational grouping.

At the bottom of the occupational structure, the declining proportion of laborers in the work force has been taken by some as an indication of a shrinking lower class. The most recent data are not consistent with this view. It is true that both the number and the proportion of laborers, both on the farms and in the cities, have declined. At the same time, however, a new category of people who do similar unskilled work for low pay and uncertain employment has been growing: cleaning workers and some others within the service category. If we grouped together all farm and nonfarm laborers, cleaning workers, private household workers, and hospital attendants, and considered them as persons doing menial labor, we would find that the number of such workers held about steady from 1950 to 1970 and then began to increase.

We have scrutinized trends at the top and the bottom of the occupational structure. What can we say about developments in the middle? On the blue-collar side, we have noted a significant contraction of craft and operative positions since 1950. The Labor Department expects this trend to continue. As these categories shrink, some of the best-paid jobs available to people without advanced education are being lost. (Of course, many unattractive, poorly remunerated jobs are also disappearing, especially among operatives.) On the white-collar side, the explosive growth of clerical employment is expected to slow, largely due to automation. Nonetheless, clerical workers will continue to be the largest single occupational group, accounting for nearly one in five workers.

The erosion in the economic and social status of lower white-collar workers, which Mills noted, continued in the decades after he wrote. The average earnings of the largely female clerical and sales labor force are far below those of the upper white-collar categories (Table 3–3). As early as 1963, the second NORC survey showed that having a white-collar job had ceased to make a contribution to occupational prestige independent of education and income (Glenn 1975: 187). As noted earlier, the very expansion of office employment contributed to the decline of clerical prestige. In a labor force that is over half nonmanual, a white collar becomes less distinctive and finally less valued. At the same time, the character of office labor is changing, and this also contributes to its declining status. Office work is being subjected to the same process of routinization for mass production purposes that marked the advance of the modern factory. Again occupational skills are being fractured into simple, monotonous operations that can be quickly mastered by untrained workers. The introduction of computerized operations has furthered this development in the office, just as the assembly line did in manufacturing (Braverman 1974: 293–358). For example, when a multi-branched bank computerized its accounting procedures, it reduced a relatively highly skilled bookkeeping staff of 600 to 150 and expanded its data-processing staff to 122. The vast majority of data-processing jobs involve low-paid, minimal-skill work such as encoding, which requires nothing more than the simple ability to operate a 10-button keyboard (Braverman 1974: 338).

The essence of this process in both factory and office is the separation of execution from conception. Initiative is transferred from factory operatives or clerical workers to administrative and technical personnel, who plan each work task in fine detail. Thus, the expansion of the high white-collar categories is directly related to the creation of low-skilled white-collar and blue-collar jobs.

Another way of looking at occupational changes is to focus on the specific occupations that are expected to account for new job creation. According to Labor Department projections, forty occupations will account for roughly half of the new job creation between 1988 and 2000. All were very large occupations by 1982, so even the ones that are growing slowly can provide large numbers of new jobs.

The forty are listed in Table 3–8. We have divided them into three broad levels on the basis of the preparation required. Low-skill occupations account for the largest proportion of new employment. For all the talk of the dawning of an "information age," more new janitors are projected than computer programmers and systems analysts combined. New hospital orderlies are expected to outnumber new physicians by more than two to one. New waiters and waitresses should outnumber new lawyers by three to one. As these examples suggest, occupations associated with the expanding service industries, from doctors to food counter workers, predominate on the list. Only a tiny fraction of the new jobs are connected with goods production.

The occupational trends associated with postindustrial society do not, then, prove that we are moving toward the more egalitarian class system illustrated by the image of an inverted pyramid. Nor do they support the notion of a consistent upgrading of occupational skills. Americans are cer-

TABLE 3–8 The Forty Largest-Growing Jobs, 1988–2000

Occupation	Number of New Jobs (thousands)	Percent
Professional and managerial	2,855	28
Registered nurses	613	
General managers and top executives	479	
Teachers, secondary school	224	
Computer systems analysts	214	
Accountants and auditors	211	
Teachers, kindergarten and elementary	208	
Lawyers	181	
Electrical and electronic engineers	176	
Food service and lodging managers	161	
Physicians	149	
Financial managers	129	
Social workers	110	
Technical and skilled	3,059	30
Secretaries, except legal and medical	385	
Truck drivers, light and heavy	378	
Receptionists and information clerks	331	
Computer programmers	250	
Licensed practical nurses	229	
Maintenance repairers, general utility	202	
Carpenters	176	
Home health aides	161	
Cooks, restaurant	156	
Teacher's aides and educational assistants	145	
Clerical supervisors and managers	136	
Blue-collar worker supervisors	133	
Electrical and electronic engineering technicians	130	
Automotive mechanics	127	
Medical secretaries	120	
Low skill	4,365	42
Salespersons, retail	730	
Janitors and cleaners	555	
Waiters and waitresses	551	
General office clerks	455	
Nursing aides, orderlies, and attendants	369	
Cashiers	304	
Guards	255	
Food counter, fountain, and related workers	240	
Food preparation workers	233	
Child care workers	186	
Gardeners and groundskeepers, except farm	183	
Stock clerks, sales floor	174	
Dining room, cafeteria attendants, and bar helpers	130	
	10,279	100

Figures are based on "moderate growth" projections.
SOURCE: U.S. Census 1990c: 392.

tainly better educated on the average than they have been in the past, but that is not to say that they are always able to use their education on the job.

Many are caught in the trap described by a job-design consultant who observed, "We may have created too many dumb jobs for the number of dumb people to fill them" (Braverman 1974: 35).

The implications of current occupational trends for the class structure are too complex to by conveyed by a single graphic image. However, what is especially striking is the tendency of a technologically sophisticated economy to create large numbers of low-skilled, low-wage jobs. Taken together with the loss of desirable blue-collar jobs, this trend implies a growing gap in the class structure between the working class (along with the working poor) and the classes above them.

CONCLUSIONS

This chapter has been devoted to tracing the evolution of the occupational structure and drawing out its implications for the class system. We have seen the country transformed from an agricultural to an industrial and finally to a postindustrial society. In the first stage of this process, the farm population dropped precipitously, while in the cities, an industrial proletariat grew and a stratum of white-collar workers appeared. In the second stage, the farm sector was reduced to numerical insignificance, the industrial proletariat began to contract, the white-collar sector swelled and became internally differentiated, and a new stratum of service-oriented menial workers developed. Both stages saw the decline of independent entrepreneurs and their replacement by salaried employees. The new class structure that took form as a result of these shifts can be described in terms of the class model we introduced at the beginning of this book: a national *capitalist class* built on corporate wealth; an *upper-middle class* of college-educated professionals and managers; a *middle class* of white-collar and some blue-collar employees, with significant job skills and relative independence on the job; a *working class* of those who work at highly routinized, closely supervised jobs; a *working-poor class* of those who work at low-wage, menial jobs; and *an underclass,* whose participation in the labor force is intermittent at best.

Our examination of recent occupational trends suggests that the working class may be shrinking while the working-poor class expands, contributing to increased polarization in the class structure. We certainly find little support for claims that our postindustrial society is evolving toward a more egalitarian class order.

There is, of course, more to a class system than the distribution of positions in an occupational hierarchy. The material we have presented here needs to be supplemented with other information—in particular, data on trends in the distribution of wealth and income and on patterns of mobility—before we can draw final conclusions about the direction of change in the class structure.

Before closing this chapter, it is appropriate to add a few observations on the political implications of the developments we have outlined. Since

Marx, observers of Western society have been keenly aware of the importance of the political role of the industrial working class. As we will observe in greater detail in Chapter 9, this class has provided the social basis for the emergence of socialist movements in many European countries and for liberal politics in the United States. However, the weight of the traditional blue-collar categories within the labor force is declining. Does this tendency foreshadow a shift to the right in the politics of postindustrial societies? The answer to this question will largely depend on the political behavior of those who work in the occupations that are growing as blue-collar employment contracts. Service and clerical workers could conceivably provide the basis for a new "working-class" political block. Professionals and managers could join the established propertied class in a new conservative alliance. Nothing is certain except the fact that occupational change will be one of the major forces reshaping American politics in the coming years.

SUGGESTED READINGS

Bell, Daniel. 1976. *The Coming of Post-Industrial Society.* New York: Basic Books.
   Broad and optimistic interpretation of postindustrial society, emphasizing changes in technology, economic organization, and occupational structure.

Braverman, Harry. 1974. *Labor and Monopoly Capital.* New York: Monthly Review Press.
   Key work on the division of labor and the transformation of work in capitalist industrial societies. Challenge to Bell.

Harrison, Bennett, and Barry Bluestone. 1988. *The Great U-Turn: The Corporate Restructuring and Polarizing of America.* New York: Basic Books.
   The transformation of the American economy in the Reagan era and its distributive impact.

Jaynes, Gerald David, and Robin M. Williams, Jr., eds. 1989. *A Common Destiny: Blacks and American Society.* Washington, D.C.: National Academy Press.
   A comprehensive study of black Americans by an outstanding panel of social scientists. Emphasis on change over the last 50 years.

Stromberg, Ann, and Shirley Harkess. 1988. *Women Working: Theories and Facts in Perspective.* 2nd ed. Mountain View, Calif.: Mayfield.
   Handy source of material on women in the labor force.

Terkel, Studs. 1974. *Working: People Talk about What They Do All Day and How They Feel about It.* New York: Avon.
   A provocative series of interviews with people, from ranking executives to unskilled laborers.

Wilson, William J. 1980. *The Declining Significance of Race.* 2nd ed. Chicago: University of Chicago Press.
   Emphasizes that changes in occupational structure may be more important for blacks than traditional forms of prejudice.

4 Wealth and Income

Let me tell you about the rich. They are different from you and me.

F. Scott Fitzgerald (1926)

Yes, they have more money.

Ernest Hemingway (1936)

Procrustes, a giant of Greek mythology, was in the habit of altering the stature of his guests to fit the length of the available bed, by either stretching them or chopping off their legs. In the next few pages, we will apply Procrustes' approach to the study of income. We have put together an imaginary parade in which the heights of the marchers are made proportional to their incomes (Pen 1971: 48–59). The parade is a convenient way of gaining an overview of the distribution of income (who gets how much?), our first concern in this chapter. Later we will take up the distribution of wealth, the causes of income inequality, and finally the personal effects of economic class position.

THE INCOME PARADE

The income procession is organized as follows: The marchers will represent the 93 million U.S. households counted by the Census Bureau. (Most of the households are family units, but about 29 percent consist of unrelated people sharing a home or individuals living alone). Each household will be counted as a single marching unit and placed in the ranks according to its total pretax money income from all sources. Marchers who receive the average (mean) household income (about $37,000 in 1990) will be of average height. Those who receive more or less than the average will, in good Procrustean fashion, be stretched or shrunk proportionally, so that those getting $75,000 will be about 12 feet tall and those earning $20,000 will be nearly three feet tall. (The parade is based on data from U.S. Census 1991a; U.S. Internal Revenue Service 1990.)

   We will have a special rule for households with more than one income earner: The chief income-earner will carry the others on his or her shoulders, so that their total height will represent their combined income and determine their place in line. Typically this will mean that husbands will be carrying their employed wives. Househusbands and housewives who are not in the labor force will be of the same height as their working spouses and walk inconspicuously by their sides. In all cases, husbands and wives will count as single marching units, and children will be left home.

One last thing: Since we want quick a impression of the distribution of income, we will make the entire procession pass by our reviewing stand at a uniform pace, in exactly one hour. This will be rough on the marchers, but it has a particular advantage for us. At any moment, we will be able to tell how much of the parade has gone and how much is to come just by looking at our watches. Let's begin the parade with the shortest marchers, the income pygmies, and work up to the giants.

The procession opens on an odd note. In the first few seconds we see nothing except a few wisps of hair moving across the horizon. It seems that the first people are marching in a deep ditch. They do not appear above ground because they are business people who have suffered income losses during the past year and have had to borrow from the bank or use up their own capital to cover them. Given the high failure rate of small businesses, we should not be surprised at this sight, however peculiar.

Five Minutes.   After they pass, we see tiny people, the size of a match or a cigarette. Five minutes into the parade, the marchers are still one foot tall, since they receive around $6,500 a year. All who have gone by so far (and many who are to come, depending on the size of their families) are "poor" by the federal government's official poverty standard. There is a notable overrepresentation of women among these dwarfs. Nearly one fifth are female heads of families who are single, divorced, widowed, or separated. A similar proportion are elderly women living alone. The majority are heavily dependent on government income-transfer programs such as public assistance or social security. Many of the marchers are malnourished, since their food needs were not reduced when we shrunk them to their present size. They could get government food stamps (which, by the way, are not counted by government statisticians as part of money income), but many of those who qualify do not receive stamps, either because they are too proud to accept them or because they are ignorant of their rights. There are also some eligible people who do not collect public assistance payments (which *are* included in income, since they are distributed in cash).

Blacks and Hispanics show up in disproportionate numbers in the first part of the parade. Nevertheless, the majority of the dwarfs who lead the procession are white and "Anglo." Actually, the single most encompassing characteristic of the dwarf marchers is that they are not employed, because they are old or disabled, temporarily out of work, studying, home with children, or not especially interested in working. Yet there is a substantial minority that does work, and among them are many people who work full-time, all year long, without exceeding dwarf height; they simply do not get paid much for their labor.

Twenty Minutes.   As the procession moves on, the marchers get taller, but only very gradually, in spite of the breakneck pace we have imposed. After twenty minutes, we are still seeing three-foot midgets who earn about $20,000 a year.

What sort of lifestyle do these midgets buy with their money? The answer depends in part on household size, since the same income will provide more comfort for a few people than for many. Families of four represented at this point in the parade lead austere lives. Most live in rented homes, since they cannot afford to buy. They own older cars, purchased used, or depend on public transportation to get around. Housing, transportation, food, and energy expenditures consume most of the available income, leaving little for discretionary extras. At best, a family vacation means a trip to visit relatives in another city. People in this part of the parade feel that their standard of living is precarious and would be seriously threatened by a few weeks of unemployment or an unexpected medical bill.

Thirty to Thirty-Five Minutes.   At half past the hour, we catch sight of little people, about four feet eight inches, who earn the median income of $30,000. (The median is the amount dividing the bottom 50 percent from the top 50 percent in any distribution.) Five minutes later, we notice marchers whom we can look in the eyes, assuming, of course, that we ourselves receive the average (mathematical mean) income of $37,000 and are therefore of average height. A careful count reveals that minorities are slightly underrepresented in this part of the parade, and female-headed households are sharply underrepresented. Many of these average-sized marchers have lower-level managerial or professional positions. Others hold one of the better-paying blue-collar jobs. A high proportion of the households represented here depend on two incomes. This is especially true of the black and Hispanic families that manage to attain this income level.

These people live substantially better than the midgets we saw fifteen minutes ago. The majority of families in this group own modest but comfortable homes (though younger families may find themselves priced out of the housing market). Late-model cars are common, and many households have two cars. Families still worry about exceeding their monthly budgets, but there is more money for luxuries such as movies, sporting events, beauty care, and annual vacations.

Forty-Five to Fifty-Five Minutes.   Ten minutes after the average-sized people pass in review, we see eight-foot marchers who might be of interest to a basketball scout. However, few of them would make the team, since the majority would have to play mounted piggyback. The same is true of the twelve-foot giants who appear ten minutes later. Although most of the heads of households marching among them hold good-paying professional and managerial jobs, families still depend on two or more incomes to reach the $80,000 required of a giant. Surprisingly, many of the heads of these households are blue-collar workers. Occasionally we see blacks or Hispanics, or even a few women who support families alone, but their representation has certainly thinned out in the last twenty minutes.

The relative affluence of the giants permits them to live much more gracefully than the average-sized people. All but a few own homes, which are larger and more elegantly furnished than the homes of shorter marchers. The household budget allows for new cars, domestic help at home, and fashionable clothing. There is a generous allowance for family recreation. Above all, household members can spend without constantly worrying that they are running up bills they will not be able to cover.

The Final Minutes.   The procession has only two minutes to run. Yet some of the most extraordinary moments lie ahead. It took nearly twenty minutes before we saw three-foot midgets and another thirty-five minutes before the twelve-foot giants appeared, but now we are looking at sixteen-foot super-giants. In about a minute and a half, we will be looking at fifty-foot Goliaths, seconds after that, 200-foot King Kongs, and then towering leviathans, thousands of feet tall. In fact, if we look down the line at the people who have yet to pass, it appears as if a steep mountain peak is advancing on our reviewing stand.

In these last minutes of the procession, the character of the marchers is changing rapidly. Instead of depending on two high salaries per household, as did most of the giants we saw a moment ago, the approaching Goliaths rarely have working spouses. The Goliaths, for whom $350,000 is a respectable income, are generally highly successful professionals (most often lawyers and doctors), corporate executives, or the owners of prosperous small enterprises. Another shift comes with the arrival of the King Kongs, who would consider $1 million a modest living. There are perhaps 65,0000 couples and individuals with incomes in excess of this amount; they rush past us in the last seconds of the parade. Many of these people hold important positions; for instance, the earnings of the top officers of the biggest corporations place them in the ranks of these mammoth marchers. However, the greater part of income at these levels does *not* come from jobs, but rather from income-producing assets. It is return on wealth, not salaries from employment, that accounts for their colossal incomes and overpowering stature.

The Final Instant.   Within the last microsecond of the parade, the leviathans we spotted earlier flit by. There are a few hundred people in this category whose incomes in excess of $20 million make the smallest of them over one-half mile tall. Who are these monsters? About one third are people who have inherited their wealth. They include such familiar names as Rockefeller, Ford, Mellon, and Du Pont. Others built their fortunes on smaller inheritances or even started with nothing and achieved leviathan status through activities ranging from the marketing of innovative products or services to real estate or financial speculation or the discovery of oil.

What do millions of dollars in income buy? Let's take a concrete case: Ray Kroc, the McDonald's magnate who put together a Big Mac fortune of several hundred million dollars. Before he died in 1984, Kroc accumulated: "a 210-acre spread in Southern California, a stadium-sized apartment on Chicago's Lake Shore Drive, a 90-foot yacht, four helicopters, and a luxurious Florida beach house whose doorbell chimed familiarly: 'You deserve a break today' to all callers" (Greene 1978: 211).

Multiple residences are fairly common among the wealthy, but other elements of Kroc's lifestyle might be a bit ostentatious for the current tastes of the established rich. But consider this: The staff required for a proper New York City townhouse, according to one of Manhattan's oldest domestic employment agencies, consists of a cook, a housekeeper, a butler, "a lady's maid," a laundress, and a chauffeur. The annual cost of these services in 1990 was approximately $165,000—an amount exceeding the income of 99 percent of American households.

LESSONS FROM THE PARADE

This chapter will elaborate some of the themes introduced by the procession. But before going on, we should pause briefly to list the general lessons that can be drawn from what we have just seen.

1. Our most general impression is one of extremely gradual increase in income levels until a break point very late in the procession. At half past the hour, we were still looking at runts four feet eight inches tall. The slow climb continued until the final minutes of the parade, when heights rose precipitously as the small numbers of Americans with very high incomes, and finally colossal incomes, strode by.

2. The parade tells us something about the relative welfare of different segments of the population. It took about twenty minutes to reach the austere and precarious living standard of the $20,000 midgets and another fifteen minutes before we saw the more secure, average-income people. Although TV sitcoms and soap operas often suggest that the lifestyles of $100,000 super-giants are typical for Americans, the super-giants did not appear until the parade was almost over.

3. The degree of participation of income units in the labor force turned out to be a critical determinant of their place in line. At the very beginning of the parade, there was a high proportion of families and individuals without any wage or salary income. Later in the parade, we noted that many of the families with relatively high incomes are dependent on the earnings of two working spouses to maintain their standard of living. This generalization applies with particular force to minority families.

4. Wages and salaries are virtually the only source of income for most people. However, in the course of the parade, we noted shifts in the relative importance of different income sources. For many of the early marchers, government transfer payments, such as social security and public assistance (welfare), are crucial. Most of the remaining marchers depend on wage or salary income from their jobs or entrepreneurial income from small businesses and professional practices. However, in the final moments of the procession, we saw marchers who are supported by wealth in the form of income-producing assets, such as stocks, bonds, and rental property. The succession of income sources toward the end of the parade is evident in Table 4-1, which is based on 1988 taxable income data supplied by the Internal Revenue Service. Note that the relative importance of wages and salary income declines above $100,000, while the significance of capitalist and small-business income grows. For those who earn above $1 million, most income is from capitalist sources.

5. Occupation, the marchers showed us, is crucial to the distribution of income, though not the overpowering cause we might have anticipated. From the reviewing stand, we saw unskilled workers gradually give way to skilled workers and then to managers and professionals. But there was considerable overlapping of occupational categories. We saw managers relatively early in the parade, and a few blue-collar workers marched with the giants. One reason for this is that a two-income blue-collar household can easily outearn a low-level manager who does not have a working spouse. Another is that two people working at the same occupation may be paid very differently.

For example, in 1987, the bottom fifth of accountants employed full-time, year-round, earned under $23,500, while the top fifth earned over $55,000. Among mechanics similarly employed, the bottom fifth earned under $10,500, while the top fifth earned over $28,000. Thus, some mechanics earned more than some accountants (U.S. Census 1989a: 160–161).

The effect of occupation on the overall income distribution depends not simply on the wages paid to particular occupations, but also on the structure of occupations within the economy. For example, the unconscionably low wages paid to farm workers will have little effect on the distribution of income, because there are very few farm workers. On the other hand, the proliferation of low-skill service jobs in recent years is likely to depress wages at the low end of the distribution.

6. Women were at a special disadvantage in the income parade. At the very beginning, when the marchers were ten-inch dwarfs, we saw many older women, typically widows. Female heads of families were a common sight among the low-income marchers. In recent years, their numbers have been increasing rapidly as a result of growing rates of marital breakup and illegitimacy. The majority of separated or divorced women do not receive alimony or child support, and few get substantial payments, even though the economic situation of men typically improves after separation

TABLE 4–1 Sources of Income, 1988

	Percent				
Income	Wage and Salary	Small Business	Capitalist	Others	Total
$1 to $20,000	69	3	11	17	100
$20,000 to $50,000	81	3	7	9	100
$50,000 to $100,000	78	6	9	7	100
$100,000 to $200,000	60	13	21	6	100
$200,000 to $500,000	52	15	29	5	100
$500,000 to $1,000,000	43	15	37	5	100
Over $1,000,000	26	19	52	3	100
Total	71	6	14	9	100

Income = adjusted gross income. The "$1 to $20,000" category excludes returns with no taxable income or returns with losses.

Small business = business or profession, "partnership and S-corporation," farms, and "sale of property other than capital assets."

Capitalist = interest, dividends, capital gains, trusts and estates, rentals and royalties, and "sales of capital assets reported on schedule D."

Others = pensions and annuities, state income tax refunds, alimony, taxable IRA distributions, unemployment compensation, and social security benefits.

SOURCE: Based on IRS data from Strudler and Ring 1990.

(Hoffman 1977). Women who enter the labor force to support their families often find themselves in the poorly remunerated pink-collar occupations discussed in Chapter 3. The median weekly earnings of women employed full-time are 70 percent of the earnings of their male counterparts (U.S. Census Bureau 1990c: 409). Even women with four years of college, employed full-time, year-round, make less than similarly employed men with only high school educations (U.S. Census 1990c: 455).

7. Blacks and Hispanics were also at a disadvantage in the parade. They were overrepresented among the early marchers, often by female heads of households. On the other hand, given their traditional position in American society, we were not prepared for the substantial number of them who marched among middle-income people. Later in this chapter, we will take a closer look at the experience of minority households.

8. The relationship between the distribution of income and the class structure is clear at the extremes but somewhat blurred in the middle. We can think of the problem in terms of the class model we introduced in Chapter 1. The little people at the very beginning of the parade, who have no employment income or work at very low-wage jobs, correspond to our underclass and working poor. The towering marchers we saw in the last two minutes of the parade belong at top of the upper-middle class or in the capitalist class. But thirty minutes into the procession, we found factory workers marching with middle-class managers. And we saw dual-income working-class marchers looking down on single-income upper-middle class marchers. In sum, the class structure as we have defined it (or, for that matter, as Coleman and Rainwater defined it for Boston and Kansas

City) does not exactly match the distribution of household income. The mismatch is greatest in the middle of the parade.

THE DISTRIBUTION OF INCOME

Although a symbolic parade of the adult population may seem like a reasonable means of expressing the national distribution of income, the data are usually published in more prosaic forms. The most common are (1) the distribution of the population among statistical income classes and (2) the distribution of aggregate or total income among various fractions of the population. The first, which was the basic source of data for our income parade, is illustrated in Table 4-2. The second, which we will examine presently, conceives of income as something akin to a national pie that has been sliced into portions, ranging in size from stingy to generous, for distribution among segments of the population.

Table 4-2 confirms our basic impression of the income parade: A large chunk of the population has very low incomes, below $15,000; three quarters of the households fall below $50,000; and an affluent few have high incomes, exceeding $100,000.

Ethnicity and family structure present variants on this basic pattern. Figure 4-1 compares the median incomes of different sorts of families. The plight of female-headed families is especially striking in this bar graph. Although half of dual-headed minority families have achieved the relative comfort of incomes over $30,000, female-headed families (both minority and Anglo) remain concentrated at the bottom of the income distribution. Of course, such families are more common in the minority population.

TABLE 4–2   Households, Families, and Individuals, by Total Money Income (1990)

Income	Percent		
	Households	Families	Individuals
Under $5,000	5	3	11
$5,000–$9,999	10	6	22
$10,000–$14,999	9	7	16
$15,000–$24,999	18	16	22
$25,000–$34,999	16	16	14
$35,000–$49,999	17	20	9
$50,000–$74,999	15	19	4
$75,000–$100,000	5	7	1
over $100,000	4	6	1
Total	100	100	100
Number (thousands)	94,312	66,322	23,590
Median income	$29,943	$35,707	$15,344

NOTE: Individuals = persons living alone. Households include families, individuals, and unrelated persons sharing housing.

SOURCE: U.S. Census 1991a.

FIGURE 4–1 Median Income by Ethnicity and Family Type, 1990

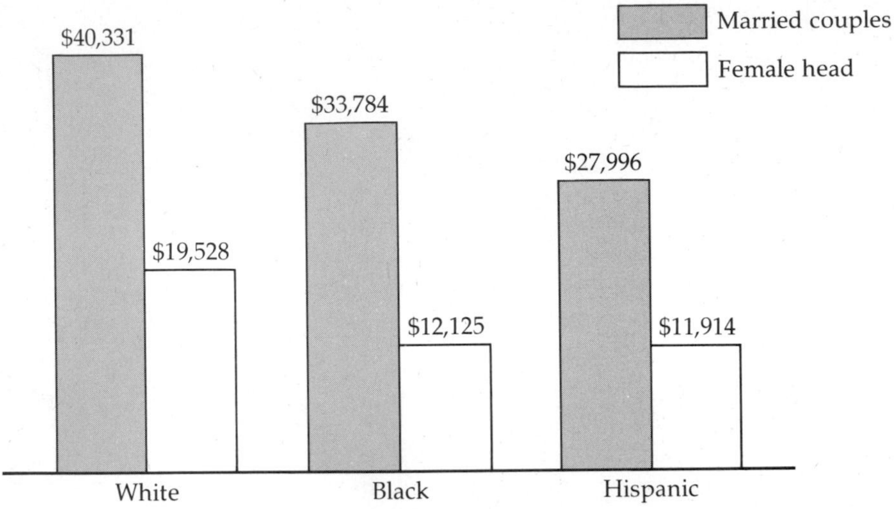

SOURCE: U.S. Census 1991a.

The slices-of-pie approach to income distribution assumes that we have divided the parade into, say, five equal intervals. We want to see what part of the total income pie goes to those in each interval. Figure 4-2 shows that marchers in the first (poorest) fifth of the parade would receive less than a 4 percent portion of such a pie. Had the pie been cut into portions of equal size, each income fifth would have received 20 percent.

The income pie illustrates the heavy concentration of income at the high end of the scale. The portion of aggregate income received by the richest fifth is more than fourteen times that obtained by the poorest fifth. In fact, more detailed data indicate that the top 5 percent alone receives almost seven times the aggregate income of the lowest 20 percent. Such comparisons belie the common notion that most income is received by the middle class. Indeed, a modern Robin Hood could transform the lives of the poor and near-poor by transferring income from the richest 5 percent of households to the poorest 20 percent until he had equalized the incomes of the two groups. In 1990, this tactic would have raised the average income of the bottom group from roughly $8,000 to $60,000 (calculated from U.S. House 1990: 24).

TAXES AND TRANSFERS: THE GOVERNMENT AS ROBIN HOOD?

But isn't the government Robin Hood? Doesn't the government use progressive taxation to collect more money from the rich than from the poor,

FIGURE 4–2   Shares of Aggregate Income Received by Fifths of Households, 1990 (pretax)

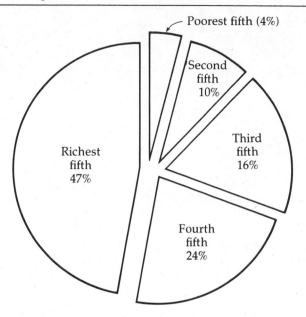

Poorest fifth (4%)

Second fifth 10%

Third fifth 16%

Richest fifth 47%

Fourth fifth 24%

SOURCE: U.S. Census 1990a: 30.

leaving us more equal after taxes than before them? The income distributions we have presented so far are all based on pretax data. What would
they look like if we took taxes into consideration? The answer is: not very
different.

We do have a national personal income tax that is somewhat progressive in its overall effect. That is, people with higher incomes pay a greater
proportion of their total income to the Internal Revenue Service than those
with lower incomes. However, other taxes operate in the opposite direction. Chief among these regressive taxes are the sales taxes, which are lev-
ied by states and localities. A sales tax imposes the same tax rate on a pair
of children's shoes whether the purchasing parent is a welfare mother or a
millionaire. But since a welfare mother spends a much higher proportion
of her income on taxable consumer items than the millionaire, she loses a
higher percentage of her income to the sales tax. Federal payroll taxes are
also regressive in their impact. For example, a worker earning $30,000 in
1990 would have paid $2,300 in payroll taxes. An executive earning
$300,000—ten times as much—would have paid only $3,900.

On the federal level, regressive taxes such as the payroll levies cancel
out the progressive effect of the personal income tax. A recent study by the
Congressional Budget Office (CBO), found that the combined impact of all
federal taxes on the distribution of income is virtually nil. According to the

CBO's estimate, no fifth of the population gained or lost as much as 2 percent of the total income pie through federal taxation in 1990. Even the extreme concentration of income in the top 5 percent of families is hardly touched by federal taxes. (Figure 4-3 presents the results of the CBO study graphically.) An earlier study that estimated the combined effect of federal, state, and local taxes reached a similar conclusion (Pechman 1985).

The government has one other way of playing Robin Hood with income, and that is through transfer payments, such as social security payments, veterans' benefits, and public assistance. Since transfer payments are counted as part of cash income, they are, unlike taxes, already reflected in the income data we saw in the last section. (Noncash benefits, such as food stamps and Medicare, are treated in Chapter 10.) The effect of government transfers is generally progressive, for the obvious reason that some are specifically designed to help the poor and the less obvious reason that over half of such payments go to the elderly, who tend to be at the low end of the income distribution (Pechman 1985: 53).

In 1989, federal transfer payments boosted the average income of the bottom 40 percent of households by about $4,000. Even the richest 20 percent of households received an average of nearly $2,000 in transfers. In general, transfer payments raised the living standard of the poorest families but did not dramatically alter the overall distribution of income. (See Figure 4-4.)

FIGURE 4–3 Income Shares Before and After Federal Taxes in 1990

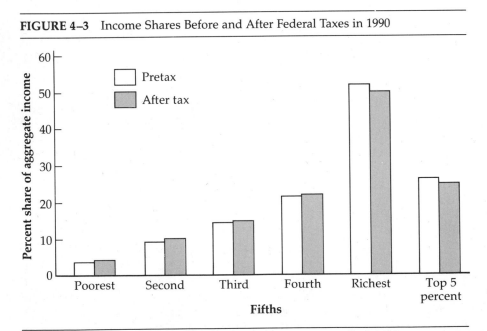

NOTE: Refers to all federal taxes and incomes of families and unrelated individuals.
SOURCE: U.S. House 1990: Tables 6 and 7.

FIGURE 4–4   Income Shares of Households Before and After Cash Transfer
             (1989)

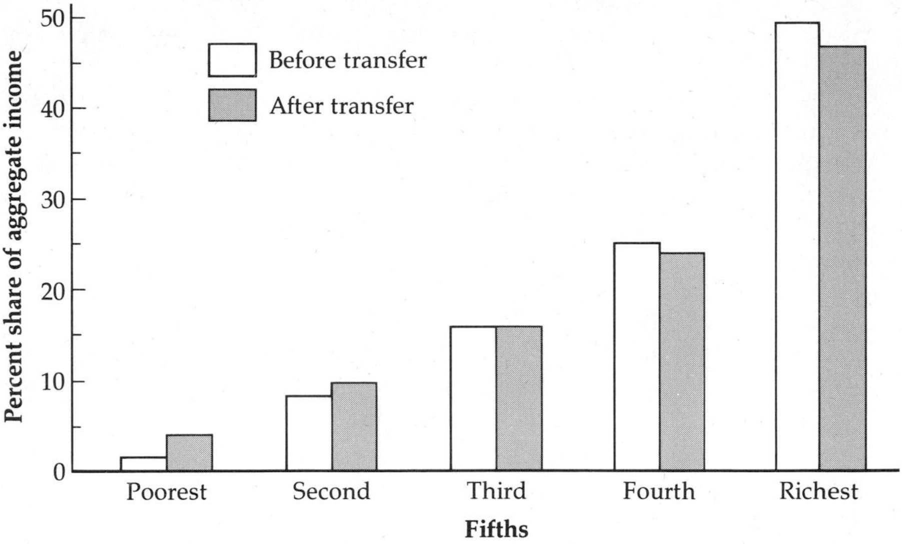

SOURCE: U.S. Census 1990d.

HOW MANY POOR?

Our discussion of income distribution and redistribution has skirted an important issue: How many people have such low incomes that they can be
considered poor? The easiest answer is based on official government statistics, which recorded 32 million poor Americans in 1990, about 13.5 percent
of the population (U.S. Census 1991b: 7). However, any count of the poor
is dependent on the standard or definition of poverty used by the counters.
Some researchers would want to adjust these figures upward or downward
because they are skeptical of the standard employed by government stati-
sticians. We will take up the problem of defining poverty in Chapter 10, so
that readers will be able to draw their own conclusions.

SOURCES OF INCOME INEQUALITY

So far, our efforts in this chapter have been largely devoted to describing
the way income is distributed among families and individuals. Now we
want to shift perspectives and ask how we might account for the existing
distribution. We have already encountered some elements of the answer.
For instance, we have observed that the concentration of wealth is the
key to explaining extremely high incomes. We know that the number of

workers in a family is crucial to its standing in the overall distribution and that female-headed families have especially low incomes. We have seen that the government is shaping the distribution when it conducts transfer programs and imposes taxes. By logical extension, all the political activity that aims to influence these policies forms part of the income-distribution system.

As these disparate observations suggest, it is no simple matter to account for the distribution of income. In fact, there is no comprehensive and viable theory of how the whole system works. What we do have are partial views of certain key aspects of the situation, usually divided into structural and individualistic approaches. The relative importance of the two approaches must be inferred by indirect analysis; what we observe and measure are the combined results as shown in the income parade. Unseen factors influence both the shape of the curve of heights in the parade and the selection of persons for the earlier or later parts of the march.

There are underlying *structural*, or institutional, factors that stem from the economic system itself and influence the opportunities that exist for earning money; they create positions or slots, which individuals compete to enter. The capitalist system, which concentrates assets into the hands of a few, is an obvious example; it permits some to collect enormous incomes without working and forces most of us to look for jobs. Furthermore, long-term trends in technology and management change the mix of different types of jobs, as shown in Chapter 3. Those who seek work do so in a preexisting labor market. The reasons why some jobs in that market provide five or ten or twenty times more earnings than others are not at all obvious; neither theories of supply and demand, nor "return on human capital," nor administrative power fully explain the differences. Some combination of theories will be needed.

The trends in the labor market create and maintain a certain proportion of low-paying and unstable jobs, and the people who have those jobs will be poor regardless of personal skills or ambitions; indeed, even if every worker had a college degree, some would be poor as long as the job structure remained the same. Similarly, owners of successful businesses will get rich even if they failed to graduate from high school.

Yet, *individual characteristics* obviously have some influence on the way young people get sorted into various career paths that lead them to higher or lower incomes. The structure may be the result of past history, but personal careers reflect talent, education, ambition, and a certain amount of luck. The relative weight of those individual characteristics will be carefully examined in Chapter 7.

THE DUAL LABOR MARKET

Dual labor market theory looks at the distribution of income from a structural perspective. The idea of dual labor markets emerged in the late 1960s from the efforts of researchers to understand the persistence of poverty in the United States. The particular question that intrigued them was this: Why is it that in a generally prosperous economy, many people work full-time without rising above the federal poverty line? The answer given by the dual labor market theorists centers on the idea that the labor market is split into two sectors, primary and secondary, and that the working poor fall on the wrong side of the divide. The sectors are distinguished by the type of jobs they offer. Primary-sector jobs are typically characterized by

> several of the following traits: high wages, good working conditions, employment stability and job security, equity and due process in the administration of work rules, and chances for advancement. . . . [The] secondary sector has jobs which, relative to those in the primary sector, are decidedly less attractive. They tend to involve low wages, poor working conditions, considerable variability in employment, harsh and often arbitrary discipline, and little opportunity to advance (Piore 1977: 93).

To understand why the working poor find themselves in the low-wage, secondary sector, we must think in terms of two separate phenomena: (1) the structure of job opportunities offered by the economy and (2) the relative standing of workers in a hypothetical "job queue," which expresses the preferences of employers among potential employees (Thurow and Lucas 1972).

A study by Barry Bluestone provides some important insights into the first aspect. Bluestone looked at the distribution of the working poor in the national economy in the mid-1960s (a period of low unemployment), when nearly a third of the families below the poverty line were headed by people who worked full-time, fifty to fifty-two weeks a year (Bluestone 1974: 140–141). He found that the poor were disproportionately concentrated in the South and were most likely to be employed by firms in certain sectors of the economy—for example, manufacturers of cotton textiles, southern sawmills and planing mills, laundries, restaurants, and limited-price variety stores. Bluestone's "character sketch" of low-wage industries emphasizes low productivity, limited capital investment, low profit margins, highly competitive markets, and product demand that is volatile with respect to price. This combination of economic characteristics makes it difficult for firms in such industries to offer their workers stable employment and a decent wage. Their inclination to pay low wages is reinforced by the absence or weakness of labor unions in these industries and federal legislation that exempts workers in some sectors from minimum wage legislation.

It should be noted that the secondary labor market is not limited to jobs in these industries. In other areas of the economy, firms sharing some of these general features are also likely to be employers of low-wage labor. Even economically secure firms may depend on such labor for certain phases of their operations, either directly or by subcontracting work to secondary-sector firms. Bluestone's analysis, then, suggests that a certain amount of poverty is built into the wage structure and that it continues to exist even in times when the economy approaches full employment. For this reason, prosperity does not trickle down to those on the bottom of the wage hierarchy.

Job seekers get channeled into primary- or secondary-sector jobs on the basis of their position in the figurative labor queue from which employers recruit workers. Sometimes trained and experienced workers match closely the needs of the available jobs, and they get hired first. But particularly for new entrants to the labor force, the employer has to do a lot of guessing about which persons in the queue might best suit the needs of the jobs. And much of the guessing concerns such qualities as punctuality, responsibility, and ability to learn simple tasks that are taught on the job. Some writers on the dual labor market believe that the significance of formal education and job skills has been overestimated. Bluestone notes that occupation-by-occupation comparisons between workers in low-wage industries and those in other industries show only slight educational differences. He argues that there is no real difference in skill levels between workers in a low-wage sawmill and those in the automobile industry. Other writers stress that most skills are acquired through on-the-job training. In fact, many skilled positions are not filled from competitive labor markets accessible to all, but rather from the restricted "internal markets" of individual firms (Doeringer and Piore 1971). For example, skilled jobs in manufacturing are typically filled through promotion, often on the basis of seniority, from the low-skilled, entry-level positions. One of the main differences between the primary and secondary sectors of the labor market is that the secondary sector offers only dead-end positions, unconnected to promotion ladders.

Behavioral expectations may be a more crucial barrier to primary-sector jobs than are job-related skills. Primary-sector employers demand greater labor discipline and are less tolerant, for example, of lateness or absenteeism. In the secondary sector, more lax employer expectations, frequent layoffs, and the absence of promotion incentives all tend to erode the work attitudes that would make an employee more attractive to primary-sector firms. Nonetheless, dual labor market researchers note that many perfectly dependable workers are trapped in the secondary labor market. Some may be victims of geography; they live in areas where better jobs are rare. Others may be excluded from more attractive positions because employers in the primary market are practicing "statistical discrimination" or stereotyping as an inexpensive screening method (Piore 1977: 94). Since it is difficult to measure the abilities and work attitudes of potential employees, such employers discriminate among job applicants on the basis of superficial resemblances to the least desirable secondary-sector employees. Thus, a high school diploma may be largely irrelevant to work on an assembly line, but an employer hiring for such a position may require one, on the assumption that high school graduates are on average more disciplined than nongraduates. Among the bases for such discrimination are race, sex (blacks and women are greatly overrepresented in the secondary sector), accent, demeanor, and scores on various sorts of tests.

Two national income studies (Wachtel and Betsey 1972; Müller 1978) have compared the effects of personal characteristics (including education, work experience, job training, race, and sex) and aspects of economic structure (including industry, region, and union membership) on workers' incomes. Both indicate that economic structure has a substantial impact on wages, *independent* of personal characteristics. In other words, two workers with precisely the same occupational abilities and backgrounds may earn different wages because of differences between the firms that employ them.

From the perspective of the dual labor market, neither an upgrading of the personal characteristics of secondary-sector workers nor the elimination of discrimination would substantially change the current situation. Their effect would be to rearrange the rank ordering of workers in the job queue or perhaps simply cause them to "bunch up" in line as they became more equal to one another. As long as a proportion of the existing opportunities in the labor market involve insecure, low-paying jobs, *somebody* will get them, and the working poor will continue to be among us.

Dual labor market theory should probably not be taken too literally: It is unlikely that there are two discrete labor markets operating in splendid isolation from one another. But the studies that have developed from the dual labor market perspective call our attention to the structural forces shaping the distribution of income. They also remind us of the limits of public policies that focus wholly on individuals, missing the character of the system that shapes the opportunities open to them.

THE DISTRIBUTION OF WEALTH

In Chapter 1, we distinguished wealth (the value of assets held at a *point* in time) from income (monetary gain over a *period* of time). One year's wages, interest, and dividends are examples of income. The real estate, bank deposits, and stock shares someone owned on, say, December 31, are examples of wealth. Typically, wealth is measured in terms of *gross assets* or *net worth*. The concept of gross assets refers to the total value of the assets someone owns; net worth is the value of assets owned less the amount of debt owed. Dependable data on wealth holding are considerably harder to come by than equivalent income statistics, which are published at regular intervals in various forms by the government. Wealth data are technically more difficult to compile, and there also appears to be considerable bureaucratic resistance to gathering them or providing relevant material to independent researchers. This state of affairs is a tribute to the political sensitivity of a topic that inevitably calls attention to the wealthy themselves.

There are two basic sources of wealth data: household surveys and estate tax returns filed with the IRS. Wealth surveys, which are occasionally conducted by the Federal Reserve System and the Census Bureau, encounter the difficulty of getting a statistically reliable sample of the very wealthy, who are few in number but control much of the existing personal wealth.

The estate tax approach, which uses the wealth of the deceased to estimate the wealth of the living, faces its own, formidable methodological problems.

In everyday speech, we reserve the term *wealth* for the rich. Most people, in fact, have only modest wealth. As Table 4-3 demonstrates, the bottom third of households are worth less than $10,000. Only a quarter of households are worth over $100,000. The table is based on a 1988 survey conducted by the Census Bureau, which also revealed that the majority of households derive most of their net worth from three asset types: home equity, car equity, and bank deposits (estimated from U.S. Census 1990f: 3–5).

In broad terms, we can distinguish three classes of wealth holders, using data drawn from the 1988 survey:

1. A *nearly propertyless class*, constituting 44 percent of the population and worth under $25,000 in 1988. Many members of this class have little or no net worth; some, in fact, owe more than they own. Others have built up precarious equities in homes and automobiles. Families worth $5,000 to $10,000 are better off than average in this class. Ninety percent of households worth $5,000 to $10,000 own cars, with an average equity of $4,400. Only 40 percent own homes, with an average equity of just $5,400. Most had several hundred dollars in a bank account.

2. A *"nest-egg" class*, including almost half of the population, with net worths from $25,000 to $250,000 in 1988. The families in this class might be described as savers rather than investors. For example, virtually all households in the $50,000 to $100,000 range have built up a cushion of safe, interest-earning assets, such as passbook savings accounts, CDs, and government bonds, worth, on average, more than $10,000. But only a quarter hold corporate stock (average value: $8,000) and just 14 percent own a small business or professional practice (average equity: $19,300). Homes and automobiles are their most substantial assets, accounting for about 70 percent of net worth.

3. An *investor class* comprising just 9 percent of families, worth in excess of $250,000 in 1988. The households in this class own most of the privately held investment assets and typically control large, diversified portfolios. The 1988 survey, which probably underestimates their holdings, indicates that nearly all own interest-bearing assets (average value: over $72,000). Over half own corporate stock (average investment: $86,300). Many have interests in small businesses, professional practices, and rental properties. On the other hand, equity in homes and autos contributes only 37 percent of net worth.

One of the clearest conclusions that can be drawn from the 1988 survey and related studies is that wealth is highly concentrated—much more concentrated, in fact, than income. For example, in 1990, the top 1 percent of in-

TABLE 4-3   Distribution of Household Net Worth, 1988

Mean Value	Percent of Households
Negative or zero	11.1
$1 to $5,000	15.1
$5,000 to $10,000	6.2
$10,000 to $25,000	11.5
$25,000 to $50,000	13.0
$50,000 to $100,000	16.7
$100,000 to $250,000	17.5
$250,000 to $500,000	6.0
$500,000 and over	2.8

SOURCE: U.S. Census 1990f.

TABLE 4-4   Concentration of Wealth, 1989–1990

	Households (by Net Worth) in Percent			
	Bottom 90%	Upper 10%	Top 1%	Top 0.5%
Gross assets	36.2	63.8	33.1	26.2
Principal residence	63.7	36.3	8.6	5.2
Other real estate	18.9	81.1	44.0	33.8
Stocks	16.0	84.0	47.0	37.5
Bonds	6.3	93.7	71.6	62.8
Trusts	10.2	89.8	52.6	32.5
Life insurance	59.0	41.0	13.2	7.7
Checking accounts	50.5	49.5	21.4	16.3
Thrift accounts	36.4	63.6	16.0	8.0
Other accounts	42.8	57.2	19.5	12.7
Small business	8.1	91.9	64.8	55.8
Automobiles	73.8	26.2	6.5	4.6
Other assets	23.3	76.7	41.9	34.4
Liabilities	62.7	37.3	14.5	11.8
Principal residence debt	76.2	23.8	3.8	2.2
Other real estate debt	19.9	80.1	43.9	37.5
Other debt	73.4	26.6	11.1	9.6
Net worth	31.1	68.7	36.7	29.0

NOTE: Figures are average of two estimates presented by the authors: one using "model-based" and the other "design-based" weighing of sample data. The differences are slight in most categories. For example, the proportion of net worth held by the top 0.5 percent is 28.8 by the first method and 29.1 by the second.

SOURCE: Kennickell and Woodbum 1992.

come earners received about 13 percent of aggregate income (U.S. House 1990: 19), while the top 1 percent of wealth holders owned 31 percent of net worth. The latter figure is from a survey of household finances conducted for the Federal Reserve Board in late 1989 and early 1990, which found that *the concentration of wealth has become so great that the net worth of the top 1 percent is greater than that of the bottom 90 percent* (see Table 4-4).

Another basic conclusion that can be drawn from studies of wealth is that ownership of investment assets, such as stocks, bonds, and rental property, is more concentrated than ownership of consumption-oriented assets, such as automobiles and owner-occupied homes. Our sketch of wealth classes reflected this fact. According to the Fed survey (Table 4-4), the assets of the wealthiest 10 percent of the population include 36.3 percent of the value of all owner-occupied residences, but 81.1 percent of other real estate and 84 percent of corporate stock. A super-rich minority, constituting 0.5 percent of the population, controls close to half of corporate stock and nearly two thirds of small-business assets; the average net worth of these households exceeds $10 million. In Chapter 8, we will consider some of the political implications of this remarkable concentration of economic power.

TRENDS IN THE DISTRIBUTION OF INCOME AND WEALTH

Normally, the distributions of wealth and income change at the millenial pace of tectonic plates shifting on the surface of the earth. But recent years have not been entirely normal.

From the 1930s until the 1970s, income inequality was gradually diminishing. Especially after World War II, the trend toward greater equality was combined with growing incomes for families at all levels (Miller 1971; U.S. Census 1989a: 42; 1990a: 5). But things changed in the early 1970s (see Figure 4-5). Overall income growth tapered off, and the fortunes of American families at different levels of income started to diverge.

While the (inflation-adjusted) real incomes of the lower three fifths of families fell or stagnated, the incomes of the richest fifth soared. The most dramatic change came in the 1980s, at the very top of the income distribution. According to a study by the Congressional Budget Office, between 1977 and 1990 the average income of the top 5 percent of the population rose from approximately $142,000 to $206,000, while the average income of the top 1 percent jumped from nearly $295,000 to $549,000 (see Table 4-5).[1]

There were other remarkable indications of widening income disparities from the mid-1970s to the late 1980s. The proportion of families with real incomes in excess of $100,000 doubled, at the same time that the poverty rate was rising. The share of the income pie reserved for the top 1 percent of households expanded by over 50 percent, while the real weekly earnings of the average worker sank. The salaries of corporate chief executive officers surged to unprecedented levels: In 1979, the average CEO earned 29 times as much as the average factory worker; by 1988, CEOs were earning 93 times as much as factory workers (Phillips 1990: 179–180).[2]

---

[1]These CBO figures and those in Table 4-7 are not strictly comparable with the U.S. Census data presented in Figure 4-5 and employed throughout this chapter. The CBO combines Census survey data with data compiled from tax forms by the IRS. The IRS data provide a more accurate picture of incomes in the top percentiles. CBO figures for 1990 are projections.

[2]The "Gini Index," a standard measure of overall income inequality, declined continuously from 1947 to 1973, then turned upward and rose steadily through the 1980s (U.S. Census 1990a, 1990b, 1990c; U.S. House 1990).

FIGURE 4–5 Mean Income Received by Fifths of Families, 1947–1989

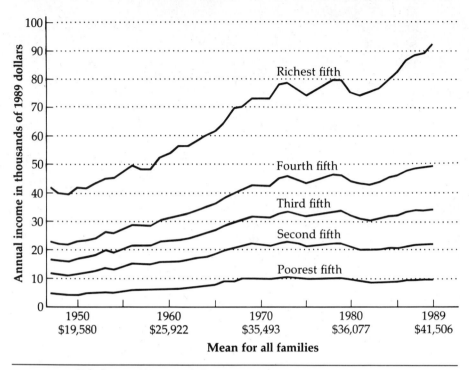

SOURCES: U.S. Census 1989a: 44; U.S. Census 1990a: 30.

TABLE 4–5 Mean Income of Households, 1977–1990, by Fifths and Top Percentiles (in 1990 Dollars)

	1977	1985	1990
Population fifths			
Poorest	$8,738	$7,109	$7,424
Second	$21,253	$18,081	$19,428
Third	$32,780	$29,263	$30,942
Fourth	$44,082	$42,208	$44,850
Richest	$82,455	$90,697	$104,248
Top 5 percent	$141,705	$171,949	$206,162
Top 1 percent	$294,874	$429,813	$548,969
Overall	$37,985	$37,559	$41,369

SOURCE: U.S. House 1990: 28; from Congressional Budget Office.

One of the most sensitive indicators of change in the distribution of income is the fate of younger families. From 1960 to 1973, the real median income of families headed by individuals 25 to 44 rose nearly 40 percent;

from 1973 to 1990, their real median income sank by approximately 7 percent—despite the rising proportion of dual-income families (U.S. Census 1991a: 52 and 1990b: 20). Declining incomes and rising home prices pushed down home ownership rates among younger families, especially in the 1980s (see Table 4-6). For example, home ownership among families with heads 30 to 34 years of age dropped from 61 to 52 percent during the decade. One the other hand, ownership rates continued to grow among older families, who had the opportunity to build up home equity or savings during the years when incomes were growing.

One reason for the deterioration of incomes at the lower end of the distribution was the proliferation of low-wage jobs, which younger workers are especially likely to hold. A recent Census Bureau study revealed a steep rise in the proportion of low-wage employment in the 1980s. The Bureau defined low earnings as yearly wages below the official poverty line for a family of four, regardless of the actual size of the worker's family. (The poverty line, which is adjusted annually for inflation, was $12,195 for a family of four in 1990—the amount a full-time worker could earn in a year at $6.10 per hour.) In the prosperous decade from 1964 to 1974, the proportion of full-time, year-round workers with low earnings shrank from 24 to 12 percent. In 1979, the proportion of such workers began to rise again, reaching 18 percent by 1990 (*Washington Post*, May 12, 1992).

The trend toward greater income inequality in the 1970s and 1980s was magnified by regressive changes in the federal tax system. The top tax rate on personal income plunged from 70 to 28 percent. Corporate and inheritance taxes, whose main effects are felt by the wealthy, were also reduced. At the same time, payroll taxes, paid largely by lower- and middle-income workers, were jacked up (U.S. House 1990). The net result of these policies can be seen in Table 4-7, which traces the evolution of *effective tax rates*— that is, the proportion of income lost to all federal taxes.[3] The story this table tells is very simple. The tax burden of the bottom 80 percent of households grew heavier from 1977 to 1990, and the tax burden of the top 20 percent became lighter. Inevitably, the top 1 percent got the biggest tax break, a reduction of nearly 10 percent.

The increasing inequality of income and the regressive tendency in federal taxation, at a time of rising stock and real estate markets, led to extraordinary accumulations of personal wealth. The number of individuals with net worth in excess of $5 million approximately doubled between 1972 and 1981 and then doubled again in three years (*Statistics of Income Bulletin* 1988: 46). In the 1980s, the combined fortunes of the people on *Forbes* magazine's list of the 400 richest Americans more than doubled in real dollar value. By

---

[3]Effective federal tax rates should not be confused with personal income tax rates, such as the 70 percent rate cited above. Personal income tax rates are legislated marginal rates. They cover only a given range of personal income and can be blunted with various loopholes and exceptions. Effective rates refer to actual outcomes and, in the current context, to all federal taxes, not just personal income taxes.

Table 4–6 Home Ownership Rates by Age of Family Head (in Percent)

Age	1973	1980	1990
Under 25	23	21	15
25–29	44	43	36
30–34	60	61	52
35–39	69	71	63
40–44	73	74	70
45–54	76	78	76
55–64	76	79	80
65–74	71	75	79
Over 74	67	68	71
Total	64	66	64

SOURCE: Joint Center for Housing Studies 1991.

TABLE 4–7 Total Federal Effective Tax Rates for All Households, 1977–1990, by Fifths and Top Percentiles

	1977	1985	1990
Population fifths			
Poorest	9.5	10.6	9.7
Second	15.6	16.1	16.7
Third	19.6	19.3	20.3
Fourth	21.9	21.7	22.5
Richest	27.1	24.0	25.8
Top 5 percent	30.5	24.5	26.7
Top 1 percent	35.4	24.9	27.2
Overall	22.8	21.7	23.0

SOURCE: U.S. House 1990: 12.

1991, the poorest of the 400 was worth $275 million; the wealthiest was worth $5.9 billion (*Forbes*, Oct. 21, 1991).

Data from Federal Reserve surveys reveal that the share of wealth held by the top 0.5 percent of households jumped in the 1980s (see Table 4-8).[4] The rise in the concentration of wealth at the top may, in fact, have been even more dramatic than the Federal Reserve surveys suggest. There is a twenty-year gap in the Fed surveys, from 1963 to 1983. Analyses of estate tax returns for that period show that the share of wealth held by the top 1 percent of households was falling (Smith 1984; Schwartz 1984–1985).[5] Taken together, the estate tax studies and the Federal Reserve surveys suggest that the long-term trend in the distribution of wealth has been U-

---

[4]Data from two Census Bureau surveys in the 1980s also point to an increase in the overall concentration of wealth. Between 1984 and 1988, according to the studies, *mean* net worth rose from $89,601 to $92,017, while *median* net worth fell from $37,012 to $35,752 (in 1988 dollars), a pattern strongly suggestive of increasing concentration (U.S. Census 1990f: 8).

[5]Unfortunately, figures from the household surveys and the estate tax surveys cannot be compared directly because of technical differences between the two methods of estimating wealth. They should, however, detect the same broad trends.

Table 4-8 Concentration of Wealth, 1963–1990

Percentile of Households	Percent of Net Worth		
	1963	1983	1989–1990
99.5 to 100	24.6	24.1	29.0
99.0 to 99.5	7.2	7.2	7.7
90.0 to 99.0	32.1	35.3	32.0
Total (top 10%)	63.9	66.6	68.7

SOURCE: Table 4-4; Avery et al. 1988: 356–361.

shaped: *Concentration fell in the late 1960s and 1970s and then climbed steeply in the 1980s.*

CONCLUSIONS

This chapter has been built around the image of an income parade. First we looked at the shape of the parade and found that as the minutes on the clock ticked by, a lot of short, poor, and almost-poor people had to pass our reviewing stand before we began to see people tall enough to live at a modest but decent standard of living. After that, the situation improved more rapidly; toward the end of the parade, the people passing by were much more affluent with every minute, until finally we saw leviathan marchers who towered over the tallest structures on the parade route.

There were not many of these monsters, but they were extraordinarily tall. That led the imagination to the idea that some Robin Hood might cut them down to size and share their incomes with the short people. Such an act of redistribution could, in theory, greatly improve the levels of living of the poorer folk. But then someone suggested that perhaps the government was already acting like Robin Hood and was producing such a redistribution. Careful examination of the data indicated that the lives of the poor were indeed being improved by government transfer payments. But most of the benefits came, not from cutting the giants down very much through progressive taxation, but by other forms of redistribution, particularly social security, which takes money from workers and gives it to those who have retired. After the government has both taxed and transferred, the very poor are less poor, but most other people's relative incomes have not been greatly altered.

After observing the income parade, we began to look at a few of the reasons why some people pass by the reviewing stand early and some arrive late. The first explanation is the ownership of wealth: The tallest people, at the end of the parade, get most of their money from return on assets in the form of interest and dividends, whether they work or not. The people in the middle of the parade get almost all their money from wages and salaries, and so the type of job they hold and the number of people in the family who have jobs determine family income and level of living. The working spouse is a particularly important element. The shorter and poorer people at the beginning of the parade either do not work or have jobs that pay no more than the minimum wage. Indeed, most of them are disabled and cannot work, or old and retired, or single women with small children to care for.

Then we began to raise questions about access to jobs: Why does it happen that among those who work, some people have good jobs that pay well and others have jobs that pay poorly? At this point, we began to hedge the answers. We noted that the question really is double-edged. (1) Why do good and bad jobs exist in the economy? (2) What selects people for the various types of jobs that do exist? The answer to the first question involves economic history, technological change, and administrative structures within a capitalist system constantly adjusting to the world trends—and this book cannot delve deeply into such issues. The second question is less structural or institutional and more individualistic, since it concerns the characteristics of individuals that give them advantages or disadvantages in competing in the existing labor market, as it is defined by institutional features. Some aspects of the selection process are obvious: people with more education tend to get better jobs, which tend to pay more money. But the details of how that works in practice turn out to be far from obvious. To take but one illustration: People with the same type of job do not always receive the same income, especially if some of them are women. That paradox will be explored later, particularly in Chapter 7.

We followed our examination of income distribution with a closer look at the distribution of wealth. Two conclusions stand out clearly from existing studies: (1) Wealth is much more concentrated than income—so concentrated that the top 1 percent of households have a bigger share of total wealth than the bottom 90 percent—and (2) investment assets, such as stocks, bonds, and income-producing real estate, are much more concentrated than other forms of wealth. These observations fit our earlier conclusions about the sources of extremely high incomes.

Finally, we examined recent trends in the distribution of wealth and income. We then discovered that the long-term drift toward greater equality had reversed itself in the early 1970s and 1980s. Both income and wealth are becoming *more* concentrated. Recently, the wealthiest 0.5 percent of households enjoyed a notable increase in net worth at the expense of the bottom 99 percent. While the incomes of the top few percent of families grew 50 percent or more, the incomes of the bottom three fifths of families fell or stagnated. The growing proportion of low-wage jobs has contributed to these changes in the pattern of in family income. Finally, the overall trend toward increased economic inequality has been reinforced by a regressive tendency in federal tax policy. Since the 1970s, the federal taxman has been acting less like Robin Hood and more like his nemesis, the sheriff of Nottingham.

SUGGESTED READINGS

Duncan, Gregg, et al. 1984. *Years of Plenty, Years of Poverty.* Ann Arbor: Institute for Survey Research, University of Michigan.
> *Summarizes findings of long-term study of economic experience of 5,000 American families, begun in 1968. Unique and valuable research.*

Edwards, Richard. 1979. *Contested Terrain: The Transformation of the Workplace in the Twentieth Century.* New York: Basic Books.
> *Development of segmented labor markets in historical perspective.*

Jaynes, Gerald David, and Robin M. Williams, Jr., eds. 1989. *A Common Destiny: Blacks and American Society.* Washington, D.C.: National Academy Press.
> *This comprehensive study includes historical data on the income of African-Americans.*

Phillips, Kevin. 1990. *The Politics of Rich and Poor: Wealth and the American Electorate in the Reagan Aftermath.* New York: Random House.
> *The trend toward economic inequality in the 1980s and its political significance for the 1990s, by one of the country's most perceptive political analysts.*

U.S. Bureau of the Census. *Money Income of Households, Families, and Persons in the United States.* Current Population Reports. Consumer Income. Series P-60.
> *Published annually, the reports in this series are the basic source of detailed information on the distribution of income. (The title varies.)*

U.S. House of Representatives, Committee on Ways and Means. *Overview of the Federal Tax System.*
> *Published annually. A clear, easy to use source of basic information on the federal tax system.*

5 Socialization, Association, Lifestyles, and Values

And this is good old Boston,
The home of the bean and the cod,
Where the Lowells talk only to the Cabots
And the Cabots talk only to God.

John Collins Bossidy, 1910

My brother down in Texas
Can't even write his name.
He signs his checks with Xs.
But they cash 'em just the same.

Irving Berlin, Annie Get Your Gun

Since Weber, many students of stratification have thought of prestige classes as "communities" characterized by more or less distinctive lifestyles and values. Sociological research has examined differences among classes in areas as diverse as consumption patterns, role relationships in marriage, sexual behavior, and language usage. Two of the variables outlined in Chapter 1 are critical for the emergence and maintenance of such differences: socialization and association. When people of similar prestige associate more often with one another than with persons of lower or higher prestige, they create identifiable strata that generate their own special subcultures. Since people tend to inherit the class position of their parents, the socialization of successive generations of the community into the patterns of thought and behavior characteristic of their elders contributes to the solidification of these subcultures.

Although this chapter will treat association and socialization as processes played out within social communities, we will find that both are shaped by occupational life and may in turn react upon it. Patterns of association through friendship, marriage, residence, and membership in formal organizations are structured by occupation. "I'm a carpenter," comments one of Coleman and Rainwater's Kansas City informants, "and I won't fit with doctors and lawyers or in country club society" (1978: 81). The professionals and business executives who do fit may be able to enhance their careers through informal contacts made in country-club society. Furthermore, research demonstrates that occupational position influences the way that parents socialize their children and that childhood socialization, in turn, cultivates attitudes and abilities that affect adult occupational achievement.

CHILDREN'S CONCEPTION OF SOCIAL CLASS

Socialization is a social learning process that prepares new members of a society for adult life. The knowledge imparted ranges from how to hold a fork to conceptions of appropriate male or female behavior. The child also absorbs adult notions about class differences. An early study of primary school students in a New England town of 15,000 showed that by the sixth grade, children had a fairly sophisticated awareness of the class significance of items such as an English riding habit, an elegantly furnished room, tattered clothing, and different occupational activities, all presented to them in pictures. And when they were asked to place their peers in one of three classes, the sixth graders agreed 70 percent of the time with adults who rated parents from the same households (Stendler 1949). Simmons and Rosenberg (1971) demonstrated that young children have a clear conception of occupational prestige differences. Even the third graders in their sample from the Baltimore city schools ordered fifteen occupations from the NORC list in a way that correlated almost perfectly with the rankings in the 1963 national survey of adults. (The children did tend to raise the relative position of their fathers' occupations, much as adults inflated the standing of their own occupations.)

Subsequent studies have concentrated on the development of conceptions of class distinctions, beginning in early childhood. Tutor (1991) showed first, fourth, and sixth graders photographs of upper-, middle-, and lower-class people. She asked the children to group the adults and children depicted into families and match them with the corresponding pictures of cars and houses. The first graders did substantially better than chance at this task; the sixth graders produced near-perfect scores; and the fourth graders were not far behind.

Naimark (1981) presented children ages 7 to 16 with "36 photographs of real Americans—depicting a wide range of occupations, activities, ages, racial/ethnic backgrounds." The children were asked to place the images on a five-point class scale ranging from "poor" to "rich." Even the youngest children achieved a significant level of consensus among themselves and with adults regarding class placements. The age-group consensus improved through the junior high years and moved closer to the adult ratings. In a follow-up study (Ramsey 1991), preschoolers were shown photographs that had consistently been placed in the top or bottom class categories and asked whether the people depicted were rich or poor. The children, ages 3 to 5, were correct almost 75 percent of the time—significantly better than chance (50 percent).

Leahy (1981; 1983) probed the thinking of children ages 7 to 17 about "poor people" and "rich people." He found that young children conceive of the rich and the poor in overt, physical terms, while older children think in terms of psychological characteristics of individuals and their positions in the society. Here some of their observations:

Joe, age 6: [Poor people have] no food. They won't have no Thanksgiving. They don't have nothing. . . . [R]ich people have crazy outfits and poor people have no outfits.

Mary, age 6: [People can become rich by going] to the store and they give you money. . . . [Or] if your husband gives you money, and your grandmother or your grandfather.

Pete, age 10: [People are rich] because they save their money and they earn it. They work as hard as they can and don't just go around and buy whatever they want.

Dean, age 12: I think that [rich and poor people] should all be the same, each have the same amount of money because then the rich people won't think they are so big.

In general, these studies show that even preschool children are aware of class differences. They suggest that as children grow older their ideas about stratification become more consistent, abstract, and "accurate." By the time they reach 12 years of age, children are not very different from adults in their thinking about class.

KOHN: CLASS AND SOCIALIZATION

While studies such as those just reviewed approach socialization through the child's developing conception of the world, a separate research tradition focuses on the parent. For several decades, studies of the latter type have recorded class differences in the way people raise their children (Bronfenbrenner 1966; Gecas 1979). The most intriguing recent work along these lines is that of Melvin Kohn and his associates at the National Institute of Mental Health in Washington, D.C. What makes Kohn's work special is his effort to isolate the particular characteristics of the lives of adults that could explain the class variation in styles of parenting. In brief, Kohn was able to show that the values parents attempt to inculcate in their children reflect the character of the parents' occupational experience. His findings suggest, moreover, that the transmission of these values plays an important role in the perpetuation of the existing class order.

Kohn directed three studies of parental values between 1956 and 1964: a survey of 421 white parents of fifth graders in Washington, D.C., a comparative survey of 861 parents in Turin, Italy, and a national survey of 3,101 men representative of all men engaged in civilian occupations in the United States (Kohn 1969: 9). In each study, he asked parents to select from a list of characteristics those they considered most desirable for a child of the same age and sex as their own child. Here are a few examples (Kohn 1969: 218):

That he is a good student.
That he is popular with other children.
That he has good manners.

That he is curious about things.
That he is happy.

Although the studies found considerable consensus across class levels about the importance of many values, others varied among levels. For example, an emphasis on consideration of other people, curiosity, responsibility, and self-control increased at successively higher class levels, while emphasis on good manners, neatness, obedience, honesty, and being a good student increased at lower class levels (Kohn 1969: 54). Table 5–1 presents class differences in mothers' responses for selected values in the Washington, D.C., study. (A smaller sample of fathers produced similar results.) In this case, parents had been asked to choose the three most important values from a list of seventeen. Households were stratified by father's class position as measured by Hollingshead's Index of Social Position, which uses education and occupation as indicators (class I is the highest). The table should be read to indicate, for example, that 61 percent of upper-class mothers considered happiness an important value. (Since mothers were permitted more than one choice, the figures in the table do not add up to 100). Although the table suggests that there are few sharp breaks between classes on parental value orientations (the breaks between classes I and II on curiosity and happiness are interesting exceptions), a clear class gradient is evident for most of the characteristics in the table.

The pattern becomes even more prominent when the occupations of employed mothers are taken into account. Kohn found that among wives of working-class men, those with white-collar jobs were closer to middle-class mothers in their responses, while those with manual jobs presented the sharpest contrast with middle-class orientations.

Kohn's interpretation of this class pattern in parental values is based on the idea that the middle-class parents who stress self-control, curiosity, and consideration are cultivating capacities for self-direction and empathetic

TABLE 5–1 Mothers' Choices of "Most Desirable" Characteristics in Child, by Class (percent)

	Class				
Characteristic	I (Upper)	II (Upper-middle)	III (Lower-middle)	IV (Upper-lower)	V (Lower-lower)
Obedience	14	19	25	35	27
Neatness, cleanliness	6	7	16	18	27
Consideration	41	37	39	25	32
Curiosity	37	12	9	7	3
Self-control	24	30	18	13	14
Happiness	61	40	40	38	30
Honesty	37	49	46	50	65

SOURCE: Kohn 1969: 30.

understanding in their children, while working-class parents who focus on obedience, neatness, and good manners are instilling behavioral conformity. The middle-class pattern, particularly in the emphasis laid on happiness, curiosity, and consideration, is oriented toward the *internal* dynamics of the person—both the child and others. The working-class pattern, on the other hand, assumes fixed *external* standards of behavior. This general difference is neatly illustrated by the contrasts between having curiosity versus being a good student, or showing consideration versus good manners. In each pair of contrasts, the first choice shows internal development, and the second shows conformity to rules.

An additional finding substantiates this interpretation. Parents were asked about the specific sorts of misbehavior for which they would discipline their children. Their responses revealed that middle-class parents were more likely to punish a child for the *intent* of his or her behavior, in contrast to working-class mothers, who were more likely to discipline for the *consequences* of behavior. For example, a middle-class mother might penalize her child for throwing a temper tantrum, while a working-class mother penalizes for boisterous play. The first suggests a loss of internal control, the second a violation of external standards.

Kohn noted that working-class and middle-class mothers also differ in their attitude toward sex-role socialization. The surveys indicate that working-class mothers are more likely to choose separate "masculine" values (dependability, ambition, school performance) for boys and "feminine" values (cleanliness, happiness, good manners) for girls. Middle-class mothers are less inclined to make such gender distinctions. Kohn argued that this difference is also related to the distinction between internal dynamics and external standards, since the working-class emphasis on traditional sex roles implies an external focus. However, this connection is certainly not so intuitively obvious as the one regarding occasions for punishment.

Kohn labeled the two underlying patterns *self-direction* and *conformity.* He demonstrated a consistent relationship between the class position of parents and the values they chose for their children. At successively higher class levels, parents value self-direction more and conformity to external standards less. But what are the roots of these differences? Kohn suggested that they reflect generalized value orientations that develop out of a specific aspect of social class: occupational experience. He reasoned, for example, that people who hold professional and managerial jobs, which are relatively unsupervised and require the considerable exercise of individual judgment and initiative, are more likely to value self-direction than those who work at highly routinized blue-collar jobs. In brief, self-direction at work should produce self-direction in values.

Evidence from both the U.S. national survey and the Italian study in Turin supports this interpretation. General value orientations in two key areas—attitudes toward authority and judgments about work—were shown to vary with social class in both countries. Authoritarian attitudes stressing "conformance to the dictates of authority and intolerance of non-conformity" (Kohn 1969: 79) became more frequent at lower-class levels. The same is true of judgments about work focusing on the *extrinsic* qualities of a job (pay, hours, etc.), whereas judgments emphasizing *intrinsic* qualities (how interesting the work is, the amount of freedom it has, the chance it affords to use abilities, etc.) are more likely at higher-class levels. Moreover, these value orientations are systematically related to the character of respondents' occupational experience. Men whose work is (1) closely supervised, (2) repetitive, and (3) oriented toward things rather than people or data are the most likely to subscribe to authoritarian values, to judge jobs on their extrinsic qualities, and, finally, to favor the conformity value pattern for children. The opposite holds for those whose work is unsupervised, unrepetitive, and oriented toward people or data.

These findings fit the causal chain that Kohn had anticipated to explain the relationship between social class and socialization patterns: Occupational experience gives rise to general value orientations, which in turn shape parental value preferences for children. Further scrutiny of the data showed that a second aspect of social class, level of education, also correlates with value orientations and parental value preferences and that this reinforces the class patterning of socialization. Kohn observed that education appears to "provide the intellectual flexibility and breadth of perspective that are essential for self-directed values . . ."(1969: 186). Education, of course, is closely related to occupation. It does not, however, "explain" the correlation between occupational experience and values. Kohn found that education and occupational conditions have independent impacts on parental values, although the effect of occupational conditions is substantially stronger.

Kohn's work has important implications for our understanding of the class system as a whole. Since Marx, sociologists have been aware that life experience, especially occupational experience, shapes social values. Kohn observed that "the essence of higher class position is the expectation that one's decisions and actions can be consequential; the essence of lower class position is the belief that one is at the mercy of forces and people beyond one's control, often, beyond one's understanding" (1969: 189). If this is true, we should expect people in top positions to learn to value self-direction and those at the bottom to learn to value conformity to authority. We might also anticipate that they will teach these values to their children. It is at this point that the larger significance of Kohn's work becomes clear. When parents inculcate values that reflect their experience of the class system, they are preparing their children to assume a class position similar to their own and, by so doing, are contributing to the long-term maintenance of the class system. But notice that these socialization values are not on the level of Marx's class consciousness. They concern the appropriate behavior of children. The connection between that behavior and the world of work is seldom recognized by the parents.

SCHOOL AND MARRIAGE

The class differences in values inculcated by parents are reinforced by the tendency of children and adolescents to associate with others of similar class background as they are growing up. Because neighborhoods tend to group people of similar economic means, the kids on the block are likely to be of the same social class. Local schools reproduce the class patterns of the neighborhoods they serve. Upper-class and many upper-middle-class families make sure that their children's classmates will be from similar households by sending them to private schools.

Although large public high schools in urban areas bring together students of diverse backgrounds, they do not necessarily mix freely. Often class differentiation within the school is institutionalized through "tracking" systems that separate students on the basis of their ability or postgraduation aspirations. Because there is a substantial correlation between social class and academic track, the system inadvertently segregates students by class (Colclough and Beck 1986: 489; Oakes 1985). Students' own preferences contribute to the class patterning of association. Studies of adolescent cliques and friendships show that students are inclined (but by no means certain) to choose friends who share their class backgrounds (Hollingshead 1949; Cohen 1979; Duncan, Haller, and Portes 1968).

Marriage choices are also shaped by social class (Hollingshead 1950; Dinitz et al. 1960; Laumann 1966; Rubin 1968; Whyte 1990). For his study of dating and marriage, sociologist Martin K. Whyte (1990) asked a sample of women in the greater Detroit metropolitan area a series of questions about their own and their husband's families when they were in high school. Given a choice of five classes (poor, working, middle, upper-middle, and upper), 58 percent of the respondents placed their parents and their in-laws in the same social class. Whyte also collected data on the occupations of respondents' fathers and fathers-in-law, which he reclassified using a system of three occupational classes (upper white-collar, lower white-collar, and manual). He found that 51 percent of respondents had married someone from the same occupational class (Whyte 1990: 105–109).

Laumann (1966), whose study of class and association we will describe in greater detail later, asked men in a Massachusetts community the occupations of their father and father-in-law at the time of the respondent's marriage. The responses, organized by five occupational prestige categories, showed that 44 percent of the respondents had married women drawn from the same stratum. Seventy-one percent had married women from the same or an immediately adjacent stratum. The inclination toward class endogamy (in-group marriage) was especially notable in the "top professional and business" grouping. Although this category constitutes a small proportion of the population surveyed, creating relatively few opportunities for endogamous marriages, 60 percent of the sons of top professionals and businessmen married women who shared their class background (Laumann 1966: 74–81). A couple of quotations from Laumann's informants neatly reveal the attitudes underlying this phenomenon (1966: 29):

> What sort of husband would a carpenter be? What sort of education would he have? My viewpoint would not jibe with a carpenter. Marriage is based on equals. I would want my daughter to marry in her own class. She would go to college and would want her husband to be educated. I would want to be able to mix with in-laws and converse with them.

> I was born into a family of great privilege—but I think I have a responsibility. Family responsibilities will be better held if the background of the members are similar. My own interests are not manual. Our family relations are very close. It would be very demanding to have an unskilled or factory worker in the family. The tradition in our family is the professions or land-owning. I would not turn out a daughter who married a machinist but—a machine operator is bound to have a modest education—probably not interested in intellectual things. Even with the best will in the world, a family relationship would be very difficult to accomplish.

The pattern of class endogamy observed by Laumann and others has an important implication for class biases in childhood socialization. It means that similar class influences are likely to be transmitted through both parents to the child.

MARRIAGE STYLES

Rainwater has studied class differences in marital patterns by looking at the role relationships of husbands and wives, "their typical ways of organizing the performance of tasks, their reciprocal expectations, their characteristic ways of communicating, and the kind of solidarity that exists between them" (1965: 28). His conclusions are based on interviews with approximately 400 husbands and wives, most living in Chicago but some living in Cincinnati and Oklahoma City.

Following Bott's (1964) earlier work on English couples, Rainwater distinguished three types of role relationships on a continuum: "jointly organized," "intermediate," and "segregated." Joint relationships focus on companionship and deemphasize the sexual division of labor. Husbands and wives with joint role relationships share the planning of family affairs, carry out many household duties interchangeably, and value common leisure activities. Even when responsibilities are parceled out by gender (wife-homemaker, husband-breadwinner), each partner is expected to take a sympathetic interest in the concerns of the other. In the segregated relationship, there is clear differentiation of concerns and responsibilities, which minimizes the husband's involvement with household matters and the wife's with the world of (the husband's) work. Husband and wife are likely to have distinct leisure pursuits and separate sets of friends. Intermediate relationships fall between these two poles.

On the basis of questions about family decision making, duties of husbands and wives, interests and activities of the partners, and the general character of the relationship, the couples surveyed were classified into one of the three categories. Of course, all couples were in some sense "intermediate"; there were no pure types. Classification was based on the preponderant tendency in each marriage. A glance at Table 5–2 will show that joint relationships predominate in the upper-middle class and segregated relationships in the lower-lower class, while the classes between them exhibit a neat gradient. Data by race for the bottom two classes (financial limitations precluded interviews with middle-class blacks) suggest a subcultural pull toward segregated relationships among blacks. However, it is not clear how precisely the black and white samples were matched, and the apparent divergence in marital role patterns may simply represent intraclass economic and educational differences between the races. In any event, the racial variance is relatively small, especially in comparison with the class differences among both blacks and whites.

The relationship between social class and marital role types is more than a matter of idle curiosity. Reported marital happiness increases with class level, especially for women (Bradburn 1969: 156), and this phenomenon is tied to the character of the organization of marital roles. In Rainwater's study, middle-class couples reported greater sexual satisfaction than lower-class couples, but the difference was largely a function of the level of role segregation. For example most lower-class wives in segregated relationships evaluated their sexual experience in marriage negatively (68 percent), while those in intermediate relationships were positive (68 percent) (Rainwater 1965: 28). In national surveys, companionship in marriage (which would appear to be similar to joint organization) is positively correlated with social class and with marital happiness (Bradburn 1969: 163).

Let's take a closer look at conjugal role types by examining how they function in upper-middle-class and working-class families. In important

---

TABLE 5–2 Social Class and Conjugal Role Relationships

| Class | Number | Role Relationships (percent) | | | |
		Joint	Intermediate	Segregated	Total
Upper-middle class	(32)	88	12	—	100
Lower-middle class	(31)	42	58	—	100
Upper-lower class					
White	(26)	19	58	23	100
Black	(25)	12	52	36	100
Lower-lower class					
White	(25)	4	24	72	100
Black	(29)	—	28	72	100

SOURCE: Rainwater 1965: 32.

ways, the very character of upper-middle-class life lends itself to the joint role relationship, since the wife's social activities are often linked to the husband's career advancement (especially if she does not have an independent career). Joint role relationships are also reinforced by the high rates of social and geographic mobility typical of this class. Isolated from kin and removed from successive sets of friends as they move from community to community and up the corporate hierarchy, husbands and wives must look to each other for support and companionship.

Among blue-collar couples, wives have no significant role to play in regard to the husband's work. Since blue-collar careers are less likely to require geographic mobility, it's easier for working-class spouses to maintain ties with kin and friends from adolescence and early adult years. Dependency on the couple's parents is intensified by the economic insecurity that is especially typical of young working-class families (Rubin 1976: 69–92). These social ties tend to draw husband and wife to separate sources of support and companionship outside the marriage. Bott's (1964) work in England showed that couples who come to a marriage with tight-knit networks of friends and kin (that is, the people each spouse knows tend to know one another) and maintain these ties are the most likely to develop segregated marital relationships. Her data suggest that social networks of that sort are least typical of professionals and most typical of manual workers. A variant on this theme can be seen in the work of Stack (1974), who found that lower-class black women in the United States develop extensive networks of kin and friends as a shield against economic vicissitudes. Such networks provide security that the erratically employed black men of this class cannot give, but they also maintain an insistent claim on the loyalties of their members, which undermines strong relationships between women and their lovers or husbands.

We have dealt with the origins of joint and segregated role relationships at the top and at the bottom of the class order. What can we say about the mix of marriage types in the middle? Two social processes seem relevant. One is social mobility: People moving up or down in the class structure may carry with them lifestyles acquired in their class of origin. Thus, the upper-middle-class origins of many lower-middle-class couples (especially younger couples) may help explain the predominance of joint relationships among them. An analogous argument can be made for the spread of segregated relationships upward. The second process, which is discussed in more detail below, is the tendency of upper-middle-class lifestyles to become generally fashionable models and filter downward. Through these processes, couples are exposed to conflicting influences, which may be resolved by the compromise implied in intermediate role relationships.

Working-class sex-role socialization is another source of differences in marital-role organization. As we noted in our discussion of Kohn's work, working-class parents are more likely than middle-class parents to hold separate sets of expectations for boys and girls. Vanfosen (1977) found that college-age daughters of working-class fathers are more likely than their middle-class peers to subscribe to traditional sex-role values as expressed in questionnaire items such as "A woman should not expect to go to exactly the same places or have the same freedom as a man." Such women are the most likely to find the segregated marital role acceptable. On the other hand, boys of all classes are taught to be more controlled, more instrumental, less emotional, and less empathetic than their sisters, but the distinction is made much more emphatically in blue-collar families (Rubin 1976: 116–117, 125–126). Rubin, who conducted lengthy interviews with working-class and upper-middle-class couples, reported:

> Not once in a professional middle-class home did I see a young boy shake his father's hand in a well-taught "manly" gesture as he bid him good night. Not once did I hear a middle-class parent scornfully—or even sympathetically—call a crying boy a sissy or in any way reprimand him for his tears. Yet, these were not uncommon observations in the working-class homes I visited. Indeed, I was impressed with the fact that, even as young as six or seven, the working-class boys seemed more emotionally controlled—more like miniature men—than those in the middle-class families (1976: 126).

Boys who are taught to be "manly" in this way are less likely as adults to feel comfortable with joint role relationships in marriage.

RUBIN AND LEMASTERS:
NEW STRAINS IN BLUE-COLLAR MARRIAGES

Rubin's study and another by E. E. LeMasters (1975) provide an intriguing postscript to Rainwater's earlier work. Despite important differences in setting, method, and the populations studied, Rubin and LeMasters reached very similar conclusions about contemporary blue-collar marriages. Rubin (1976) conducted intensive, wide-ranging interviews with fifty blue-collar couples and a smaller control group of twenty-five upper-middle-class couples in their homes in the San Francisco Bay Area. In all cases, the wife was under forty and the couple had at least one child under twelve. LeMasters spent five years in a participant observation study of approximately fifty men and women, regular patrons of a "family-type," working-class tavern (The Oasis) in Wisconsin. Virtually all the men in LeMasters' study were skilled construction workers. On average, they seem to have been older and more prosperous (he calls them "blue-collar aristocrats") than Rubin's respondents. However, both Rubin and LeMasters concluded that marital norms filtering down from the upper-middle class are creating enormous strains in working-class marriages.

Studies of working-class family life from the period of Rainwater's research (for example, Komarovsky 1962) depicted unmistakably segregated marital relationships. One of LeMaster's informants, a woman married for thirty years, bitterly described the traditional pattern:

The men go to work while the wife stays home with the kids—it's a long day with no other adult to talk to. That's what drives mothers to the soap operas—stupid as they are.

Then the husband stops at some tavern to have a few with his buddies from the job—not having seen them since they left to drive home ten minutes ago. The poor guy is lonely and thirsty and needs to relax before the rigors of another evening before the television set. Meanwhile the little woman has supper ready and is trying to hold the kids off "until Daddy gets home so we can all eat together." After a while, she gives up this little dream and eats with the kids while the food is still eatable. About seven o'clock, Daddy rolls in, feeling no pain, eats a few bites of the overcooked food, sits down in front on the TV set, and falls asleep.

This little drama is repeated several thousand times until they have their twenty-fifth wedding anniversary and then everybody tells them how happy they have been. And you know what? By now they are both so damn punch drunk neither one of them knows whether their marriage has been a success or not (LeMasters 1975: 42).

Now this pattern is being challenged by notions of intimacy, companionship, sharing, and equality that are received from above. The problem is that these ideals do not appeal equally to wives and husbands. Women have been prepared for them by their socialization and in many cases by contact with a middle-class world through white-collar employment. They are exposed to the new thinking through women's magazines and television soap operas. Men, on the other hand, are quite satisfied with traditional role relationships, which they regard as part of the natural order of things:

I couldn't stand being home every day, taking care of the home, or sick kids, or stuff like that. But that's because I'm a man. Men aren't supposed to do things like that, but it's what women are supposed to be doing. It's natural for them so they don't mind it (Rubin 1976: 105).

Men sense, of course, that many women do "mind it," but they are inclined to think that women's complaints are groundless. From The Oasis:

What is the hell are they complaining about? My wife has an automatic washer in the kitchen, a dryer, a dishwasher, a garbage disposal, a car of her own—hell, I even bought her a portable TV so she can watch the goddamn soap operas right in the kitchen. What more can she want? (LeMasters 1975: 85).

But behind the bluff there is fear. From the less "macho" setting of his living room, one of Rubin's informants phrased the problem differently:

I swear I don't know what she wants. She keeps saying that we have to talk, and when we do, it always turns out I'm saying the wrong thing. I get scared sometimes. I always thought I had to think things to myself; you know, not tell her about it. Now she says that's not good. But it's hard. You know, I think it comes down to that I like things the way they are, and I'm afraid I'll say or do something that'll really shake things up. So I get worried about it, and I don't say anything (Rubin 1976: 121).

For their part, working-class women in these studies are very dissatisfied but also frightened and confused and occasionally given to wondering whether asking a man to be more than a conscientious provider is indeed asking too much.

> I'm not sure what I want. I keep talking to him about communication, and he says, "Okay, so we're talking, now what do you want?" And I don't know what to say then, but I know it's not what I mean. I sometimes get worried because I think maybe I want too much. He's a good husband; he works hard; he takes car of me and the kids. He could go out and find another woman who would be very happy to have a man like that and who wouldn't be all the time complaining at him because he doesn't feel things and get close (Rubin 1976: 120).

There is a second aspect of blue-collar marriage that is being subjected to strain as middle-class norms filter down: sexual adjustment. Problems in this area are not new. For instance, husbands and wives have long clashed over the desirable frequency of sexual intercourse. LeMasters heard this complaint among patrons of The Oasis (1975: 101). However, difficulties of more recent origin, deriving from the sexual revolution of the 1960s and 1970s, are evident in the comments of the younger couples interviewed by Rubin.

Comparative evidence from the 1940s and 1970s, including Rubin's own research, suggests that working-class sexual behavior is moving closer to middle-class norms. For example, working-class couples have nearly caught up with middle-class couples in their willingness to engage in once exotic sexual variants such as cunnilingus and fellatio; working-class men have become similar to middle-class men in their concern for their wives' sexual satisfaction (Rubin 1976: 134–135, 137–148). But change is not without psychological costs. Again, differential receptivity to new standards—in this case, husbands are the more open—creates stress for working-class marriages. The blue-collar men whom Ruin interviewed wanted freer, more expressive, more mutually satisfying sexual relationships with their wives, as their remarks show:

> I think sex should be that you enjoy each other's bodies. Judy doesn't care for touching and feeling each other, though.
>
> She thinks there's just one right position and one right way—in the dark with her eyes closed tight. Anything that varies from that makes her upset.
>
> It's just not enjoyable if she doesn't have a climax, too. She says she doesn't mind, but I do (Rubin 1976: 136).

But their wives respond warily. This comment from a woman married twelve years is typical:

> He says I'm old fashioned about sex and maybe I am. But I was brought up that there's just one way you're supposed to do it. I still believe that way, even though he keeps trying to convince me of his way. How can I change when I wasn't brought up that way; [*with a painful sigh*] I wish I could make him understand (Rubin 1976: 138).

The emphasis here is on "how I was brought up." Working-class women have been socialized to fear or deny their own sexuality and to expect men to divide women into "bad girls" whom they use and "good girls" whom they marry. Thus, even married women pressed by their husbands for freer sexual expression are inhibited by the fear voiced by this woman after ten years of marriage: "How do I know in the end he won't think I'm cheap?" (Rubin 1976: 141). If working-class men find change easier to assimilate, it is because they have less at stake. Nonetheless, Rubin finds that some of the husbands are, in fact, still ambivalent in their own attitudes toward female sexuality and probably convey that ambivalence to their wives.

Rubin reports a different situation among her upper-middle-class couples. The working-class woman had typically remained at home until they were married, but the middle-class women had the more open atmosphere of their college experience (typically in the 1960s) to free themselves psychologically from traditional sexual constraints. Nearly all the couples in both groups had had premarital intercourse, but the middle-class wives expressed less guilt about having done so (1976: 60, 62). The distinct character of the upper-middle subculture in sexual matters is reflected in two of Rubin's observations. The middle-class men were less likely than their working-class counterparts to issue ambivalent messages in sexual matters and less given to "good girl" versus "bad girl" conceptions. And the middle-class women were more likely to feel guilty about their sexual inhibitions than about their sexual behavior.

INFORMAL ASSOCIATION AMONG ADULTS

Warner, whose classic Yankee City study we examined in Chapter 2, considered patterns of association so critical to understanding the class system that he sometimes appeared to define class in terms of association. A social class, he suggested, is a group of people who belong to the same social cliques, intermarry, dine in each other's homes, and belong to the same organizations. Warner defined a social clique as "an intimate nonkin group," with no more than 30 members. He and his associates collected elaborate data on the clique membership of families in Yankee City. They found that most cliques joined people of the same or adjacent classes (Warner and Lunt 1941: 110–111, 350–355). But there was enough clique formation across class boundaries that we consider it impractical to define class in terms of exclusive patterns of association.

No recent study has attempted the thorough mapping of social cliques carried out in Yankee City. Such a research task would be all but impossible in a larger urban setting, but sampling methods can be used to gauge the influence of social class on adult patterns of informal association. Edward O. Laumann conducted two surveys of adult association patterns in metropolitan areas. We have already referred to the initial study, published in

1966, which was based on a poll of 422 white male residents of Cambridge and Belmont, Massachusetts, both part of the greater Boston metropolitan area. (Areas of Cambridge close to Harvard, MIT, and Radcliffe were not included in the survey because of their large student populations.) The second survey, conducted three years later, covered approximately 1,000 white male residents of greater Detroit (Laumann 1973). The two studies reached similar conclusions, but we will focus on the Cambridge-Belmont results, which are reported in a more accessible form.

In both studies, respondents were asked the occupations of their three closest friends. The distribution of respondents from the Cambridge-Belmont survey is given in Table 5–3. In the percentage distribution (the middle part of the table) it is evident that men at all levels are most likely to choose intimate friends from among the members of their own, or an adjacent, occupationally defined class. The tendency is strongest at the extremes. For instance, unskilled and semiskilled workers found more than three quarters of their friends close to their own group and relatively few within any of the white-collar categories. Members of the top professional and business class drew approximately three quarters of their friends from their own grouping. This tendency toward self-selection appears weakest in the middle-status categories, particularly clerical–small business. Nonetheless, the pattern does not indicate the disappearance of the distinction between middle class and working class. Most clerical–small business friends are drawn from the white-collar categories (65 percent), and most friends of skilled workers are themselves manual workers (62 percent). Overall, fewer than one quarter of all friendships reported cross the manual-nonmanual line.

Again, it is worth quoting Laumann's informants for the attitudes behind associational choices. The following remarks are in response to questions eliciting subjective reactions to association with people in occupations ranging from janitor to physician (Laumann 1966: 28–29). Unfortunately, Laumann does not indicate the speaker's own occupation, though it can usually be inferred.

> I am not a snob, but it is a fact of life that in most of these occupations we would have nothing in common. There must be an intellectual ground for my associations.
>
> I have had experiences with factory workers as a class of people and they're rough. They tend to go on the rough side of living—"hurray for me and the hell with the other guy" is their attitude. . . . Top executives tend to, when that high up, to get standoffish and independent, and I can't afford to have independent friends.
>
> The doctor has his trade; the machine operator also has his trade. As long as a man puts his mind into his trade, it is OK. A tradesman is not much of a big man compared to a top executive. The top executive would try to be uppity—would try to lord it over me. I have a little bit of this in the family already. This guy (my relative) is too big for his own family. His wife is not good enough for him any more. Just because you are a top executive does not mean that you can act uppity.

TABLE 5–3 Cambridge-Belmont Study: "Best Friends'" Occupations by
Respondent's Occupation

Friends' Occupations	Respondent's Occupation					
	Top Professional, Business	Semiprofessional, Middle Business	Clerical, Small Business	Skilled	Semiskilled, Unskilled	Total
Frequency distribution						
Top professional, business	178	54	34	19	14	299
Semiprofessional, middle business	41	113	47	34	40	275
Clerical, small business	6	31	35	28	44	144
Skilled	9	21	39	73	82	224
Semiskilled, unskilled	6	17	24	56	180	283
Total	240	236	179	210	360	1225
Percent distribution						
Top professional, business	74.2	22.9	19.0	9.0	3.9	24.4
Semiprofessional, middle business	17.1	47.9	26.3	16.2	11.1	22.4
Clerical, small business	2.5	13.1	19.6	13.1	12.2	11.8
Skilled	3.8	8.9	21.8	34.8	22.8	18.3
Semiskilled, unskilled	2.5	7.2	13.4	26.7	50.0	23.1
Total	100.0	100.0	100.0	100.0	100.0	100.0
Ratio of observed to expected frequencies						
Top professional, business	3.0	0.9	0.8	0.4	0.2	
Semiprofessional, middle business	0.8	2.1	1.2	0.7	0.7	
Clerical, small business	0.2	1.1	1.7	1.1	1.0	
Skilled	0.2	0.5	1.2	1.9	1.2	
Semiskilled, unskilled	0.1	0.3	0.6	1.2	2.2	

SOURCE: Laumann 1966: 65.

Not all informants accepted class segregation in friendships as desirable. One man commented:

> I just don't believe it is possible to know anything about a man just by knowing his occupation. If he is a good guy, I don't care what he does . . . The job isn't important—it's what he's like himself.

Laumann's analysis deals systematically with a problem that appears to be ignored by Warner (Warner and Lunt 1941: 350–355) and is only alluded to by Hollingshead (1949: 214) in their discussions of cliques: Classes, by most definitions, vary in size. Thus, the statistical opportunities that an individual has to associate with members of his or her own class vary with the size of the class. If there are more low-skilled, blue-collar workers than clerical workers in the local population, it will be easier for the blue-collar workers than for the clerical workers to develop friendships with peers. There is a related "ceiling" and "floor" problem, which is especially significant in the top and bottom categories. Members of the top class cannot choose superiors as friends but have a broad opportunity to select status inferiors. The reverse holds true for those in the lowest class. These tendencies in the data reflect a structured social reality, but they may also derive from the system of categories chosen by the investigator to present the data. Fewer and broader categories will tend to exaggerate the degree to which peers associate by masking selection upward and downward within categories.

The bottom part of Laumann's table measures the extent to which the tendency toward differential association goes beyond statistical or random opportunities. It is based on the assumption that, if men selected their friends without regard to class, then we would expect the distribution of friendships reported by the respondents in each occupational grouping to be roughly the same as for the sample as a whole. Thus, since 23.1 percent of all friends reported were low-skilled workers, we would expect the same proportion of low-skilled workers among the friends of top professionals and businessmen. The actual figure (from the second part of the table) is 2.5 percent, approximately one tenth that amount. Ratios between the expected and actual frequencies are given in the corresponding cells of the third section of the table; a ratio equal to the random probability is indicated by 1.0. Thus top professionals and businessmen were three times more likely than chance to choose friends from their own class, and ten times less likely to choose unskilled workers.

These ratios confirm the impression given by the percent distribution that friendship choices are structured by social class. For most categories, the probability of self-selection is approximately twice chance. Moreover, the probabilities fall off in a regular pattern with increasing prestige distance. In both senses, the self-isolation of the top professional and business elite is greater than for any other grouping, including the bottom blue-collar category. On the other hand, the friendship patterns of clerical workers and small-business owners show a weaker class bias than other groups.

FORMAL ASSOCIATIONS

Studies of formal associations—large groups or organizations with explicit purposes and rules of membership—show the same class patterning that we have observed for cliques and friendships. In Jonesville, Warner and his associate Marcia Meeker (Warner et al. 1949a: chapter 9) found that most associations draw their membership from only one or two strata, and the prestige of the association matches that of its members. Furthermore, each stratum favors different types of organization. The general pattern they observed would seem to fit most American communities (Hodges 1964: 105–115), with the exception that larger cities typically have special upper-class men's clubs such as New York's Links or Boston's Somerset (Domhoff 1967: 19). (In Chapter 8, we will treat the political and economic significance of these organizations.)

Jonesville's upper class patronized the country club, three exclusive women's clubs, and certain professional and business groups. The women's groups were partly social, partly charitable in function. Actually, these organizations had as many upper-middle- as upper-class members, but the latter dominated them and set the tone. This tone was well described for the women's clubs:

> The members of the upper class profess an interest in travel, a knowledge of foreign lands; they value objects associated with tradition and antiquity; they are preoccupied with leisure-time pursuits, with activities which have dubious economic value, but which are considered worthy, noble, and honorable. The upper-class woman who has both sufficient money and sufficient leisure time (with servants to take care of her home) is expected to become a patron of the arts.
>
> Through these associations, the members of the upper class not only express their common interests, but also perform activities of high value to the total society. They preserve relics and other symbols of prestige which are valued by the community. They patronize and sponsor the arts, through lectures and discussions of books, music, and the theater. They perform charitable activities by contributing to needy individuals and institutions (Warner et al. 1949a: 133).

Many upper-middle-class people belonged to the upper-class organizations but stood somewhat aside in deference to their superiors. (The small size of the upper class in Jonesville forced them to mix with those of slightly lower rank.) The upper-middle class also belonged to many groups that reached far below them in status, such as lodges. But their major energy and leadership aspirations went into civic clubs—Rotary, the women's clubs, and educational and health groups. Instead of emphasizing graceful leisure pursuits, the civic clubs were busy promoting community improvement. Most characteristic, probably, were the male luncheon clubs, which featured conviviality and ritual expressions of community solidarity, guest speakers with messages of uplift, and committees that promoted betterment projects of various sorts.

> There is, in general, a sharp break between the upper-middle and lower-middle classes with respect to the kind and amount of participation in associations. While the members of the upper and upper-middle classes participate together, there is little participation between the upper-middle and the lower-middle classes. . . . While the differences between the upper-middle and the lower-middle classes are quite clear, it is difficult to distinguish between associational behavior of the lower-middle class and that of the upper-lower (Warner 1949a: 138).

This gap in participation above the lower-middle group, and the continuity between that group and the upper-lower, parallel their behavior in informal cliques, as discussed above. But Warner and Meeker reported that this vast "common-man" group displayed two different types of activity: the fifth who were white-collar workers tended to participate either in the civic clubs already mentioned or in others that were modeled after them but catered to the "little fellows." Thus, they were either quiet members of Rotary or more active members of Lions. They also belonged to the more ritualized lodges like the Masons and the Eastern Star. Their wives participated in the ladies' auxiliaries of the lodges, emphasizing a more rigid sex separation than was the case with higher-status families. These white-collar families were usually very active church members (probably more so than any other group) and participated extensively in the clubs sponsored by their churches.

The four fifths of the common-man group who were foremen and skilled workers and steady semiskilled operatives were much less active in formal organizations. They might belong to a church or a lodge or a labor union, but in general they preferred to stay at home.

The organizations of the common man stressed inclusiveness—they said that they were friendly folk and that anybody could belong. Thus, their clubs had a wider spread of rank within them than did the exclusive groups of the top people in town. Warner and Meeker added:

> The ideologies of the lower-middle and the upper-lower associations express the ideologies of the people—patriotism, brotherhood, democracy, equality—and the symbols are those which provoke common interest on the part of the "common man." This preoccupation with equality gives satisfaction to individuals of low status, for it minimizes, or overlooks, status distinctions and gives them a sense of similarity with other individuals regardless of their position in the community. It reflects the attitudes of those people who are reluctant to accept an inferior status and declare, "We haven't any classes here, we're all equal" (Warner et al. 1949a: 142).

Members of the lower-lower class, feeling hostile and suspicious toward the "snotty" folk above them, seldom join anything, and they attempt to reduce interaction with people outside their own level to an absolute minimum. They withdraw to escape being snubbed.

Even churches—institutions supposedly rejoicing in the common brotherhood of a common Father—are class typed. In most American towns, the people of higher status belong to those Protestant denominations that feature services of quiet dignity and restrained emotion, such as the Episcopal or Unitarian groups. The common men are more often seen at the Methodist and Baptist churches, where the services are more vigorous, or in Catholic churches (reflecting their origins as part of the "new" immigration from southern and eastern Europe). Those lower-class individuals who go to church at all are most likely to join revivalistic and fundamentalist sects (Demerath 1965; Laumann 1966: 55; Smith 1980: 514).

CLASS DIFFERENCES IN SOCIAL PARTICIPATION

We have demonstrated a strong tendency for people to associate with class equals in their friendships, marriages, organizational memberships, and residential choices. To this generalization we can add a second observation about the relationship between social participation and class: The *amount* of formal and informal association in which people engage is directly related to their class level. For example, a study of the friendships of 199 men in Cambridge, Massachusetts, conducted by Kahl and Davis, found that the percentage of people who live without close friends (defined as those with whom one exchanges house visits once a month or more) increases continuously as one descends the occupational hierarchy. The proportion of such social isolates ranged from 10 percent among top professionals and managers to 30 percent among unskilled workers (Kahl 1957: 138). These findings were supported by Hodges (1964) for the San Francisco area. Curtis and Jackson (1977: 169) studied six communities, varying in population from 5,000 to 700,000, and found in each a significant correlation between higher class position and the frequency of men's visits to friends.

The strong correlation between class and the amount of informal association may help explain one of the findings in a recent Gallup Poll survey on friendship: Higher-income people are less likely to feel lonely. Seventy-three percent of respondents with household incomes over $50,000 said that they "never" or "seldom" feel lonely. Among those with incomes under $20,000, only 53 percent gave these same responses (DeStefano 1990: 31).

The patterns of participation in formal organizations parallel those for informal association: People of higher prestige status belong to more voluntary formal organizations and devote more time to them—this despite the fact that many organizations are specially designed for lower-status people. Furthermore, when an organization has members of various prestige backgrounds, the higher-status persons are more likely to be the leaders (Smith 1980; Curtis and Jackson 1977: chapter 8; Reissman 1954; Warner et al. 1949b: chapter 9; Warner and Lunt 1941: chapter 16).

CONTRASTING LIFESTYLES

Let's explore a little further the contrast suggested above between a socially active upper-middle class and a relatively inactive working class. Many middle-class people combine their social and business lives. In a crude sense, we may say that they use social opportunities to make contacts that have a business or professional function, in that they provide new customers or clients or opportunities for bureaucratic advancement. But in a broader sense, it is better to say that these people simply merge the two spheres and do not see them as separate compartments of living. The husband meets people through his business life, entertains them, and they become friends; or his wife meets people in the community who become business contacts. It is a basic part of the job of most businessmen to be at ease in their contacts with new people; they come to value sociability as an end in itself, and it cannot be expected that they will suddenly reverse themselves at 5 P.M.

Business people live in neighborhoods filled with others like themselves, who share the values of sociability. Their homes are not overcrowded, and they can entertain without strain. They become activists in club work and are likely to run into the same townspeople at the parent-teacher association, the Rotary Club, or the country club. Particularly in small towns, the number of such active business families is small enough for them all to become acquainted. Even their children form a single, extensive web of contacts.

In the past, business connections were primarily based on buying and selling: The insurance man, the doctor, and the owners of the retail stores and the small factories all were tied together in a mutual exchange of goods and services. But the more recent tendency is for these men and women to turn into executives in large corporations, for the local stores and factories have often become branches of national chains. This change means that the executives are less-permanent members of their local communities, for their companies shift them around from one part of the country to another. And it means that the relationships among them have changed flavor somewhat, for instead of independent entrepreneurs who both cooperate and compete with each other in the open market, they have become incumbents within hierarchies who are more likely to seek personal esteem and advancement than a sale.

William H. Whyte, Jr., of *Fortune* magazine, wrote a penetrating account of the clique behavior of the male executive of 1952 in *Is Anybody Listening?* He got most of his information from their wives but also interviewed many junior executives, as well as some of their bosses. The senior men told him, "with a remarkable uniformity of phrasing," that the social activities of junior executives and their wives were extremely important for business success and that the good wife is "(1) . . . highly adaptable, (2) highly gregarious, (3) realizes her husband belongs to the corporation" (Whyte 1952: 146). Here are a few examples of descriptions of the ideal wife:

Executive:	She should do enough reading to be a good conversationalist. . . . Even if she doesn't like opera she should know something about it, so if the conversation goes that way she can hold her own. . . .
Executive:	The hallmark of the good wife is the ability to put people at their ease.
Wife:	The most important thing for an executive's wife is to know everybody's name and something about their family so you

> can talk to them—also, you've got to be able to put people at their ease (Whyte 1952: 152–154).

There are patterns of behavior that are appropriate for each level of the hierarchy, and no ambitious junior executive dares to get far out of line. He cannot drive a Cadillac before he has passed through the Buick stage, or he will be thought "pushy." On the other hand, he must not drive an old Ford, or he will look like a failure. One wife, speaking of this pattern, told Whyte:

> It makes me laugh. If we were the kind to follow The Pattern, I'll tell you just what we would do. First, in a couple of years, we'd move out of Fern-crest Village (it's really pretty tacky there, you know). We wouldn't go straight to Eastmere Hills—that would look pushy at this stage of the game; we'd go to the hilly section of Scrubbs Mill Pike. About that time, we'd start going to the Fortnightly's—it would be a different group entirely. Then about ten years later, we'd finally build in Eastmere Hills (Whyte 1952: 154–155).

Whyte (who wrote in a time when almost all business executives were male) emphasized that the rule was to keep up with the Joneses but never to get far ahead of them. The timing of each shift had to be carefully calculated. Similarly, the executive must not have odd personal tastes that made him too "different." An "intellectual" or "aesthete" was mistrusted because he was different, and different people could not be understood, so their actions could not be anticipated. It was because competition within corporate hierarchies depended so much on the interpersonal relations among executives (equals and immediate superiors) that their total way of life, which included their wives and children, became important. If you dealt with a man in a market situation, you dealt primarily with the quality of his goods or services. But if you had to work with him every day, his total personality was relevant. Therefore, corporations thought of themselves as "one big happy family," and most families are intolerant of too much "difference" among members, which might threaten the smooth operation of joint living.

A similar picture emerges from Rosabeth Kanter's more recent *Men and Women of the Corporation* (1977), a study of a single multinational corporation that she calls "Idsco." The similarity is ironic because Idsco, in contrast to at least one of its competitors, maintained a "libertarian" policy that officially denied any claim on the private lives of its employees and avoided involving spouses in company-related activities. Nevertheless, the social lives of the company's executives and their families were inextricably bound to corporate concerns. Although a few wives with serious careers of their own insisted on their total independence from Idsco, the kind of complaint Kanter heard most frequently from the wives of Idsco executives revolved around the contradiction between the company's official denial of their existence and the "strong demands to be gracious, charming

hostesses and social creatures, supporting their husbands' careers and mo-
tivating their achievements, with the boundaries of their own life choices
set by the company. . . ." (Kanter 1977: 108). By the time Idsco men reached
middle-management positions, their wives, like the women Whyte
interviewed,

> realized that friendships were no longer a personal matter but had business
> implications. Social professionalism set in. The political implications of
> what had formerly been personal or sentimental choices became clear. Old
> friendships might have to be put aside because the organizational system
> makes them inappropriate. . . . The public consequences of relationships
> made it difficult for some wives to have anything but a superficial friendship
> with anyone in the corporate social network. Yet since so much of their time
> was consumed by company-related entertainment, they had little chance for
> other friendships and reported considerable loneliness. A few wives com-
> plained about other costs to instrumentalism in relationships; having to en-
> tertain in their homes people they did not like and would not otherwise
> have invited, the need to be consistently cheerful and ready to be on display.
> Duplicity in relationships was one result (Kanter 1977: 116).

Whyte and Kanter, and other students of upper-middle class social life,
portray what has been dubbed the "two-person career"—a pattern of advancement dependent on a wife's unpaid talents. Less studied and still relatively uncommon is the "dual-career" family, in which both husband and
wife pursue demanding professional or managerial careers. A survey of
1,000 working-age women in Chicago conducted by Lopata and associates
(1980) found that dual-career couples are as likely as single-career couples
to mix social and professional life. About 60 percent of wives employed as
managers or professionals reported that their husbands helped them with
career-related entertaining at home. According to the respondents, husbands in the same occupational categories were somewhat more likely to
receive such help from their wives. (It made little difference whether the
wife was employed.) On the other hand, women employed in blue-collar
jobs or married to men in such occupations reported little job-related
entertaining.

The Chicago survey confirms the contrast indicated by earlier studies
between the merged recreational and business life of the career elite and
the separated work and family life of blue-collar workers. A study that il-
luminates this difference was done by Floyd Dotson in New Haven, Connecticut, in 1948. He interviewed fifty working-class couples in their
homes. They no longer lived in the ethnic districts of their parents and
were typical of the "new" working class who were reared in American cities rather than on farms or in Europe. Most of the husbands held semiskilled jobs in the factories. Most of the wives stayed home and did not
work for pay (Dotson 1950; 1951).

Dotson found that in some respects, the way of life of these people
approximated middle-class norms: They emphasized the importance of the
nuclear (as compared to the extended) family and preferred not to live with

relatives nor to depend upon them much for exchange of such services as baby-sitting. The notion of "permissiveness" in child rearing was widespread. It appeared to Dotson that as these young couples had learned American ways different from those of their parents, they had learned family patterns that were not far divergent from middle-class standards.

However, there were two striking differences: (1) These working-class people had very few intimate friends other than kinfolk; (2) they interacted frequently with siblings. It must be remembered that because they came from working-class homes, they tended to come from large families and had many brothers and sisters available. Furthermore, a sizable proportion of their siblings lived in or near New Haven, as working-class people tend to stay put if employment opportunities are available. About two fifths of the families had no intimate friends other than kin; and almost equal number belonged to loose friendship groups that included husbands and wives, but exchange of visits was not frequent. In only two instances (out of fifty!) did the husband and wife participate in a tight clique of the middle-class style. In six instances, the wife continued to interact with a group growing out of her school contacts, and in five the husband continued to spend much time with his boyhood friends (usually via an athletic club). Rarely did a man interact in evening hours with friends from work.

Why did these people not participate in much clique activity? The sheer availability of siblings was one reason, for they offered all the companionship many of the couples desired. Furthermore, they were easy to interact with—they all understood one another, lived in the same way, did the same things. It was simple to bundle up the kids, take them to sister Jane's house, put them to sleep on any available bed, and then play cards and talk and drink beer. In those few instances where the respondents had a sibling who had climbed into the middle class, visits were rare. Several couples reported receiving invitations to Christmas dinner from successful siblings, but they declined because they would not feel natural and relaxed in the more opulent homes.

Similarly, a few of the respondents found it easy to maintain contact with some old school chums. But as the number of small children in the house multiplied, interaction rates went down. Money was one reason: It was expensive to hire a baby-sitter, and they felt less free in taking infants on visits with nonrelatives. Furthermore, the emphasis on family life led many people to prefer the company of their children: "Sure, kids are a hell of a lot of expense and trouble, but what would a home be without them? If you want the honest-to-God lowdown on me, you can say that my real life is right here with my family. We're family people; we find most of our fun in life right here at home" (Dotson 1950: 122).

Through the years, these factors (plus movement from one part of town to another) broke up most of the adolescent cliques that these people had once had. Dotson expected new friendships formed from the immediate neighborhood or from work to replace them and was surprised to find so little interaction. Although the informants verbalized about the ideal of neighborliness, mostly they meant only that is was proper to smile and say hello. Many were quite reluctant to form close ties with neighbors. One man said: "I've found out you can't really be friends with most people because pretty soon they'll do you dirt" (Dotson 1950: 161). Nonetheless, some women did have one or two close friends among immediate neighbors (about two fifths of the respondents). And interestingly enough, most of the couples remembered with pleasant nostalgia the more intimate atmosphere that existed in the ethnic neighborhoods of their youth and regretted that their current area was "colder." Yet they made it cold by their own behavior.

No recent study gives as systematic a picture of working-class association patterns as Dotson's work, but its general outlines are confirmed by later research. Curtis and Jackson's survey of six communities, although it does not deal separately with working-class people, shows that lower-prestige individuals are more likely to value visits with relatives over those with friends, are more likely to have relatives in the same community, and (largely for that reason) visit their relatives more frequently (1977: 161–178). Lillian Rubin, whose study of working-class families in the San Francisco Bay Area we examined above, found that the extended family is still "at the heart of working-class social life" (1976: 197). Visits with parents and siblings are frequent. Holidays are typically celebrated with large family dinners, which may include aunts, uncles, and cousins. Most important, kin are those "whose lives are shared both emotionally and socially . . . with whom intimacies are maintained, who can be trusted with the care of young children on the rare occasion when a couple takes an evening out alone . . ." (Rubin 1976: 197).

Rubin believes that working-class couples are more likely to exchange visits with nonkin than they were in the past, but the practice is still infrequent. The dinner party, "that backbone of middle-class social life" (Rubin 1976: 195) is virtually unknown. Couples with young children are the most likely to invite friends to the house—typically late Saturday night, when the kids are asleep, for dancing or cards:

> We don't have money to go out hardly at all, but I get lonely to see some friends without the kids around sometimes. So once in a while, we invite them over, and we play some cards and have a little beer and a snack (Rubin 1976: 196).

Women who are not in the labor force may see old friends during the day and develop casual relationships with female neighbors. Employed men and women find new friends at work, and men often "stop off for a beer with the guys" after work (a custom much resented by their wives). But the guys are seldom invited home, and the neighbor has her well-defined place in the housewife's daily routine; so these new relationships do not change the essentially kin-oriented character of the working-class pattern of informal association.

The Lynds remarked about Middletown that working-class people were taught to use their hands and middle-class people their personali-ties—that the one group sold physical labor, the other social skill. It may well be that social skill is a trait of personality that, on the average, gets people into middle-class occupations. Or it may be that the cultural tradi-tions of urban working-class life do not teach people to make friends easily or to trust them too fully. At any rate, the evidence indicates a marked quantitative and qualitative difference between the two classes in social behavior.

CONCLUSIONS

This chapter has moved back and forth between the economic circum-stances of people and certain aspects of their values and lifestyles. We noted that parents are influenced by the characteristics of their jobs and the sizes of their incomes; they learn to adjust attitudes and behaviors to eco-nomic reality. Oversimplifying, we can say that most working-class parents tend to see their lives as limited to routine and subordination, and they train themselves and their children to conform. By contrast, most middle-class people see the world as more open to control, and they encourage themselves and their children to develop active personalities that seek to shape the environment rather than adjust to it.

These characteristics of personality are linked during adolescence to behaviors at school and interactions among school friends. Attitudes to-ward books, sports, sex, work, and authority all reflect these class-related perspectives. Young people sort themselves out so as to associate with oth-ers like themselves; this facilitates friendships and shared activities, but it also tends to reinforce and sharpen the original subcultural perspectives by reducing contact with people who are different and might teach new ideas. The residential segregation of large cities adds to the separation; people with the same incomes live near one another and send their children to local schools. The net result is to close the circle: Ideas that seem appro-priate to the types of jobs parents have and to their ways of handling given amounts of money for family consumption—ideas, in other words, that stem from material or economic circumstances—shape the thinking and behavior of children in ways that lead them to recreate the same economic circumstances when they are adults. Material facts and human feelings about those facts influence each other in a reciprocal fashion.

However, in a society like that of the United States, all of these state-ments are no more than loose trends. The trends are real, as shown by statistical comparisons of attitudes studied in opinion polls, in counts of who interacts with whom, and more dramatically revealed in the conver-sations of men and women when they are talking freely about their lives. Some aspects of the reciprocity between circumstances and ideas they see clearly and can articulate; other aspects are more subtle and less conscious and must be inferred by the observer. But our social classes are not sharply defined groups completely segregated from one another, and our society does not stand still long enough for subcultural patterns to grow entirely consistent within themselves and then be passed on from one generation to the next with little change. The economy keeps transforming itself—from farming to small business to corporate enterprise; large numbers of people move from one part of the country to another; immigrant groups get assimilated and lose much of their ethnic tradition; individuals climb or fall from one social level to another in their own lifetimes, or experience such mobility between generations. Consequently, there are always people at any social level who show values and behaviors that are typical of another level—from which they came or toward which they aim. And people of all levels learn standardized ideas and lifestyles from the pervasive mass media, even if they lack the jobs or the money to allow them fully to play out the appropriate roles.

This constant state of flux brings tensions. People are not always sure how they want to behave, or they lack the economic means to live as they desire. Husbands, wives, and children in the same family may have different goals. To some degree, an open society allows people to reorganize their lives: Children move away from the world of their parents; spouses separate and seek new partners who are more appropriate to changing lifestyles; people change careers in midlife, especially housewives who go to work outside the home. The tensions from change bring more change, which sometimes solves problems. But for many people, life remains, as so often portrayed in contemporary literature, a matter of accepting quiet desperation. The novelist tends to heighten individual idiosyncrasy; the sociologist looks for the common patterns—especially those linked to the facts of life shared within each prestige class—that help explain why so many lives are indeed similar to one another. These similarities in private lives get projected through formal organizations into public influences; each prestige class formulates goals and policies and creates appropriate organizations to promote its interests or solve its problems, and these differing policies become the battlegrounds of political action, as we shall see in more detail in a later chapter on class consciousness and ideology.

But first, let us study in more detail the issue of social mobility: What proportion of the population moves up or down? Are the rates changing through time? What factors cause mobility? The next two chapters examine these questions.

SUGGESTED READINGS

Cookson, Peter, and Caroline Hodges Persell. 1985. *Preparing for Power: America's Elite Boarding Schools.* New York: Basic Books.
   How the upper class is educated.
Fussell, Paul. 1983. *Class: A Guide Through the American Status System.* New York: Ballantine.

One man's view of the class system: idiosyncratic, witty, perceptive, arrogant. Emphasizes class differences in lifestyle, covering such areas as dress, speech, recreation, and home interiors.

Gecas, Victor. 1979. " The Influence of Social Class on Socialization." In *Contemporary Theories about the Family,* edited by W. R. Burr et al. New York: Free Press.
Systematic critical review of the literature.

Hollingshead, August B. 1949. *Elmtown's Youth.* New York: John Wiley.
This classic work from the Warner era includes a comprehensive study of patterns of association among high school students. Demonstrates strong correlation between social class and social cliques.

Kanter, Rosabeth. 1977. *Men and Women of the Corporation.* New York: Basic Books.
A rich ethnographic work whose full range we could only sample in this chapter.

Kohn, Melvin, and Carmi Schooler. 1983. *Work and Personality: An Inquiry into the Impact of Social Stratification.* Norwood, N.J.: Ablex.
Synthesis of their research.

Miller, S. M., and Frank Reissman. 1961. "The Working Class Sub-Culture: A New View." *Social Problems* 9:86–97.
Excellent synthesis of the literature

Rubin, Lillian Breslow. 1976. *Worlds of Pain: Life in the Working-Class Family.* New York: Basic Books.
Sensitive portrayal of working-class life.

6 Succession and Mobility: Structural Opportunities

Some people's money is merited
And other people's is inherited.

Ogden Nash, The Terrible People

As each generation succeeds its predecessor, there occurs a vast sifting process that places individuals into class levels. If a society were completely "open," the forces of pure competition would sort people according to their native talents and the efforts with which they used those talents. Individuals would get neither help nor hindrance from the social positions of their parents in the competition for worldly success; the correlation coefficient between the class positions of parents and those of adult sons and daughters would be low (reflecting purely genetic influences).

Reality does not match the abstract model of the completely open society. Not only is talent partly inherited in the genes, but in addition, families greatly influence the *motivations* of children, through both deliberate and unnoticed patterns of socialization that shape ambitions and the drive for success. Parents also teach *skills* of personality, verbal fluency, and technical adroitness in various tasks that affect later life chances.

Furthermore, the free play of competition in the market has sharp limits, which bias the distribution of rewards. Education is necessary before talent can be properly exploited, and it is expensive in tuition and in the less obvious but often higher costs of supporting students during the years of study and of replacing the income that might otherwise be earned during the same years. Once out of school, a young person gains advantages from parents with lots of money or good connections or even a distinguished name in the community. There are limits on free competition because the stratification variables reinforce one another and because occupation, money, connections, and values are interdependent and are partly passed on from one generation to the next. This is sufficient reason, even if there were no other, to speak of a class system instead of a collection of separate stratification variables.

No society is completely open. In order even to imagine a high degree of openness in his ideal republic, Plato had to suggest that children be separated from parents, sorted into homogeneous groups by potential talent, and then reared by the state in ways that would maximize the development of talent—some were trained for war, others for politics, others for labor. But neither is it possible to have a society that is completely closed, with a system in which all children inherit their parents' positions and no movement up or down the scale ever occurs. The latter type is most closely approached in caste systems, where occupation is normally inherited and marriage is usually confined to caste equals, but even in traditional India, some mobility existed. The correlation coefficient between the positions of parents and children is never equal to 1.0.

In order to understand the degree of openness that exists in the United States today, we must approach the problem in stages. Our first task is to measure the amount of inheritance of class position that occurs from one generation to the next. Inheritance of position will be called *succession*, and movement to a different level will be called *mobility*. The total amount of mobility that occurs in our society is a complex result of many past influences on our occupational structure. Once we understand the structure itself and its changes through time, we can focus attention on the sorting of particular types of individuals into their different careers, which is called the *process of status attainment*.

PROBLEMS OF PERSPECTIVE AND MEASUREMENT

The distinction made above suggests two perspectives that can be applied to the overall question of succession and mobility, and it is important to keep them separate in both theory and measurement. The first concerns the shape of the stratification structure and the trends that influence it. Here we are concerned with the relative proportions of positions at high, middle, and low levels of the system, and also tendencies that are changing those proportions. For instance, an aristocratic society that is made up of a mass of peasants dominated by a handful of large landowners and their administrators will show a shape like a pyramid that is distorted by a very high peak and a narrow middle. When such a society industrializes, the landed aristocrats are replaced by industrial and commercial capitalists at the top, probably in greater numbers than the landlords they push aside. The middle sectors grow, with new engineers, teachers, and skilled workers, who are needed in profusion to make the system function with efficiency. The pyramid gets more squat in shape.

A hierarchy with few positions at the top allows only a few people to move up, regardless of how fair the system is, but a hierarchy with many positions at the top is likely to generate more movement. Particularly during the period of transition from the one type of society to the other, the very expansion of new positions, which are disproportionately placed in the middle and upper levels, creates mobility. That is, many sons and daughters of peasants have a chance to climb into newly created jobs and thus raise themselves above their parents. In such a situation, it is possible to have a lot of upward mobility without much downward slippage. Thus, changes in the structure can increase opportunities for everybody or, to phrase it another way, can enhance life chances throughout the system. Children of people near the top can stay there, and at the same time, many children from below get a chance to improve their position, because the upper strata are expanding.

Once we have described the structure and its trends, we can turn our attention to the second perspective on succession and mobility, that of the particular factors that influence the careers of individuals. Within a given structure, we still might wonder why particular individuals manage to take advantage of the existing opportunities and others do not. Here we pay attention to the many stages of careers and try to understand what is happening at each decisive step, and overall, to determine the relative weight of factors that combine to cause final status placement as adults. For instance, how much difference does a father's occupation make for a son's career, once we control for the son's education? Or does an intelligent son have a good chance to become successful even if his parents are poor? Or do the careers of daughters follow the same model as those of sons? Or do black people suffer from discrimination to such an extent that they cannot get good jobs even if they have a good education?

In order to simplify our story, we start with a discussion of the gross amounts of succession and mobility in the United States and of the structural facts and trends that explain them. Furthermore, we will begin by comparing the careers of fathers and sons. Since women have entered the labor force in massive proportions only in recent years, the basic patterns of careers, especially in the past, have been mainly studied by concentrating on men. Once male patterns are understood, we will ask questions about women. We will first concentrate on structure, turning to the process of individual status attainment in the next chapter.

It would be useful to compare a large sample of fathers and sons on their overall status placement, their scores on some standard SES (socioeconomic status) index. But in practice, that is impossible, because we never have enough data on both the fathers and the sons to compute a good composite index for each generation. To get a large sample, we must start with sons and ask about their fathers. Because many of the fathers are retired or dead, the most reliable information the sons can give about them concerns occupation; sons rarely know much about the details of the incomes of their fathers. Thus, it is useful to begin with a simple comparison of just the occupations of fathers and sons, and most large-scale studies have done so.

Whenever we work with occupational data, we must decide how to group the hundreds of specific occupational titles used by our respondents into some form of ranking that approximates the hierarchy of socioeconomic status. Many different schemes are used, but in essence they reflect only two options: (1) discrete categories taken directly from or somewhat adapted from those used by the U.S. Census or (2) a continuous scale that assigns to each occupational title a numerical value that has been shown to approximate its prestige value in the minds of the public. Both procedures have been explained in Chapters 2 and 3. Many researchers have shown that the relative prestige ranks of occupations change slowly, so for practical purposes, we can compare fathers and sons by assuming similar ranks in each generation.

When using the census categories, two facts should be remembered. First, the classification is a crude one and obscures many shades of meaning. For instance, in the category of proprietors, managers, and officials, the scheme puts all sorts of men together, ranging from the president of General Motors to the man who runs a peanut stand. The occupational classification treats these two gentlemen as equals—which they may be before God and the law but are not in the stratification system. Second, it should be noted that the smaller the size and the larger the number of categories, the more the mobility, assuming mobility is defined as the number of men born in one category who ended up in another. By creating more categories, one creates more places to move and thus more mobility.

HOW MUCH MOBILITY?

How much social mobility is there? Unfortunately, no one has done a full-scale national mobility survey since 1973. But we persuaded Professor Robert Hauser, one of the top authorities in the field, to estimate mobility in the 1970s and 1980s, using data from the annual General Social Survey (GSS). The GSS, which is produced by the National Opinion Research Center, asks respondents their own occupation and the occupation of their father (or other family head). Using this information, Hauser could answer questions like these: What chance does the son of an unskilled worker have of attaining a professional or managerial position? What are the social origins of the people in high-status occupations?

The GSS is a small-scale, national poll. To obtain dependable results, Hauser had to aggregate the data on working-age men in successive GSS surveys. He also had to employ a simplified, five-step occupational hierarchy, instead of the 17-step hierarchy employed in earlier, landmark studies (Blau and Duncan 1967; Featherman and Hauser 1978). The data cover all men, aged 25 to 64, who are employed or who have work experience and are looking for a job.

Table 6–1 is an "outflow" table that starts with fathers and shows what happens to their sons. It is drawn from Hauser's analysis of GSS polls from 1982 through 1990. To answer the question about the sons of unskilled workers, find "lower manual" under "father's occupation" on the left side of the table and read across. The percentage breakdown shows that many of these sons (27 percent) attained "upper white-collar" (professional or

TABLE 6–1 Outflow from Father's Occupation to Son's Occupation, 1982–1990

Father's Occupation	Son's Occupation (in percent)					
	Upper White-Collar	Lower White-Collar	Upper Manual	Lower Manual	Farm	Total
Upper white-collar	60	14	11	14	1	100
Lower white-collar	39	18	19	23	1	100
Upper manual	32	10	33	24	2	100
Lower manual	27	11	25	37	1	100
Farm	19	10	22	33	16	100
Total (N = 4,632)	36	12	23	27	3	100

	Up	Stable	Down	Total
	44	36	20	100

Occupational categories are: upper white-collar—professionals, managers, officials, nonretail sales workers; lower white-collar—proprietors, clerical workers, and retail sales workers; upper manual—craftsmen and foremen; lower manual—service workers, operatives, and nonfarm laborers; farm—farmers and farm workers.

SOURCE: Data from Davis and Smith 1990; analyzed for this volume by Robert Hauser.

managerial) positions, but the largest group (37 percent) followed their fathers into unskilled "lower manual" jobs. (See the notes at the bottom of the table for a fuller description of the five occupational categories.)

Several general conclusions can be drawn from these data. (1) There is a high level of occupational succession, sons following fathers, in American society. This is especially true at the top: Well over half of the sons of managers and professionals hold similar positions. (2) But there is also considerable movement up and down the occupational ladder from one generation to the next. The figures at the bottom of table sum up the results. Forty-four percent of sons moved up, about half as many slipped downward, and the rest remained at the same level as their fathers. (3) There is a barrier, albeit a fairly porous one, between manual and nonmanual jobs (i.e., between the bottom three categories and the top two). Roughly two thirds of those born on either side of this divide will remain there. The other third moved up or down across the divide (calculated from the original data).

If Hauser had been able to use a more detailed occupational hierarchy, he would have uncovered a large volume of short-distance occupational mobility: the son of an unskilled construction laborer who becomes a semiskilled factory worker, or the son of a salesman who becomes a social worker. Both moves would go undetected by the five-step occupational hierarchy we are using here. Large-scale studies in 1962 and 1973 suggest that mobility often takes this form—a few steps up or down the ladder (Blau and Duncan 1967; Featherman and Hauser 1978).

Table 6–2 presents the same data in a form that allows us to answer a different set of questions. It is an "inflow" table, which groups the survey respondents according to their *own jobs* and asks *where they came from.* (The

TABLE 6–2 Inflow to Son's Occupation from Father's Occupation, 1982–1990

Father's Occupation	Upper White-Collar	Lower White-Collar	Upper Manual	Lower Manual	Farm	Total
Upper white-collar	36	25	11	11	4	21
Lower white-collar	14	19	10	11	5	12
Upper manual	22	20	35	22	11	24
Lower manual	20	24	29	37	5	27
Farm	8	12	15	19	75	15
Total (N = 4,632)	100	100	100	100	100	100

See Table 6–1 for source and definitions.

previous "outflow" table grouped people according to their *fathers' jobs* and asked *where they ended up*). The difference is evident in the way the tables are percentaged. The "inflow" table calculates percentages down the columns, instead of across the rows.

For example, if we read down the "farm" column on the right side of Table 6–2, we find, to no one's surprise, that most farmers are the sons of farmers. It is similarly unremarkable that most manual workers are the sons of manual workers. But reading down the upper-white-collar column, we discover that the people in these better occupations (who constitute about one third of the sons, according to Table 6–1) are of very diverse social origins. In fact, nearly half are from farm or manual homes. We get this result, in spite of the high level of father-to-son succession at the top, because the relative number of professional and managerial jobs has been growing, opening opportunities for advancement from below.

The mixed social origins of people at the upper end of the occupational structure bears on a question we have raised at various points in this book: the degree of consistency or clarity in the class system. Social mobility appears to reinforce consistency toward the bottom, while undermining it at the top. In practice, this might limit the homogeneity of values or political outlook among the people in the upper occupations.

SOCIAL MOBILITY OF WOMEN

Studies of social mobility have focused almost entirely on men. Until recently, this emphasis seemed justified by the fact that most women did not hold jobs. But it also reflected the increasingly dubious notion, which sociologists have shared with much of the public, that women's work did not matter, since so many women held part-time or sometime jobs just to supplement family income.

Studying women's mobility presents a special set of problems. For one, with whom do we want to compare women workers—their mothers or their fathers? Of course, many mothers in previous generations did not work or worked only part-time or intermittently; they were not the economic mainstay or foundation of social status for their households. If the comparison is with fathers, we face the problem that women workers are distributed across occupations in a very different fashion than men. Thus, if we see that many daughters of lower-manual men have moved "up" to white-collar work, are we looking at a difference between generations or between genders? If we find that the daughters of upper-white-collar men are concentrated in upper-white-collar positions, can we assume that they hold the same sort of jobs as their fathers? Probably not, given the segregation of women workers into a relatively small number of occupations. We know that the women in this category are much more likely than men to work as teachers, nurses, and social workers.

Table 6–3 does not resolve these problems (a task beyond the scope of this book), but it does give us a preliminary picture of women's mobility. The data employed were aggregated from successive GSS surveys in the same fashion as the data on men we just examined. The one conclusion that we can confidently draw from this table is that women's occupational achievement, like men's, is powerfully influenced by their occupational origins. For example, over half of the daughters of upper-white-collar men, but only 24 percent of the daughters of lower-manual workers, hold upper-white-collar positions.[1] Few daughters of white-collar workers hold blue-collar jobs, but nearly 40 percent of the daughters of lower-manual men hold such positions.

CHANGING CAUSES OF MOBILITY

We have been looking at the pattern of intergenerational movement up and down the occupational hierarchy. But what are the causes of this mobility? To simplify the problem, imagine a completely closed society of male workers in which all sons replicate the positions of their fathers. Consider the factors that could open it up. There are four basic ones:

1. *Circulation or exchange mobility.* Some sons slip down the scale and thereby make room for others to climb up. In this instance, mobility is a two-way street: There must be some who lose for others to gain.

2. *Occupational or structural mobility.* If technological and organizational changes are occurring in an asymmetric way that creates jobs at a faster rate in the middle and upper levels of the hierarchy than in the lower levels, then some sons will have the chance to climb into the new positions without displacing any-

---

[1] By referring to daughters and fathers in this discussion, we are, of course, using the same shorthand we used in our exposition of men's mobility. In the GSS surveys, occupational origin refers to father or, in a minority of cases, other family head—such as mother or grandfather.

TABLE 6–3 Outflow from Father's Occupation to Daughter's Occupation, 1976–1985

Father's Occupation	Daughter's Occupation (in percent)					
	Upper White-Collar	Lower White-Collar	Upper Manual	Lower Manual	Farm	Total
Upper white-collar	51	36	2	11	0	100
Lower white-collar	39	43	3	15	0	100
Upper manual	27	45	3	25	0	100
Lower manual	24	38	3	35	0	100
Farm	21	32	3	44	1	100
Total ($N = 2,462$)	32	39	3	27	0	100

See Table 6–1 for definitions.
SOURCE: Data from Hout 1988: Appendix.

body. In this instance, mobility is a one-way street, and every-body gains. (Of course, occupational change could also reduce opportunities and force some people down.)

3. *Reproductive mobility.* If men in upper levels of the system have fewer children than those in lower levels, then the system will reproduce itself in an asymmetric way that allows some sons from the bottom to climb into higher positions without forcing anybody to decline. This is another one-way street.

4. *Immigration mobility.* If many immigrants enter the system at the lower levels, they make it possible for the system as a whole to grow in ways that give native-born sons more chance to move into higher positions than they would otherwise have.

Complexities of data and measurement make it impossible to give pre-cise answers on how much mobility at any given moment comes from each of the four causes, but some partial estimates can be made and some rough trend lines drawn (Duncan 1966). One of the most obstinate complexities could be called the *multiplier effect:* If a new job is created near the top of the hierarchy, it is likely to be filled by someone from the middle, and that person's job in turn is taken by someone from below. Thus, one new open-ing can create two or more moves in a step-by-step progression. Another problem is caused by the fact that many men change jobs during their careers; therefore, intergenerational and intragenerational mobility get confounded.

Of course, a man who has advanced in the world as compared to his father has no idea which of the causes made it possible for him to get ahead, and he probably would not care if he were told. Thus, analyses of these separate factors cannot be used in a direct way to explain class con-sciousness or other subjective phenomena. But a student who looks at the system as a whole to measure the amount of mobility that exists wants to understand the main causes of that mobility and current tendencies among them.

Let us begin with occupational or structural change. We have already discussed the trends in Chapter 3, where it was shown that some occupational categories have been expanding faster than others. For example, in the decades from 1940 to 1990, professional and technical jobs for men increased over 300 percent, whereas farm jobs decreased by 75 percent. The average growth for the whole male labor force was 64 percent, as shown in Table 6–4. (Remember that in earlier discussions, we dealt with the full labor force, including women, but at the moment, we are focusing on fathers and sons.)

Concentrating on the nonmanual jobs, most of which put men into the middle and upper classes of society, we note that the categories of professional and technical, managers and proprietors, and clerical and sales expanded in absolute numbers from about 10.5 to 29.6 million positions in three decades, or a growth of 182 percent. If those positions had expanded at the same rate as the average for the labor force as a whole, there would have been only 17.1 million men in those jobs in 1970. The difference of the actual over the "expected" growth was 12.5 million positions, and that many men had a chance to move up in the system to fill the newly created jobs. Those men constituted 19 percent of the male labor force in 1990. And this estimate is an absolute minimum, since we have no way of adding in those men who moved up as a result of the multiplier effect described above. Also, we are not counting here the sizable expansion in jobs at the craftsman and foreman level, which is often called the "aristocracy" of manual labor.

Similar calculations for the three decades preceding 1940 show that the average growth for the male labor force was 31 percent, but the growth of the nonmanual jobs was 73 percent. The relatively faster growth of the better jobs allowed at least 9 percent of the labor force to experience upward

---

TABLE 6–4 Male Occupational Distributions, 1940 and 1990

	In Millions		Percent Change
	1940	1990	
Total	39.2	64.4	+64
Professional/technical	2.3	9.7	+322
Managers	3.4	8.9	+162
Sales/clerical	4.8	11.0	+129
Craftsmen	6.1	12.5	+105
Operatives	7.1	9.2	+30
Service workers	2.4	6.3	+163
Laborers, except farm	4.7	4.0	−15
Farm occupations	8.5	2.1	−75

Because of modifications in Census Bureau occupational categories and the shift from 14 to 16 as the minimum age for which occupational data are typically tabulated, these 1940 and 1990 distributions are not strictly compatible (see Table 3–4, especially 1972 tabulations). But they are an appropriate basis for the rough comparisons made here.

SOURCES: U.S. Census 1975; U.S. Labor 1991.

mobility, a somewhat smaller percentage than occurred in more recent decades. Consequently, the rate of mobility caused by occupational or structural change appears to have speeded up during the middle years of the century.

We have only limited information about the relative reproduction rates of men at different levels of the occupational hierarchy. But we do know that before World War II, the families of white-collar workers were much smaller than those of blue-collar workers, especially farmers. To take a specific example to illustrate the difference, it was estimated that 1,000 professional men had 870 sons, whereas 1,000 farmers had 1,520 sons (Kahl 1957: 257). Thus, the sons of professionals were too few in number to take over their fathers' occupational positions, to say nothing of filling the expanding number of professional jobs in the system. It was calculated that if fathers at all status levels had generated families of equal size, the mobility of blue-collar sons would have been reduced by some 7 percent in the decades before the war. Since then the differences in reproduction rates among the strata have declined markedly, thus reducing the mobility opportunities arising from this particular cause.

Until recent decades, the natural growth of the American population produced by an excess of births over deaths was not sufficient to meet the needs for labor in an expanding economy. In the eighteenth and nineteenth centuries, we had a wilderness to conquer, and millions of farmers were enticed or forced to come from Europe and Africa to clear and cultivate the land. Around the beginning of the twentieth century, the farm population began to stabilize, and the new demand for workers was in the mines and urban factories. "In 1907, the peak year of the pre–World War I immigrant inflow, one third of the nearly one million immigrants who reported an occupation upon arrival were classified as farm laborers and nearly three fourths of the total were classified as either farm or nonfarm laborers, as domestics, or as other service workers" (Wool 1976: 63). In fact, most of those who arrived as farmers took jobs in the cities. In other words, the expanding needs for unskilled and semiskilled labor were still being partly met by immigrants. Their presence was essential for economic growth and permitted native-born citizens (including the sons of foreign-born parents) to climb into the ranks of skilled manual and nonmanual workers.

The First World War was a major turning point. After it was over, modern methods of farming increasingly replaced men with machines, and the farm population began to decline, both relative to the total and also in absolute numbers: sons of farmers moved to the city seeking work. In 1910, there were 10.4 million male farm owners and laborers; in 1940, there were 8.5 million, and by 1970, only 2.2 million. These "internal migrants" were somewhat more prepared for city life than the average immigrant: They spoke English, most were literate, and indeed, some were well educated. But the majority of them moved (at least at first) into the lower levels of skill in the city labor force. This was particularly true of the blacks who left the farms of the South (where 90 percent had lived) and streamed toward the cities of both North and South. The new availability of more-than-ample numbers of internal migrants collided with the new technology in the cities that produced more goods with fewer men. The expansion of unskilled jobs began to slow just as these flows of rural-to-urban migration peaked.

The political response to these combined trends was a drastic change in our immigration laws, which cut the flow from almost a million a year before the First World War to less than half that number after it. In proportion to our now larger population, the cut seems even more drastic (Higham 1963). Furthermore, the newer immigrants in the 1950s and 1960s were more skilled and entered the labor force at all levels, so they had less overall effect on the mobility chances of the native born.

The situation keeps changing. By the late 1970s, the pool of excess farmers had dried up (including Puerto Ricans), and the flow of rural-to-urban migration had become a trickle. Simultaneously, a new phenomenon has appeared: large numbers of illegal or undocumented immigrants, two thirds or more from neighboring Mexico. They slip into the country against the wishes of the authorities, and nobody knows for sure how many there are, although estimates indicate several million of them. At first, they were mainly temporary agricultural workers who came for a few months during harvest time in Texas and California. But now even those jobs tend more and more to be done by machines, and many undocumented workers are found in Los Angeles and Chicago and New York working in tasks that do not demand high skills. They are especially useful to marginal entrepreneurs (for example, in the garment or restaurant industries) who need cheap workers willing to toil long hours under unsavory conditions without complaining to official inspectors. Native-born workers have higher expectations and tend to refuse such jobs; if unemployed, they can subsist through the work of other members of the family or by turning to the welfare system. Thus, it is possible to have illegal immigrants working while the unemployment rate among citizens remains high (U.S. Labor 1978: chapter 5).

It is difficult to quantify the exact impact of immigration on the mobility chances of native-born citizens, but the trend line is clear enough. In the earlier years of this century, a significant number of Americans experienced mobility into higher jobs, because many lower ones were filled by willing immigrants. In recent years, the overall effect of legal immigration on mobility has become minor, and the effect of illegal immigration is not known for sure but is probably less than the impact of the migration before World War I.

TRENDS IN SOCIAL MOBILITY

In earlier chapters, we examined trends in income, wealth, and occupational structure. What can we say about trends in social mobility? Is there more or less upward mobility than there was in the past? How has the pattern of mobility changed?

We have two landmark studies of mobility: *The American Occupational Structure* (1967) by Peter Blau and Otis Dudley Duncan, and *Opportunity and Change* (1978) by David Featherman and Robert Hauser. These books present sophisticated analyses of social mobility, based on massive national surveys conducted in 1962 and 1973. In the next chapter, we will look at their conclusions about the determinants of individual mobility, but here we will focus on their measurement of the overall pattern of mobility. Table 6–5 presents the gross mobility results from these two surveys, together with Hauser's estimates for the 1970s and the 1980s compiled for this book. Unfortunately, data from the first two are not perfectly compatible with the data from the GSS polls used by Hauser. One problem is that Hauser's estimates refer to men aged 25 to 64 at the time of the survey, whereas the earlier surveys cover men 21 to 64. We will have to be cautious about any comparisons between the first two and the second two.

Moving from period to period in the table, we find that the sons of lower-manual men experienced declining succession and increasing upward mobility. The proportion of farm sons following their fathers, already low in 1962, plunged further by 1973. It also appears that there was a big leap in the proportion of sons following fathers into upper-white-collar professions in the 1970s. But the difference between the 1973 and 1972–1980 figures is probably the result of the exclusion of men in their early 20s from the data for the latter period. These young men were still completing the schooling or early work experience that would allow them to move into upper-white-collar jobs.

For each period, the bottom row of numbers (Total) gives us the occupational distribution of all sons. These figures, traced from panel to panel, provide a way of tracing structural mobility. They indicate—as we might expect—that more sons are finding upper-white-collar jobs, and fewer are working on the farm. This shift in occupational structure creates what we might describe as a mobility updraft, drawing people into the higher positions. On the other hand, this phenomenon is playing itself out as the proportion of farm fathers declines and the proportion of fathers in the top stratum increases. (The rightmost column gives the distribution of respondents' fathers).

Judging from the summary statistics at the bottom of the table, the overall pattern of mobility has not changed dramatically over the last three decades. The most notable change seems to be between the 1970s and 1980s, when upward mobility dropped off, downward mobility correspondingly increased, and movement both ways across the blue-collar/white-collar line increased by a few percent.

Since the biggest differences are the most recent, we should take a closer look at the last two periods. One way to do this is to concentrate on workers in the first half of their careers. These younger workers are the most likely to feel the effects of change. Focusing on them reduces the

TABLE 6–5 Outflow from Father's Occupation to Son's Occupation, 1962–1990

Father's Occupation	Son's Occupation (in percent)						
	Upper White-Collar	Lower White-Collar	Upper Manual	Lower Manual	Farm	Total	Row %
1962							
Upper white-collar	54	17	13	15	1	100	17
Lower white-collar	46	20	14	18	2	100	8
Upper manual	28	13	28	30	1	100	19
Lower manual	20	12	22	44	2	100	27
Farm	16	7	19	36	22	100	29
Total	28	12	20	32	8	100	100
1973							
Upper white-collar	52	16	14	17	1	100	18
Lower white-collar	42	20	15	22	1	100	9
Upper manual	30	13	27	29	1	100	21
Lower manual	23	12	24	40	1	100	29
Farm	18	8	23	36	15	100	23
Total	30	13	22	31	4	100	100
1972–1980							
Upper white-collar	65	11	11	12	0	100	14
Lower white-collar	45	22	15	18	0	100	12
Upper manual	29	11	33	26	1	100	24
Lower manual	25	11	23	40	1	100	28
Farm	16	9	24	34	18	100	23
Total	32	12	23	27	5	100	100
1982–1990							
Upper white-collar	60	14	11	14	1	100	21
Lower white-collar	39	18	19	23	1	100	12
Upper manual	32	10	33	24	2	100	24
Lower manual	27	11	25	37	1	100	27
Farm	19	10	22	33	17	100	15
Total	36	12	23	27	3	100	100

	Up	Stable	Down	Total
1962	49	34	17	100
1973	49	32	19	100
1972–1980	50	35	16	100
1982–1990	44	36	20	100

	Blue-Collar to White-Collar (As % of Sons with Farm or Manual Fathers)	White-Collar to Blue-Collar (As % of Sons with White-Collar Fathers)
1962	31	31
1973	35	34
1972–1980	34	28
1982–1990	37	32

See Table 6–1 for definitions of occupational categories.

SOURCES: Featherman and Hauser 1978 and GSS (Davis and Smith 1990).

birth-cohort overlap that dilutes the contrast between successive surveys. (For example, the cohort of men who were 45 to 54 in 1980 would be 55 to 64 in 1990 and therefore be represented in surveys both years.)

Using the same GSS data he employed for the general mobility analysis, Hauser looked at the experience of men, 25 to 44, in the 1970s and 1980s. The results are reproduced in Table 6–6. Two differences are immediately apparent. First, in the 1980s, upward mobility was less common for young men, and downward mobility was more common. Second, young men found it harder to move into the top occupational category. Look down the upper-white-collar columns for the two periods. In virtually every case, a smaller proportion of sons reached this level. Even the sons of upper-white-collar fathers had a tougher time retaining their fathers' occupational status. The sons of men in both white-collar categories were more likely to be bounced down into lower positions than they were in the 1970s.

What about the mobility of younger workers across the blue-collar/white-collar line? In the 1980s, the proportion of white-collar jobs increased slightly. Nonetheless, upward mobility across the line *decreased*, and downward mobility *increased*.

The 1980s drop in overall mobility was, in good measure, the result of the increasing status of fathers. From the 1970s to the 1980s, the proportion of fathers in the top two categories rose, and the proportion in the bottom three fell—thereby reducing the possibilities for upward movement by sons.[2] This brings us back to the question of structural mobility. How has the changing occupational structure affected the overall level of intergenerational mobility?

In the last section, we looked at this question using national occupational statistics going back to the early twentieth century. In Table 6–7, we chart an index of structural mobility based on the difference between the occupational distributions of sons and fathers. The index is the minimum percentage of sons who have moved because of changes in occupational structure between generations.[3] The fall of the index from 1962 to the 1980s suggests that structural mobility has been declining, especially in recent years. The trend is clearest among younger workers. We may conclude that shrinking structural opportunities for mobility were contributing to the decline in overall mobility during the 1980s.

---

[2]If, hypothetically, the sons of the 1980s had the fathers of the 1970s (that is, had the same social origins as workers in the 1970s), overall mobility in the second decade would have looked like this: 45.0% up; 34.2% stable; and 20.7% down. These figures were calculated by taking the numbers of men in each category of occupational origin in the 1970s and assigning them destinations in the proportions recorded in the 1980s.

[3]The index is calculated by subtracting the percentage of fathers from the percentage of sons for each occupational category in which the latter exceeds the former. The sum of these differences is the value of the index.

TABLE 6–6   Mobility of Younger Workers (25–44): Outflow, 1972–1990

Father's Occupation	Son's Occupation (in percent)						
	Upper White-Collar	Lower White-Collar	Upper Manual	Lower Manual	Farm	Total	Row %
1972–1980							
Upper white-collar	61	10	15	14	0	100	24
Lower white-collar	47	26	12	15	1	100	7
Upper manual	32	10	32	26	1	100	25
Lower manual	28	9	23	39	1	100	28
Farm	15	8	27	33	17	100	16
Total ($N = 1{,}780$)	36	11	23	27	3	100	100
1982–1990							
Upper white-collar	54	11	15	19	1	100	31
Lower white-collar	40	25	13	21	1	100	9
Upper manual	30	9	34	26	2	100	24
Lower manual	26	11	26	37	1	100	26
Farm	16	9	22	33	20	100	11
Total ($N = 2{,}282$)	36	12	23	27	3	100	100

	Up	Stable	Down	Total
1972–1980	47	36	18	100
1982–1990	38	37	24	100

	Blue-Collar to White-Collar (As % of Sons with Farm or Manual Fathers)	White-Collar to Blue-Collar (As % of Sons with White-Collar Fathers)
1972–1980	45	19
1982–1990	40	29

See Table 6–1 for source and definitions.

TRENDS IN BLACK MOBILITY

The 1962 and 1973 mobility surveys we have been examining were large enough to include a significant sample of African-American men. The general results reveal a remarkable change over a relatively short period (Table 6–8). Upward mobility increased, especially mobility from blue-collar to white-collar positions. Slippage from white-collar to blue-collar became much less common. Data broken down by age show that even larger gains were made by men 45 or younger (Featherman and Hauser 1978: 326).

In the 1973 survey, fathers in higher occupational categories were much more likely to pass their advantage on to their sons. In 1962, for example, only 13 percent of the sons of upper-white-collar men had similar occupations. By 1973, 43 percent did. Nonetheless, the mobility chances of black men still lagged behind those of whites, even when they started at similar levels. For example, in 1973, at all five levels of occupational origin, white men had a better chance than African-American men of attaining upper-

TABLE 6–7 Index of Structural Mobility, 1962–1990 (in percent)

	All Sons	Sons 25–44
1962	22	—
1973	19	—
1972–1980	19	18
1982–1990	14	10

SOURCE: See Table 6–5.

white-collar status (compare the right-hand columns for 1973 in Tables 6–5 and 6–8).

Unfortunately, no mobility data are available for African-Americans in the 1980s. It would be especially interesting to see how blacks who entered the labor force after the civil rights revolution of the 1960s have fared. However, the occupational data that we examined in Chapter 3 suggest continuing improvement in mobility prospects for African-Americans.

CONCLUSION

Four basic factors described in this chapter influence the amount of social mobility:

1. *Circulation or exchange.* Some sons slip, creating opportunities for others to rise.

TABLE 6–8 Mobility of African-American Men: Outflow, 1962 and 1973

Father's Occupation	Son's Occupation (in percent)						
	Upper White-Collar	Lower White-Collar	Upper Manual	Lower Manual	Farm	Total	Row %
1962							
Upper white-collar	13	10	14	63	0	100	4
Lower white-collar	8	14	14	64	0	100	3
Upper manual	8	11	11	67	3	100	9
Lower manual	7	9	11	71	2	100	37
Farm	1	5	7	66	20	100	48
Total ($N = 1,122$)	5	8	9	68	11	100	100
1973							
Upper white-collar	44	12	8	36	0	100	4
Lower white-collar	20	21	13	46	1	100	4
Upper manual	16	14	16	54	0	100	10
Lower manual	12	12	14	61	1	100	46
Farm	5	7	17	63	8	100	36
Total ($N = 3,493$)	12	11	15	59	4	100	100

SOURCE: Featherman and Hauser 1978: 326.

2. *Occupational or structural change.* As the economy evolves, the occupational structure is altered, opening and closing opportunities.
3. *Differential reproduction.* Class differentials in fertility can create room at the top.
4. *Immigration.* New immigrants take bottom positions as the native born move up.

Mobility data from the 1980s reveal a high level of father-to-son succession. Nonetheless, about two thirds of sons have moved up or down from their fathers' occupational level, and upward mobility is far more common than downward mobility. Daughters' mobility cannot be described in quite the same terms as sons', but similar data show that father's occupation is a strong influence on daughter's occupational destination. We do not have recent data on the mobility of African-American men. But we do know that in the early 1970s, they still had slimmer mobility chances than white men from similar occupational backgrounds. Nonetheless, their situation had improved from what it was in the 1960s.

Over the last generation, the overall pattern of mobility has been relatively stable. Upward mobility, for example, has consistently outweighed downward mobility. Succession to father's status is little different now than it was in 1962. Upward mobility has declined, but only slightly. Movement either way across the blue-collar/white-collar divide has remained close to one third of the workers in the surveys.

But when we sharpen our focus and look at the experience of younger men in recent years, we find evidence of stagnating opportunities. Significantly more workers are downwardly mobile, fewer are moving from blue-collar backgrounds to white-collar careers, and it is becoming harder to move into the upper-white-collar stratum. Furthermore, it is evident that the structural change that has fueled mobility in the past (#2 in our list above) is waning. These findings regarding the experience of younger workers are our best window on the future. What they suggest is not encouraging.

These trends could have powerful political implications. Already many young workers fear that they will not have the same opportunities for advancement that their parents enjoyed and that they may not even be able to maintain the living standard that they grew up with. Especially in periods of economic stagnation, such feelings could turn them against politicians they identify with the status quo.

SUGGESTED READINGS

Erikson, Robert, and John Goldthorpe. 1985. "Are American Rates of Social Mobility Exceptionally High? New Evidence on an Old Issue." *European Sociological Review* 1: 1–22.

Comparison with U.K. and Sweden. References to other important literature on this topic.

Featherman, David, and Robert Hauser. 1978. *Opportunity and Change.* New York: Academic Press.

Detailed report on the last major national mobility survey. Comparisons with 1962 survey.

Hout, Michael. 1988. "More Universalism, Less Structural Mobility: The American Occupational Structure in the 1980s." *American Journal of Sociology* 93: 1358–1400.

Trends in mobility of men and women from 1972 to 1985.

Newman, Katherine S. 1988. *Falling From Grace: The Experience of Downward Mobility in the American Middle Class.* New York: Free Press.

How ex-managers, laid-off skilled workers, unemployed professionals, divorced women, and their families respond to their fall from economic grace.

7 Family, Education, and Career

The transmission of property from generation to generation, in the same name, raised up a distinct set of families, who, being privileged by law in the perpetuation of their wealth, were thus formed into a Patrician order, distinguishable by the splendor and luxury of their establishments. From this order, too, the king habitually selected his counsellors of State; the hope of which distinction devoted the whole corps to the interests and will of the crown. To annul this privilege, and instead of an aristocracy of wealth, of more harm and danger, than benefit to society, to make an opening for the aristocracy of virtue and talent, which nature has wisely provided for the direction of the interests of society, and scattered with equal hand through all its conditions, was deemed essential to a well-ordered republic.

Thomas Jefferson (1821:38)

Jefferson proposed to do away with the aristocracy of wealth by changing the laws of inheritance to encourage men to divide their landed estates equally among all their children and thus eventually arrive at small holdings. He proposed to establish the aristocracy of virtue and talent, characteristics he assumed were "scattered with equal hand" among all strata, by creating a public school system in the state of Virginia that would provide primary education for all citizens and then select the best students for further training in high schools and universities. In his later years, he founded the University of Virginia as the capstone to this system and was so proud of it that he ordered that his tombstone should record his two greatest accomplishments: the writing of the Declaration of Independence and the founding of the University of Virginia.

From Thomas Jefferson forward, American political leaders have endorsed high rates of mobility. In doing so, they have reflected the general values held by most Americans, which accept some inequality in society, believing that people should get different rewards for different kinds of work, but also believing that each generation should start fresh and compete in a "fair" way for those rewards. In other words, we believe that young people ought to have careers based on their own talents and desires rather than have their lives determined by the class positions of their parents; they should have "equality of opportunity" but not "equality of result." Accepting the fact that some aspects of the talent and desires of the children are inherited from or shaped by the parents, we nevertheless believe that a high degree of equality of opportunity can be achieved through the school system. If all children have access to good schools at all levels, *regardless* of the financial resources of their families, then the graduates should be able to compete on a reasonably fair basis.

The practical means for achieving this end began in our early history with free elementary schools in every village and town, supported at first by churches (more interested in equal access to the Bible than to economic success) but later by local tax money. Later, free high schools were added, and in most small towns, they are now the most prominent buildings to be seen. In the mid-nineteenth century, free or inexpensive state universities were founded with federal aid (Jefferson was a generation ahead of his time). In recent decades, a variety of new programs have been created, ranging from Head Start classes for poor youngsters before first grade, through vocational and job training, to college scholarships and other forms of affirmative action for bright but economically disadvantaged students. Through more schools and more financial support programs, it was thought, most children would be given access to training appropriate to their talents and aspirations, thus increasing equality of opportunity for all.

However, this approach was based on the assumption that the problem could be solved by various government programs that would keep disadvantaged youngsters in school longer, in combination with financial help to poorer schools that would make them "better." But behind that assumption were three beliefs about the way schools worked in the real world:

1.   Good schools can overcome handicaps from other social forces, particularly family background.
2.   Bad schools can be made better by spending more money on them.
3.   There is a close connection between the amount of education a youngster gets and the kind of career she or he will have.

To the degree that those beliefs are not correct, government attempts to increase equality of opportunity through changing the schools to make career competition more "fair" will fall short of their goals.

Recent research has cast partial doubt on all three beliefs. The massive work of James Coleman and his associates in the mid-1960s showed that the money available to particular primary and high schools did not much influence the amount the students knew. Furthermore, what the students did know was influenced more by family background than by anything else the researchers could measure. Indeed, the longer they stayed in school, the more the children of higher-strata families moved *ahead* in their test scores while the children of lower-strata families slipped *behind*. In other words, instead of removing family handicaps, the schools often appeared to accentuate them (Coleman et al. 1966; Averch et al. 1972; Hodgson 1975).

A few years later, the work of Christopher Jencks and his associates reported that the income earned by a man was only loosely related to the number of years he stayed in school (although that was a better predictor than any other single variable that could be measured). A lot of other factors intervened that shaped careers once people left school and started working—so many factors, and so hard to pin down, that Jencks at one point dismissed them as "luck." The implication was that even if we equalized schooling, we would not equalize career outcomes very much (Jencks et al. 1972 and 1979).

Both of these researches have become highly controversial. They stimulated new studies and even more debates about just what the studies meant. The issues raised are so essential to the understanding of both succession and mobility, and are so important to policy decisions about how the schools should be managed, that we must examine the discussion in considerable detail (for a cogent critique, see Aaron 1978: chapter 3).

Perhaps the main reason why the connections between family background, education, and career outcome are so debatable is the obvious fact that so many things influence one another: It is hard to separate out just what determines what at each stage of a person's development. To take a few examples: Some aspects of intelligence may be transmitted through the genes, but since they do not show up until they interact with environmental forces that shape a child's behavior, we cannot be sure what portion of observed and measured intelligence is genetic and what portion is environmental. (And since the subject involves politically volatile issues connected with race and ethnic background, the research is often designed to prove a point more than to provide scientific enlightenment.) Indeed, we do not fully understand exactly *how* the early social environment—ranging from nonverbal interaction between mothers and babies, through family styles of speech and reading, to contact with other children in the local neighborhood and the school—shapes those aspects of intelligence that we can measure through tests. (Only through standardized tests, with all their faults, can we study large samples and come to any general conclusions.) And we do not know very much about how other personality factors, such as energy level, sense of discipline to aim at long-term goals, and ability to deal with other people, combine with intelligence as measured by the tests to influence the outcome of schooling and the impact of schooling on occupational career. Nor do we fully understand how the subtle parts of the school environment, such as the way the teachers stereotype the children into those who are "good learners" and "bad learners," begin to form the self-images of the children and thus influence their ambitions and their behavior. Those stereotypes can be related to sex, to race, or to socioeconomic status: Girls are often supposed to be good at poetry, boys at mathematics, blacks at basketball, the poor at auto mechanics, and the rich at medicine, and teachers sometimes tell them so.

Many things in life are going on at once and influencing one another, but systematic research must make some choices and concentrate on a few of them. We will emphasize the standard variables of stratification research and only mention superficially the most relevant aspects of personality and school environment. We will do so by following a formal model that is based on the idea of a sequence of events in which several measurable traits of earlier years are correlated with outcomes in later years. Instead of just relating father's occupation with son's occupation, as we did in the last chapter, we will follow the various pathways through which the former does and does not influence the latter. In the lingo of mobility models, this sequence is often called the *process of status attainment*. The inequalities among individuals in occupational prestige, income, and education that are the subject matter of these models are typically referred to as differences in *socio-economic status* or SES.

A warning to the reader is important at this point. By committing ourselves to the type of formal model we shall use, we help untangle some of the main variables that are at work. However, to handle the large samples we need in order to follow several variables over long periods of time, we automatically oversimplify the real world. There is a great temptation to jump from the model to reality, to forget that when we speak of "education," for instance, all we mean is the number of years of school attended (regardless of the type of school or the style of teaching). Each variable is measured by a particular operational definition that may be picking up a lot of "noise"—many aspects of life connected with that definition that we are not aware of. And it may imply to the reader still other characteristics not intended by the researcher that may or may not actually exist. The conclusions stated are based on the statistical manipulation of the variables as defined. The fit between this type of model and the real world is never as good as we would like it to be.

STATUS ATTAINMENT: BLAU AND DUNCAN

We begin by returning to the classic study of mobility by Peter Blau and Otis Dudley Duncan (1967) that we referred to in the last chapter. Their analysis of mobility tables was a refinement on earlier work but not a major change. However, the second part of their book was a creative advance that set off a flurry of follow-up studies in this country and abroad. In that new analysis, they turned their attention away from gross amounts of succession and mobility and toward the specific sequence of events that shaped the careers of individual men. They did so not with tables of cross-classification, which would have become unwieldy, but with a form of regression analysis that produced path diagrams. Much of this analysis was worked out by Duncan, and we shall refer to it by his name.

The simplest form of regression analysis computes the well-known correlation coefficient; it gives an indication of the degree to which one variable is connected with another. If you can completely predict everybody's score on the second, or *dependent*, variable from a knowledge of their score on the first, or *independent*, variable, then the coefficient is 1.00. If there is no relation between the scores of persons on the independent and the dependent variables, the coefficient is 0.00. On most matters that concern social science, there is an in-between relationship, and the correlation coefficients are intermediate.

For example, we can take our group of sons and give each person two scores using the Duncan index of occupational prestige discussed in

Chapter 2: one for the son's occupation and one for that of his father. The correlation coefficient between the scores for fathers and those for sons will be about 0.41. Strictly speaking, this allows us to predict partially the scores of sons from the scores of fathers, and we can convey the size or power of that prediction by indicating the percentage of the total variation in the positions of sons that can be accounted for by our knowledge about their fathers.[1] We get this percentage by squaring the correlation coefficient: 17 percent of the variance is explained. That percentage may seem low and suggest that our nation comes close to being an open society, in which sons find their own careers without much influence from their fathers, but the data we examined in Chapter 6 showed that in concrete terms, there is a clear influence from fathers to sons.

The goal of Duncan was much more ambitious than to measure the relationship between the occupations of fathers and those of their sons: He wanted to trace out the specific pathways of influence. Just *how* did the fathers influence their sons? The model he developed was based on sequence in time, with the assumption that earlier events accounted for later events. Thus, he measured the father's education as an early background factor. Then he measured the father's occupation when the son was sixteen years of age, just at the point of deciding about education and career. Then he indicated the son's education by counting the number of years he stayed in school. Then he computed the status index of the son's first job, and finally the job he held at the time of the interview in 1962. Thus, he put five variables into his model.

The first step in building up the model was to calculate the simple or zero-order correlation coefficients among the five variables, and they are shown in Table 7–1. Inspection of the table begins to show the pattern: The most important correlate of a son's current job is his own education; after that, in descending order, come his first job, his father's job, and his father's education. About a third of the variation in the prestige of men's jobs for the whole sample is predicted or explained by the amount of their education; that relationship is the highest in the table, yet still leaves a large residual of unexplained variance.

But then we run into a new puzzle. A man's education is predicted to some degree by his father's occupation (explained variance is about one fifth). So we might well wonder: how important as an independent cause is a son's education? Maybe there is a *chain of causes* at work: father's occupation (which serves as a proxy for all his SES variables) "causes" son's education, which "causes" son's job. If the links in that chain are very tight, then the son's education is merely a mediating or intervening variable that explains the process through which the father influences the son's career

---

[1] We often tend to substitute *caused by* for *predicted by*, but doing so is dangerous, since we need a clear theoretical understanding of the chains of linkage before we can safely speak of causation. Sometimes correlations are coincidences or are caused by a complex chain of influences not revealed by the data at hand.

TABLE 7–1 Simple Correlations among Five Variables, Fathers and Sons

| | Variable | | | |
Variable	First Job, Sons	Education, Sons	Fathers' Occupation	Fathers' Education
1962 occupational status of sons	.54	.60	.41	.32
First-job status of sons	—	.54	.42	.33
Education of sons		—	.44	.45
Fathers' occupational status			—	.52
Fathers' education				—

SOURCE: Blau and Duncan 1967: 169.

but has little independent effect of its own. That is, high-status fathers keep their sons in school and thus assure their success in life. On the other hand, if the linkages between the first two variables are weak, and the last two linkages are strong, then the son's education could have a lot of *independent* influence and indeed could be the key to a position *different* from that of the father. That is, many poor but bright sons use the schools to achieve mobility.

A little further thought produces another complication. Maybe the father's status position influences the son's career more than once. It certainly has some impact on the education the son gets, but perhaps it has an additional impact later in life, helping the son to get a good first job and then advance in his career. After all, some craft unions give preference to the sons of union members in getting admitted, and people in some businesses and professions can benefit a lot from family connections. Simple or zero-order correlation coefficients merge these two influences of father on son into a single number, the correlation between the occupations of fathers and sons.

The only way to sort out these influences is to create a flow chart in which one follows the son's career through several stages and recalculates the impact of earlier forces on that career at each stage. The goal is to separate out the *special* influence of *each* variable, *independent* of the other variables in the model (that is, to assume that they are held constant). If we are sure of the sequence because we have a theory that tells us what comes first and what comes later, eliminating worry over the direction of causality connecting the variables, we can organize the correlations into a special form that produces a path diagram. The model, introduced to sociology by Duncan, is shown as Figure 7–1, and since the events mostly follow each other in time, there can be little argument about the sequence.

Note that most variables are connected by a straight arrow; the coefficient on the arrow shows the relative influence of the first variable on the second, independent of all the other paths in the diagram. For example: Father's occupation is linked to son's education with a coefficient of 0.28. That shows the "direct" connection. Furthermore, son's education is connected directly with his current occupation in 1962 with a coefficient of

FIGURE 7–1 Path Coefficients among Five Variables, Fathers and Sons

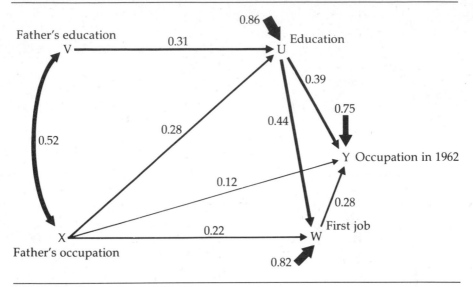

SOURCE: Blau and Duncan 1967: 170.

0.39. Therefore, the "indirect" effect of father's occupation on son's job via the intervening variable of son's education must be a combination of those two coefficients (although not a simple arithmetic sum). But in addition to the indirect influence through education, the father's occupation also influences the son's occupation directly, with a coefficient of 0.12, which shows some small influence that goes through routes *other* than education, such as the father's help to the son's career that comes after schooling is over.

What do these numbers mean? One can read the chart by simply comparing the sizes of the coefficients (or in this version, the thickness of the arrows) in order to get an intuitive grasp of the *relative* strength and weakness of different paths, always remembering the distinction between a simple direct path and a complex indirect path that cumulates influences. One can visualize how the overall relationship between the early variables of parental status and the final dependent variable, son's career, can be decomposed into chains of events.

There are rules for cumulating various coefficients to make statements about how the total variance is decomposed into pathways.[2] Following

---

[2] The gross effect of father's occupation on son's occupation, direct and indirect, is shown by the simple correlation coefficient between X and Y, or 0.41, which explains about 17 percent of the variance (since the variables are given in standardized form, an increase of 1 standard deviation in X produces an increase of 0.4 of a standard deviation in Y). Of that gross effect, about a quarter (0.12 divided by 0.41) is produced by the direct connection. Another quarter is produced by the indirect connection through son's education (0.39 times 0.28, which equals 0.11 divided by 0.41)—the indirect coefficients are multiplied to obtain their cumulative effect. For details, see Blau and Duncan 1967: 166–167, and Jencks 1972: 354–356.

those rules, we can say that of the total relationship between father's occupation and son's occupation, about one quarter is produced by the direct connection, and another quarter is produced by the indirect connection via education. The rest comes from other indirect connections (such as the impacts of the father's occupation and the respondent's education on the first job, which in turn influences later jobs).

The reader should note the short arrows that come into the chart without any previous ties: Each indicates the residual coefficient measuring unexplained influences up to that point. For instance, the residual coefficient for occupation at the time of the interview in 1962 was 0.75; its squared value is 0.56, which means that 56 percent of the variance in the occupational prestige scores of sons remains unexplained after all the independent variables in the model are cumulated, both directly and indirectly. The model "explains" about half of what is going on but leaves the other half unexplained. That is one way of saying in numbers what we have previously said in words: There is a lot of mobility in American society, and the occupational positions of men cannot be predicted with great precision from a knowledge of their family backgrounds, *even if you add knowledge about their educations.* That in turn implies that schooling may be important for career but does not completely determine the outcome.

Blau and Duncan warn us that the large unexplained variance does not necessarily mean that the model is weak, since the residual paths are "standing for all other influences on the variable in question, including causes not recognized or measured, errors of measurement, and departures of the true relationships from additivity and linearity, properties that are assumed throughout the analysis. . . . The relevant question about the residual is not really its size at all, but whether the unobserved factors it stands for are properly represented as being uncorrelated with the measured antecedent variables" (Blau and Duncan 1967: 171, 175). Subsequent research does suggest that improvement in measurement techniques to reduce error would increase the explained variance, so the model is somewhat stronger than the numbers suggest (Bielby et al. 1977; Corcoran 1978). Beyond such improvements, the usual temptation is to build in more redundancy (for example, adding the mother's education). Since much of the impact of a redundant variable is already included in some variable in the system with which it is correlated (such as father's education), adding the new variable will not greatly change the pattern but will certainly increase confusion to the eye. And it is the pattern that mainly intrigues us: the *relative* weight of the variables. Nevertheless, the unexplained variance is always a challenge to the next investigator, who tries to do better.

Let us summarize the results of the Blau and Duncan model. Using simple correlations, they indicated that 17 percent of the variance of the occupational status of sons could be accounted for by the occupational status of their fathers. Furthermore, they showed that the fathers' status explained 19 percent of the variance in the length of education achieved by the sons, which led to the suspicion that part of the function of education was to promote succession rather than mobility: middle-class sons stayed in school longer than working-class sons, and that allowed them to get better jobs (a point strongly emphasized by Bowles and Gintis 1976). The path coefficients showed that the fathers' education and occupation had about equal influence on the sons' education, and that the sons' education (including the part of it explained by family background and the part that was independent of family background) was the biggest single influence on sons' eventual occupation, but that there was a small additional influence on career that the family background could continue to assert even after the sons had graduated.

JENCKS ON INEQUALITY: PART ONE

The new challenges to the conventional wisdom about the equalizing power of the schools, particularly the work of James Coleman and his colleagues, led a group at the Harvard School of Education, under the leadership of Christopher Jencks, to try to integrate most of what was known into one grand model of status attainment that would include more variables than Blau and Duncan had studied. The new book was called *Inequality: A Reassessment of the Effect of Family and Schooling in America.* It was published in 1972 and created quite a stir.

Jencks did not collect original data but rather accepted the Duncan results, with two useful additions. (1) He interpolated a lot of material from other studies. (2) He organized it into a series of discrete stages corresponding to points in the life cycle, instead of relying on a single path diagram that summed up the whole process.

Jencks added income as the final dependent variable in the chain rather than occupation, and reported that there is only a moderate zero-order correlation, 0.44, between occupational status and income (using "estimated true" coefficients, which adjust the figures upward to compensate for measurement errors, and limiting the sample to native white males of nonfarm background, aged 25 to 64 in 1962). Obviously, men whose occupational titles are the same often earn different amounts of money. There are many reasons for that: varying pay scales in different parts of the country; the increases that come from seniority on the job; variations that come from the size and profitability of the firm or the strength of the union; and the fact that the occupational title itself is only an abstraction that covers different tasks and different amounts of authority. (See Kalleberg and Griffin 1980.)

Jencks also found a correlation of only 0.35 between respondents' education and income and only 0.29 between fathers' occupation and son's income. Every time we add another variable into a chain of causes and effects, we weaken the connection between the earliest predictors and the final outcome, because at each step in the process, other factors are at work (or "luck" keeps reappearing). Consequently, this new model, which uses income as the final dependent variable instead of occupational prestige, reduced the measured impact of education on life chances.

The Harvard team also wanted to measure somewhat more sharply the "pure" effect of years of schooling, independent of the qualities that students bring with them to the school building. To do so, they added another new variable to the chain: the son's IQ at age 11, on the assumption that it is the best available measure of the talent or "cognitive ability" of the child that reacts with the stimuli of the school. (That got them into the issue of genes versus early socialization experiences in the home as "causes" of IQ, but since schools cannot affect those proportions, we shall pay no attention to the issue here.)

In organizing the vast statistical material from many different studies that is integrated in *Inequality,* Jencks tended to stress a polemic point. Economic (occupational and income) inequality in the United States is very large, and simply improving the quality of bad schools and reducing the differences among individuals in the number of years they attend school will not go very far in eliminating the economic differences. He estimated that most—indeed close to three quarters—of the variation in the incomes of adult men cannot be explained by cumulating all the usual predictors: family background, IQ score, years of education, and even job title. He wrote:

> Economic success seems to depend on varieties of luck and on-the-job competence that are only moderately related to family background, schooling, or scores on standardized tests. . . . The fact that we cannot equalize luck or competence does *not* mean that economic inequality is inevitable. Still less does it imply that we cannot eliminate what has traditionally been defined as poverty. It only implies that we must tackle these problems in a different way. Instead of trying to reduce people's capacity to gain a competitive advantage on one another, we would have to change the rules of the game so as to reduce the rewards of competitive success and the costs of failure (Jencks et al. 1972: 8).

In other words, if we really propose to equalize incomes, we must do something that *directly* alters the operation of the labor market to reduce the range of incomes: Corporation presidents should receive less, and people who fry hamburgers should receive more. The amount of government intervention required to produce such a result is often call socialism, and Jencks was not afraid of adopting the word. He implied that our national preoccupation with changing the schools was a distraction from the real issue.

JENCKS ON INEQUALITY: PART TWO

The reaction to Jencks' conclusions was strong; he was not faulted on technical grounds, but many critics were unhappy about the implications of the words *luck* and *socialism.* Jencks and colleagues set out on a second round of research, including the integration of several new surveys of careers (including those of brothers) that had since become available. The second book, *Who Gets Ahead?*, published in 1979, was much less controversial, for at least two reasons. First, it gave up the noble attempt of the earlier volume to communicate with the general public and instead offered dry and statistical prose in a highly technical style. Second, it avoided the discussion of policy implications for the most part. The authors no longer emphasized the difficulty of creating more equality of result; they simply reported on the amount of variance in schooling, job, and income that can be explained by different background characteristics.

These policy weaknesses of the book were offset by a prodigious amount of new work, using many new sources of data, that certainly increased the accuracy of the calculations, although it did not dramatically alter the conclusions. Let us report a few of the more significant analyses.

Who Gets Ahead? made a number of changes in procedure, but the most interesting for our purposes comes from a recognition that the standard SES measures on parents are crude measures of the total effect of family background. Although it gave those measures, it also added, where possible, a measure based on the similarity of the careers of brothers. Jencks assembled some new data of his own as well as material available in other studies in order to increase knowledge about what happens to boys reared in the same home, although the difficulty of tracing both brothers keeps the samples much smaller than one would like. The resemblance between brothers is a total measure of family influence that is useful but not very specific: It covers shared genes, shared environment in the home and the neighborhood, and the possible influence of one brother on the other. It predicts career outcome somewhat better than that conventional SES indicators, thus reducing the size of that bothersome residual variance.

Jencks and his colleagues also rounded up a number of scores that went beyond the usual IQ scores, including measures of personality. They found that no one of the additional measures was particularly powerful, but a combination of them predicted career outcome about as well as the IQ score by itself and seemed to be adding something new to the mix.

Let us start with the overall conclusion and then break it down into a series of sequences that follow one another. *The resemblance between brothers, or total family background, explained almost half the variance in the occupational statuses of men and a little less than a third of the variance in their incomes.* (If we were able to measure the lifetime earnings of men instead of using only one year as the dependent variable, the predictability would be even higher, since temporary ups and downs would be averaged out.) Obviously, this result is much more powerful than that of Blau and Duncan and suggests more inheritance of position than was previously believed.

The most important single indicator of total family background was the father's occupational status, but that accounted for only a third of the resemblance between brothers. Adding in a string of other demographic variables (education of both father and mother, income of both, family size, and race) adds another third. The remaining third of the variance "is presumably due to unmeasured social, psychological, or genetic factors that vary within demographic groups" (p. 214).

If the schools are to be equalizers, then talented children of poor families must stay in school as long as talented children from rich families. Otherwise, the schools are just mediators or the transmission belts used by privileged families to obtain privileged careers for their children. If we measure talent at a relatively early age and then follow the subsequent paths taken by the children, we can estimate how much of school attainment is a consequence of talent (as measured by the tests), regardless of the socioeconomic background of the students' families. Jencks and his colleagues reported that early cognitive test scores explain about 40 percent of the variance in ultimate educational attainment; about two thirds of that is *independent* of the connection between the IQ scores themselves and family background. This implies that there is a lot of variation of talent picked up on the tests, even among families of similar social status (indeed, even among brothers), and the variation in talent has a significant amount of influence on educational attainment. However, test scores have far less effect on educational attainment than a completely meritocratic system would require, and they have even less effect on ultimate earnings, predicting only about 5 to 10 percent of their variance (all else being equal). The schools are already more "fair" than the labor market, according to these particular measures.

Who Gets Ahead? spelled out some of the details as follows:

> The fact that demographic background affects cognitive skills and educational attainment explains more than half of its effect on occupational status and earnings. . . . (p. 82)
>
> We turn now to explaining *how* ability exercises its influence on an individual's schooling. Higher-scoring individuals [with all other variables constant] are treated differently than lower-scoring individuals, especially in school. Adolescents with high scores are more likely to be in a college curriculum, more likely to receive high grades, more likely to report that their parents want them to attend college, more likely to say their friends plan to attend college, more likely to discuss college with teachers, and more likely to have ambitious educational and occupational plans . . . (p. 106).
>
> A man's ability in sixth to eleventh grade has important effects on his later occupational status, but 60 to 80 percent of the effect is explained by the amount of schooling he gets (p. 115).

That last paragraph indicates that employers are more interested in the years of schooling than in the cognitive ability itself.

The earlier book reported that once a cohort of youths is in high school, variations in the quality of the schools (measured by expenditure per pupil or degrees held by teachers) do not explain much variation in the eventual educational attainment of the students, independent of their cognitive ability. That suggests that efforts in the courts and legislatures to equalize spending in different school districts will not make much difference in the impact of education on careers, although it might make life pleasanter for the students and the teachers. This conclusion does not deny that some particular schools, or programs within schools, can change the motivations and performance of some students, but it does assert that differences in the quality of schools are subtle matters of leadership and social climate, which cannot be measured by the size of the school budget or the formal training of teachers.

What about the influence on students of their peers in the same school? There are two types of segregation, often intermixed in practice, that tend to make high schools different from one another: segregation by SES and segregation by race. Although earlier research tended to conclude that segregation by SES raised the aspirations and the attainment (such as college attendance) of pupils who went to predominantly middle-class schools, more careful studies have disputed that result and often find the opposite pattern. In *Inequality,* Jencks evaluated the evidence and found two conflicting forces in the "better," or middle-class, schools. (1) Because most of the students are thinking about going to college, aspirations are raised even among the minority of students from poorer families not accustomed to college. (2) The better schools have higher academic standards, so the grades of many students are lowered in the overall competition, which tends to lower their aspirations for college. These two forces may cancel out each other.

Concerning racial segregation, Jencks almost threw up his hands in despair. Most of the research compared white and black and mixed schools at a given date. But schools are desegregated under different circumstances, and the effects of those differences on the students is important; consequently, results varied by year and by location. Furthermore, very few studies actually followed groups of black students who moved to predominantly white schools to see how the particular change affected their aspirations and performance. So Jencks wrote, "the most reasonable assumption at present is that racial desegregation [in high schools] makes little or no difference to students' college prospects" (Jencks et al. 1972: 155).

Entering a college that has a selective admissions policy provides some career advantage over a nonselective college, but additional variations in the quality of the colleges do not add to that effect. However, choice of curriculum makes a difference: People preparing to study law or engineering will end up with higher incomes than those who are studying music or poetry (Alba and Lavin 1981; Velez 1985).

In a general sense, everybody knows that staying in school pays off— that is, people with more education have higher-status jobs and earn more money. But *how much* difference does it make? And is the difference linear, so that an extra year of high school is worth about the same as an extra year in college? Indeed, is the last year in college, which provides the coveted bachelor's degree, worth something extra?

Studies done after Jencks's first book was written indicate a slightly higher correlation between education and earnings than he first reported, probably due more to methodological improvements than to actual trends in society. The various surveys indicated an average correlation of about 0.40 between years of education and earnings in dollars, without control-

ling for other influences. The second book contained a chapter on education written by Michael Olneck (Jencks et al. 1979: chapter 6), which shows
that the relationship is not linear: Years in college are worth more than
years in high school, particularly if one stays long enough to pick up a
degree. Four years of high school bring an increase in dollar earnings over
elementary school graduates of about 40 percent; four years of college
bring an increase of almost 50 percent over high school graduates. Thus,
teachers appear to be correct when they tell students to stay in school because doing so will improve their careers.

But are they really so accurate? We know that family background affects both talent and the length of schooling; maybe it is the family influence that counts, not the impact of the schooling by itself. Olneck estimated that if we control for both family background and IQ scores, the
"pure" effect of a high school diploma is reduced by half, to about a 20
percent increase in earnings, and the effect of a college degree is reduced
to an additional 35 percent in earnings. Notice that the reduction for the
controls is greater for the high school years than for the college years. There
appears to be a gradual reduction in the impact of background variables as
one gets older. Thus, *once young men get into college,* they are on their own
to a greater extent, and the payoff of higher education (especially for those
who stay long enough to get the degree) is almost as great for those from
poorer families and for those with less measured talent as it is for the others. In the earlier years, education is as much a reflector of family background as an equalizer that offsets family background, but in the later
years, education has more independent influence.

These measured effects of education fit with our discussion in Chapter
3, which emphasized a rather sharp break in the occupational system, dividing men with college degrees from all others. Employers screen applicants through the simple device of academic credentials, and jobs at the
managerial level, even for young people just starting to work who are
being selected and trained for such jobs, are given to those with college
degrees in hand. Over past decades, we have dramatically increased the
level of education for all segments of the population: Almost everybody
now finishes elementary school, and over 80 percent finish high school.
Among those who go no further, there is a large pool of workers, and the
jobs they get and the earnings they receive depend more on family background, individual talent, and the oddities of work experience than on a
few years more or less of schooling. But at the upper levels of the occupational system, the good jobs go primarily to those who have received a
complete college education—about a quarter of young men and women.
Within that select group, the further differential impact of family background or even test scores is rather small in terms of jobs received or dollars earned: It is the degree that counts. Since background has a lot to do
with the chance of getting the degree in the first place, we reach a double
conclusion: *College degrees both protect the privileges of people born into upper-
status families and permit many from lower-status families to climb into the elite.*

In the 1960s and 1970s, some observers noted that the rising numbers of new college graduates seemed to be reducing the value of the degree in increased income. But in the 1980s, the income gap between high school graduates and college graduates was again increasing. It seems that we will have to wait a while before these trends can be fully understood (Freeman 1976; Jencks et al. 1979: 188–190; U.S. Census 1990b: 74).

Olneck commented:

> Our findings place a number of widespread presumptions in doubt. The most significant of these is that high school dropouts are economically dis-advantaged because they fail to finish school. Our results suggest that the apparent advantages enjoyed by high school graduates derive to a signifi-cant extent from their prior characteristics, not from their schooling. Unless high school attendance is followed by a college education, its economic value appears quite modest. . . .
>
> Furthermore, while the distribution of elementary and secondary educa-tion has become considerably more equal, the distribution of higher educa-tion has become somewhat less equal.[3] Since higher education has more impact on earnings than elementary or secondary education, at least for those who complete college, the increasingly unequal distribution of higher education may more than offset the effects of equalizing the distribution of elementary and secondary education (Jencks et al. 1979: 189–190).

Careful reading of those paragraphs allows one to understand what seems to be a paradox. In recent decades in the United States (and Europe), the general population's average level of education has improved markedly, and educational inequality has been reduced. That is, the gap between the poorly educated and the average has declined. Yet the inequality in earn-ings persists as before. More equal schooling has not produced more equal earnings. Reforming the educational system by itself will not reform the economic system. Employers raise their standards and insist on higher cre-dentials for the same jobs that used to go to people with less education. The relative earning power of a college degree remains high, compared to those without a degree. And the relative impact of one's family of origin on eventual placement in society remains about the same as before (for a com-parative analysis that includes Europe, see Boudon 1974).

SCHOOLS, STUDENTS, AND STATISTICS

One of the main reasons why school reforms have less overall impact on mobility chances than anticipated is the "creaming process." If new high schools or new branches of the state university are built to make education more accessible, the best students from the most advantaged families who

---

[3]That is, more or less equal in terms of the impact of family socioeconomic status on years of schooling.

are waiting in line will be the ones most likely to enroll. That is, before the change, there were many students who could not get into the existing schools, and they formed a pool of prospective new students. Once the new facilities are available, the most ambitious and talented, and probably the ones from the highest SES families in the previously excluded pool, will take most advantage of the new opportunities. Thus, the average schooling for the community goes up, but the correlation between family background and education does not necessarily change. The new schools get "the cream of the crop" from among those who were previously waiting in line, and the most underprivileged remain outside the door (for one study, see Lavin et al. 1981).

The other main factor that weakens the effect of school reforms on mobility chances is the looseness of fit between years of school (all else held constant) and the final outcome in the form of earned income. The impact of family background plus the unexpected consequences of what happens on the job considerably weaken the pure influence of education.

Those new high schools and colleges do in fact change the lives of many individuals, so from the practical point of view they are certainly useful. They increase the career chances of the new students (particularly those from disadvantaged groups), give them a sense of real progress compared to their parents, and convince them that the system is more fair than it was. Moreover, those who receive a fuller education are typically better informed, more flexible, and more tolerant—desirable qualities in themselves, whatever their effect on mobility. But on the average for the whole society, those new schools do not change the *relative* life chances of many young people from different backgrounds, at least not enough to make major changes in the coefficients of the path diagrams. By emphasizing those coefficients and deemphasizing the number of young people whose lives have been changed, much of the research literature gives a partial impression of social trends that may lead many citizens to the wrong conclusion, such as a vote against school taxes.

The statistical model Jencks and others have adopted from Duncan concentrates on the relative chances of individuals from different backgrounds, and it is perfectly legitimate to do so when discussing the stratification system as a whole or for comparing one such system with another. But the model is an abstraction, and we should never forget the gap between it and the reality experienced by the particular graduates of the new schools. Even in terms of improving the relative chances of some groups compared to the average, certain programs can make a difference. Some small (but expensive) experiments have raised the IQ scores of preschool children from impoverished backgrounds by twenty to thirty points, and some scholarship programs (such as the GI Bill of Rights or special aid for minority students) have increased considerably the college attendance of bright students from poor homes. Indeed, changes can affect groups (such as blacks) and still not significantly alter coefficients for the aggregate of all individuals (Jencks et al. 1972: 14, 358).

In recent years, the study of the status attainment process may have gone too far in its emphasis on this one statistical model. Those who are intrigued with computer manipulation of many variables for large samples of people find path diagrams and similar statistical tools to be an elegant way of presenting lots of output to sophisticated readers. They even like to call it "causal analysis," because it separates out the relative influence of different variables on the final outcome. But from a sociological perspective, it is more descriptive than causal: It maps out the routes different individuals take within a social structure that is assumed to be constant. It tells us little about the causes of that structure or of changes in it. Its guidance for citizens and policymakers is limited by the fact that most of the diagrams cover the whole range of the population, instead of concentrating on those particular groups where a change in government programs could in fact alter the lives of underprivileged children. (See the debate between Hauser 1976 and Boudon 1976; also Horan 1978.)

The aim of most practical reform is to improve the opportunities of some special groups—often small ones with limited impact on national, aggregate statistics—that have previously been unfairly handicapped. Path diagrams can help spot those groups and diagnose the problem. But research designs based on tabular analysis similar to our earlier discussion of mobility tables are also likely to be helpful. With them, we can separate out of the total population those particular groups we wish to help, we can choose the point in the life cycle we hope to influence (such as the decisions made in the last years of high school), and we can study the two or three variables most amenable to policy influence. Furthermore, legislators, school board members, and parents will be able to understand and participate in the discussion. More complex styles of research are likely to lose readers in this larger audience, and they are the ones who pay the costs of the investigations and have ultimate responsibility for the application of the results. So we should either learn to present complex research in simpler ways or keep the citizens in mind when we set up the original research designs.

As an example of how research can be organized in a way that communicates a more vivid sense of practical reality, let us turn to a study that concentrates on a specific point in the life cycle that shapes the careers of so many young people.

WHO GOES TO COLLEGE?

Not all people with college degrees have outstanding careers, but few people (other than athletes and entertainers) achieve important positions without them. We already know that family SES, some additional influences from family background as shown by similarities among brothers, and measured IQ all influence the number of years of school achieved. But how can we use these same variables to explain more precisely why some people go to college and others do not?

William H. Sewell and his associates took up this challenge in a massive study that followed the careers of 9,000 high school seniors for nearly two decades after their graduation (Sewell and Hauser 1975; Sewell and Shah 1977). All the respondents graduated from Wisconsin high schools in 1957. By 1964, nearly 40 percent had entered college, but Sewell found large differences in college attendance by class, gender, and IQ. For example, 44 percent of men began college, but only 31 percent of women. Here are some further breakdowns of data for young men (the pattern for women was similar, although the percentages going to college were lower):

1. Men in top IQ quartile:
   a. Highest SES: 91 percent started college.
   b. Lowest SES: 52 percent started college.
2. Men in bottom IQ quartile:
   a. Highest SES: 39 percent started college.
   b. Lowest SES: 6 percent started college.

Clearly, both ability and social status influenced the college attendance of male high school graduates. Indeed, high social status was even powerful enough to get many rich dullards into college, but high intelligence was only strong enough to get half of the bright boys from poor homes into college. Of course, these same two variables influenced the drop-out rates in high school, thus shaping in advance Sewell's sample of high school graduates.

Have things changed since the 1960s? In some ways yes, in others no. Overall, the proportion of high school graduates who enter college has risen significantly. The large gender gap in college attendance has all but disappeared (U.S. Census 1990c: 151). But the class gap has persisted and even widened in recent years. Figure 7–2 shows college attendance rates for recent high school graduates by family income quartile. In 1989, nearly 80 percent of unmarried graduates in the top income class, but only 45 percent of those in the bottom income class, had attended or were in college. The disparity had shrunk in the 1970s but grown again in the 1980s. As a result, the class gap in college attendance was wider in 1989 than it had been in 1970 (Mortenson and Wu 1990).

The earlier Wisconsin study, with its cross-classification of college attendance by both IQ and SES, showed one intriguing fact: About half of those with high ability from low-status homes did make it to college. Unlike some other groups whose paths were almost completely determined by background (smart students from rich homes, or dull students from poor homes), this one special group had an open fifty-fifty chance. Can we go beyond the statistics and explain why some bright but underprivileged youths could make it, while others could not? What additional factors of motivation can we understand that shape individual choices in a group that seems to have so much leeway for personal decision?

The Wisconsin study, along with others, indicated that *within* any given status level, some families are more ambitious for their children than others (or, to be more precise, that the children perceive such differences and can

FIGURE 7–2 College Participation Rates of Young Adults by Income Quartiles, 1970–1989

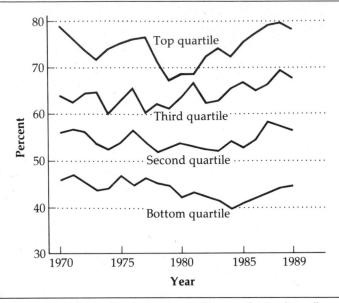

Refers to unmarried high school graduates, ages 18–24, currently attending college or previously enrolled for at least one year.

SOURCE: Mortenson and Wu 1990: viii. Based on U.S. Census, *Current Population Survey*, Series P–20.

report them on a questionnaire). The ambition is transmitted to the child through encouragement to succeed in school as a means of getting a better job later on. Children with such encouragement are likely to do a little better than others, particularly if they are talented; the approval and encouragement they receive from teachers and the higher grades they get in school are rewarding and begin to shape self-perceptions as persons who are "college material." That process in turn helps the motivated students to get into college preparatory courses, and such participation in turn increases the chance of making friends with other college-bound students. These encouraging factors of parental ambition, school performance, and influence of peers and teachers all lead to the formulation of college aspirations and plans. For upper-status students, these pressures toward college are almost automatic, and only those with severe handicaps in intelligence or personality fail to succumb to them. Thus, for upper-status students, the factors described are the "mediating" variables that translate the abstraction of socioeconomic status into particular realities that stimulate and guide behavior. But within lower-status groups, there is enough openness of choice to allow some students, particularly those of high talent with ambitious parents, to aim toward college.

Actually, the factors we are discussing can be observed even before the end of high school (Spenner and Featherman 1978; and articles by Kitsuse and Rist in Karabel and Halsey 1977). For example, an earlier study conducted in the Boston metropolitan area reconstructed the school history of a sample of boys from the first grade through the end of high school (Kahl 1953). Boys with high IQ scores usually had good marks starting with the first grade, but even more, those with low IQ scores had poor marks. Father's occupation did not affect school performance in the earlier grades, but it began to take effect in the fourth grade and, by the time of junior high school, was slightly more important than IQ in predicting performance.

However, some of the bright boys from lower-status homes kept up their motivation and aimed for college. The Boston research indicated that the main reason they did so was the special encouragement they got from home. These were usually homes headed by skilled workers or routine white-collar workers, not by men at the very bottom of the occupational scale. The boys themselves could be divided into two groups, each reflecting to a remarkable degree the values about life held by their parents: those who believed in "getting by" and those who believed in "getting ahead." This basic split showed in their more specific attitudes toward the details of schoolwork, after-school recreation, and jobs. The boys who believed in just getting by were generally bored with school, anticipated some sort of low-level job, and found peer-group activity to be the most important thing in life. The boys who were striving to get ahead took schoolwork more seriously than recreational affairs. Both groups noticed the differences: The nonstrivers said the others "didn't know how to have any fun," whereas the strivers said the others were "irresponsible; didn't know what was good for them."

Two quotations from interviews with parents will show how these contrasting attitudes of the boys were reflections of the values about life, work, and school held by their parents (Kahl 1953: 194–196):

Case A   The father is a bread salesman; he has five children. He is a high school graduate.

> I was never a bright one myself, I must say. The one thing I've had in mind is making enough to live on from day to day; I've never had much hope of a lot of it piling up. However, I'd rather see my son make an improvement over what I'm doing and I'm peddling bread. . . . I think he's lazy. Maybe I am too, but I gotta get out and hustle. . . . I don't keep after him. I have five kiddies. When you have a flock like that, it is quite a job to keep your finger on this and the other thing. . . . I really don't know what he would like to do. Of course, no matter what I would like him to do, it isn't my job to say so as he may not be qualified. I tried to tell him where he isn't going to be a doctor or lawyer or anything like that, I told him he should learn English and learn to meet people. Then he could go out and sell something worthwhile where a sale would amount to something for him. That is the only suggestion that I'd make to him. . . . I suppose there are some kinds who set their mind to some goal and plug at it, but the majority of kids I have talked to take what comes. Just get along. . . . I don't think a high school

diploma is so important. I mean only insofar as you might apply for a job and if you can say, "I have a diploma," it might help get the job, but other than that I don't see it ever did me any good.

Case B The father is a foreman in a factory with about twenty men under him. He had three years of high school and is convinced that he would have gotten further ahead if he had had more education.

> Down at the shop, we see a lot of men come in and try to make their way. The ones with the college education seem to succeed better. They seem better able to handle jobs of different sorts. They may not know any more than the other fellows, but they know how to learn. Somehow they've learned how to learn more easily. . . . If they get a job and see that they aren't going to get anywhere, they know enough to get out of it or to switch. They know enough to quit. . . . So that's why I hope my boy will go to college. . . . The college men seem also better able to handle themselves socially. They seem smoother in getting along with people and more adaptable to new situations. I think that I would have gotten along a lot better myself if I had had that sort of education.

In most instances, the parents who believed in getting ahead were somewhat frustrated: The father had not reached the place in the hierarchy that he had expected and desired. He blamed his failure on insufficient education and was determined that his son would do better. On occasion, the father was satisfied, but the mother was not; she then thought of her husband as a failure and tried to influence the son to outshine him (for further research, see Kohn 1977).

The two basic factors of class and IQ not only influence the chance of continuing education after high school; they also influence the type of school chosen and the chance of graduating. About one third of all college students start at two-year community colleges. These schools' curricula usually permit some students to transfer later to four-year colleges, but many concentrate on technical courses that prepare students directly for the world of work in positions that are somewhere in the middle of the occupational status range: dental assistants, automobile mechanics, bookkeepers, and the like. Indeed, the whole system of higher education is stratified according to the quality of the education provided and the particular career preparation emphasized, and that hierarchy is paralleled by the stratification of students' families. Two-year community colleges tend to draw their students from the lower half of the income distribution. Private colleges and universities recruit disproportionate numbers of students from high-income families. The admissions profile of public four-year colleges and universities places them somewhere in between (Berg 1970; Karabel in Karabel and Halsey 1977).

These studies of college attendance and graduation can be used to give partial support to both of two opposing ideological positions. (1) The American system of higher education is so big and so open that it provides major opportunities for talented youths from poor families to prepare themselves for successful careers that raise them above the level of their parents. (2) The American system of higher education is sufficiently stratified that its main function is to reproduce for each generation of children the status positions held by their parents. Does it promote mobility or succession? The answer is: both.

Sewell actually calculated the "loss" of potential talent through the effect of the biases against poorer families and against girls. Using the Wisconsin data from the late 1950s, he estimated that if all children of both sexes had the same chance to go to college as did boys of high SES (that is, if their selection was based solely on talent as shown by the relation between IQ and schooling among high-SES males), then 32 percent more young people would have gotten some training beyond high school, 43 percent more would have entered a college, and 47 percent more would have graduated. Those are dramatic figures that indicate the potentials for further improvement in our educational system. (See further: Levine and Bane 1975; Sewell et al. 1976; Karabel and Halsey 1977; Persell 1977; Hurn 1978.)

COLLEGE AND THE CAREERS OF WOMEN AND MINORITIES

In recent decades, the ethnic and gender disparities in college attendance have narrowed, even though the class gap has not. Figure 7–3 is what Thomas Mortenson (1991), a researcher with the American College Testing Program (ACT), calls an "equity scorecard." He compares people aged 25 to 29 in each group with a reference population of the same age: females with males; minorities with whites; and the bottom income quartile with the top quartile. The "equity scores" he charts since 1940 are the ratios between the college completion rates of each group and the corresponding reference population. A score of 100 means parity with the reference population in the percentage of people who have received college degrees.

The abrupt gyrations of the black and female trend lines for the 1940s correspond to the reduction of the male college population during World War II and the generous college financial aid provided to returning soldiers under the G.I. bill. African-American veterans were quick to respond to this opportunity, which was not available to the next generation of black high school graduates.

In the 1960s and early 1970s, the black-to-white ratio oscillated around 40 percent. It then jumped to roughly 50 percent, where it has remained. The data for Hispanics, though less complete, suggest a similar pattern, starting 10 points lower. Given the lag time required to push up the numbers of 25- to 29-year-old college graduates, these quantum leaps must represent the college freshmen of the 1960s, when federal aid to higher education was increasing.

The progress of females relative to males has been steady since the 1950s. In the 1990s, the female-to-male ratio tracked here can be expected

FIGURE 7–3 Equity Scorecard on Attainment of College Degrees: Disadvantaged Populations Compared to Reference Groups, 1940–1989

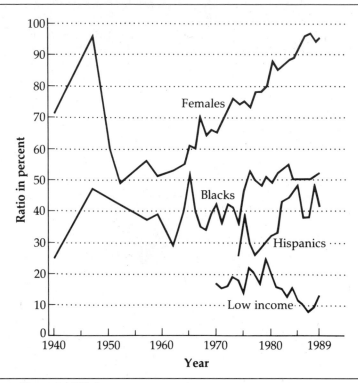

SOURCE: Mortenson 1991: ix.

to exceed 100—that is, a higher proportion of women than men in this age group will hold college degrees. The ratio of the lowest to the top income quartile has turned in the opposite direction. Although some progress was made in the 1970s, due to improved financial aid, the ratio fell sharply in the 1980s.

Even so valuable a credential as a college degree is not automatically converted into career success. In general, education yields greater economic returns to white males than to females or African-Americans. (See Coverman 1988; Treiman and Roos 1983; Schiller 1989: 120–126; Jaynes and Williams 1989: 299–301.) Both blacks and women face discrimination in hiring and promotion. Women at all levels of education tend, as we have seen, to be segregated into comparatively low-paying occupations. Women's earnings are further undercut by their tendency to move in and out of the labor force for family reasons. However, research has shown that gender differences in labor force attachment, absenteeism, and self-imposed limitations on work hours account for only a small fraction of the earnings gap between men and women; the rest is apparently the result of discrimination (Coverman 1988; Duncan et al. 1984: 152–172; Treiman and Roos 1983).

In 1990, the average yearly earnings of college-educated African-American men were 73 percent of the earnings of their white counterparts (U.S. Census 1991a: Table 29). This was far from parity, but an improvement over 1949, when the ratio was just 52 percent (Jaynes and Williams 1989). One way to get a fix on the effect on earning power of education for disadvantaged groups is to make comparisons of this sort using populations with carefully matched characteristics. For example, we can compare black and white men with the same age, education, and work pattern, as in Table 7.2.

The table makes race and gender comparisons of median income for full-time, year-round workers, matched for age and years of education. People with exactly four years of high school or exactly four years of college are singled out, as are those 25 to 34 years old. We focus on these younger workers because their careers should most reflect the changes promoted by the civil rights revolution of the 1960s and the women's movement that began in the 1970s.

Not evident from the table, which is based on relative comparisons, is the fact that education pays an absolute dividend for all categories of workers. For every group of workers (females, white females, young black men, etc.), a college degree is worth roughly $10,000 more in annual earnings than a high school diploma. Two encouraging findings can be seen in the table. (1) A college degree consistently moves black earnings closer to white earnings for corresponding groups, and likewise brings women's earnings closer to men's. (2) Among younger workers (25–34) with college degrees, race and gender disparities are generally smaller than they are among their older counterparts. Young, college-educated blacks have closed most of the gap between themselves and whites. These generaliza-

TABLE 7–2 Earnings Equity Ratios, 1990 (Full-Time, Year-Round Workers, by Education and Age, in Percent)

| | Black/White | | Female/Male |
	Males	Females	All
All educational levels	72	91	69
All educational levels, 25–34	75	86	78
High school	76	94	67
High school, 25–34	75	86	73
College	80	100	72
College, 25–34	88	97	83

Based on median annual earnings of workers over 25. Includes wage, salary, and self-employment income, but not transfer payments or interest, dividends, etc.

SOURCE: U.S. Census 1991a: Table 29.

tions reinforce the idea that a college education pays off for disadvantaged groups and suggest that it will do so more in the future than it has in the past.

The gender gap in earnings, according to these data, is larger than the race gap. This fact of social life undercuts the encouraging black/white ratios among females. If black women earn close to what white women earn, that still leaves them well behind men. For example, the average young man with four years of college, fully employed in 1990, earned $31,320 if he was white and $27,626 if he was black. A similar female graduate earned $25,733 if white and $25,019 if black. A fully employed white male high school graduate was not far behind at $22,985. But his twin sister earned just $16,481 (U.S. Census 1991a: Table 29).

We can conclude that education, especially college education, is an equalizer—but with two caveats: It is less a gender equalizer than a race equalizer, and it is only an equalizer if people have good access to education. We have seen that access for women is virtually equal to that of men, but access for blacks and Hispanics is well behind that of whites, and low-income young people have suffered a deterioration in their chances of getting a college degree.

CONCLUSIONS

The previous chapter focused on measuring the gross amount of intergenerational mobility and explaining the broad social factors that create mobility opportunities. In this chapter, we shifted perspective and concentrated on the determinants of individual success within the framework of gross mobility. Two factors—family background and education—are strongly correlated with career success. But measuring their separate influence is complicated by the fact that the two are correlated with one another. In general, people from privileged backgrounds get a double boost for their careers: a direct advantage (e.g., father's connections help son get job) and an indirect advantage (son gets more education). But education also exercises a strong independent influence. In effect, this means that the son of a blue-collar family who manages to get a college education is likely to do well.

Jencks used comparisons between brothers to measure the effect of family background. Building on Blau and Duncan's work, he concluded that background variables (including father's occupation, parents' education, income, and race) account for nearly half the variance in occupational attainment. (This figure includes family influence on years of schooling.) If Jencks' estimate is correct, we can make a good guess at a boy's odds of occupational success on the day he is born. A good guess, but not a sure thing: There is another 50 percent to be determined by factors ranging from individual initiative to pure luck.

The literature on status attainment suggests that the influence of education on occupational success is of similar magnitude to the influence of family background. Jencks' research improved on earlier studies that had simply measured the cumulative effect of years spent in school. He showed that the value of time spent in the classroom, measured by gains in lifetime income, varies considerably. A year of college is worth more than a year of high school. The final year of college, if it results in a degree, is worth much more than any of the preceding years. The power of the degree is such that among those who manage to graduate, the influence of family background is greatly reduced.

But if college graduates are equalized in this sense, access to college remains very unequal. There is, for example, an enormous gap in the college participation rates of young adults from high- and low-income families. The gap has actually increased in recent years. This finding is consistent with our conclusion in Chapter 6 that mobility among younger workers seems to be slowing.

What would happen if recent trends were reversed—if there were a significant increase in the proportion of high school graduates going on to college? Would this reduce the influence of family background on career chances? It might. But it could also contribute to "degree inflation" similar to what happened with a high school diploma—once a valuable credential, now almost universal and therefore devalued. If a high percentage of the population earned bachelor's degrees, the best jobs would probably be reserved (as they increasingly are) for people with postgraduate degrees, who would likely come disproportionately from privileged class backgrounds. The outcome, of course, would not simply depend on the supply of workers with education credentials, but also on the demand created by economic growth and the changing shape of the occupational structure. As we have seen, individual ambitions, abilities, and family advantages are only half of the mobility picture.

SUGGESTED READINGS

Bowles, Samuel, and Herbert Gintis. 1976. *Schooling in Capitalist America.* New York: Basic Books.
> *How schools meet the needs of the economy and the class system.*

Jaynes, Gerald David, and Robin M. Williams, Jr., eds. 1989. *A Common Destiny: Blacks and American Society.* Washington, D.C.: National Academy Press.
> *Basic source on African-Americans. See chapter 7, "The Schooling of Black Americans."*

McLeod, Jay. 1987. *Ain't No Makin' It: Leveled Aspirations in a Low-Income Neighborhood.* Boulder, Colo.: Westview.
> *Engaging portrait of black and white teens in a public housing project. Shows how their occupational aspirations are shaped by peer group, family, and school.*

8 Elites, the Capitalist Class, and Political Power

Those who hold and those who are without property have ever formed distinct interests in society.

James Madison (1787: 79)

Although we are all of us within history we do not all possess equal powers to make history.

C. Wright Mills (1956: 22)

Our interest in this chapter and the one that follows will be the relationship between the class structure and the political system. These chapters deal successively with two aspects of that relationship: political power and class consciousness. The treatment of power focuses largely on those who occupy positions at the top of the class structure and the apex of the political system—an approach that adheres to the emphases in the existing literature and appears to be guided by the inherent logic of the subject. After all, the people at the top are the most likely to occupy themselves with the manipulation of instruments of power. Yet the people beneath them are far from powerless, especially—and here is where class consciousness comes in—when they are united by a shared awareness of common goals. Thus, we see these two chapters as complementary and hope that readers will do the same.

This chapter covers power structures in the community and the nation and examines empirical evidence developed from three competing theoretical perspectives: elite, class, and pluralist. The elite perspective makes a sharp distinction between an organized minority (the elite) that rules and an unorganized majority that is ruled. The class perspective, which is rooted in Marxist theory, also emphasizes the power of a minority, but it is more specific about the identity of the rulers and the structure that creates them: They are the owners of productive wealth, or the capitalist class. The pluralist perspective denies that power is concentrated in one group. It maintains that in democratic societies, there are multiple bases of power representing the interests of competing groups, and no minority can hope to impose its will. Obviously, the first two approaches have much more in common with one another than with the third.

HUNTER: POWER IN ATLANTA

Two books published in the 1950s ignited a major controversy among social scientists concerning the degree of concentration of political power in the United States. C. Wright Mills' *The Power Elite* (1956), which we will take up later in this chapter, concluded that national decision making is being concentrated in the hands of a small, unrepresentative elite of corporate, political, and military leaders. Floyd Hunter's *Community Power Structure* (1953) suggested that a parallel tendency is affecting local communities. Those who defended the Mills-Hunter position came to be known as *elite theorists*, in contradistinction to the opposing *pluralists*, who contended that power was so diffused among competing groups that no single group could dominate the others.

Hunter's book is a case study of the structure of power in Atlanta, Georgia (which he called Regional City). Using a new methodology, which we will describe shortly, Hunter concluded that a small circle of men, most of them executives in leading banks and corporations or lawyers with corporate practices, made the key decisions in the community. The members of this elite filled the first two levels of a four-tier power-ranking system, which Hunter described as follows:

First rate: Industrial, commercial, financial owners and top executives of large enterprises.

Second rate: Operations officials, bank vice presidents, public-relations men, small businessmen (owners), top-ranking public officials, corporation attorneys, contractors.

Third rate: Civic organization personnel, civic agency boards of directors, newspaper columnists, radio commentators, petty public officials, selected organization executives.

Fourth rate: Professionals such as ministers, teachers, social workers, personnel directors, and such persons as small business managers, higher paid accountants, and the like (Hunter 1953: 109).

Note that Hunter placed the top executives above even the highest-ranking public officials (such as the mayor). The text leaves no doubt that he relegated most public officials to the third tier. As Hunter described the decision-making process in Regional City, important public initiatives originated out of public view, in the offices and private clubs of the city's corporate elite. From there, they filtered down to the second and third levels of the power system. By the time they became public knowledge, the key decisions had been made.

How did Regional City's power structure relate to its class structure? Hunter did not use *social class* or related terms, but he was obviously suggesting that the city's upper class (represented by wealthy businessmen) stood at the top of the power structure. The role of other classes is left ambiguous. Hunter did attribute some influence over the city's affairs to certain labor and black community leaders. But in general, he appeared to be suggesting that middle- and lower-class people in Regional City were so isolated from the exercise of power that their influence was of minor significance.

THE REPUTATIONAL METHOD AND ITS CRITICS

Hunter's book was influential in two ways, both of which gained it friends and enemies. One was its critical attitude toward American institutions. If Hunter is right, political democracy is weak in the face of economic inequality; middle- and lower-class citizens cannot hope to match the power of the wealthy. The other source of influence was the book's innovative technique for identifying community leaders. The new procedure, which came to be known as the "reputational method," was widely imitated and just as frequently attacked. Taking a clue from earlier community studies, Hunter systematically recorded the opinion of community members about power standings, just as Warner and Hollingshead had done for prestige standings.

To determine who were the most influential people in Regional City, Hunter asked 40 of the city's ranking public officials, civic leaders, businessmen, and professionals whom they would choose "if a project were before the community that required *decision* by a group of leaders—leaders that nearly everyone would accept" (Hunter 1953: 62). This poll yielded a select group of approximately twenty leaders who were regarded as influential by the leaders themselves.

The reputational technique provided a quick way to identify community leaders, which contributed to its popularity. However, pluralist critics raise several objections to the procedure. They charge that Hunter and his followers:

1. *Assume what needs to be proven.* By asking their informants *who* the decision makers are, researchers are implicitly assuming the existence of a small, decision-making elite. According to political scientist Nelson Polsby, the proper question is not Who runs the community? but Does *anyone* at all run the community? (Polsby 1970: 297).

2. *Equate a reputation for power with power itself.* The information that researchers using Hunter's technique receive from informants could well be inaccurate because informants misperceive the power structure.

3. *Assume that a single elite deals with a broad range of issues.* Decision makers may well be specialized in their interests and powers, thus forming a system of multiple elites, perhaps representing different segments of the community.

4. *Assume that leaders form a cohesive elite.* The identification of leaders does not tell us whether leaders agree on issues and exert their power in a concerted fashion.

5. *Downplay the role of formal political institutions* by largely ignoring the role of city government, political parties, and elections.
6. *Base research on an asymmetrical (one-sided) conception of power,* which is virtually implicit in a narrow focus on the identity of leaders. Policy makers may, in fact, find themselves compelled to take the probable reactions of other groups into account as they make decisions. If this is the case, power is in some sense reciprocal.

Hunter's critics were formally less concerned with his conclusions than with how he had reached them. Yet many of their criticisms implied that community power is dispersed rather than concentrated and that the reputational approach is defective because it is incapable of detecting such dispersion. Pluralist researchers proposed a methodological alternative designed to meet these objections. Instead of focusing on the identification of putative *decision makers,* they insist that students of community power should be studying *decision making,* by examining how representative community issues are resolved. If it can be shown that a well-defined group consistently triumphs over opposition in these matters, the existence of a dominant elite has been demonstrated. Otherwise, the pluralists contend, it must be assumed that power is divided among competing groups and shifts from one to another as the issue that is the focus of debate changes. (For criticism of the reputational approach see Dahl 1967; Polsby 1970; Aiken and Mott 1970.)

DAHL: POWER IN NEW HAVEN

The best-known piece of research employing a decisional methodology is Robert Dahl's study of New Haven, Connecticut, *Who Governs?* (1961), which was presented as a pluralist refutation of Hunter's work as well as Mills's examination of the national power structure. The core of Dahl's book is an examination of community decision making in three issue areas: (1) party nominations for political office, (2) urban renewal planning, and (3) public education. Dahl was interested in establishing who the decision makers were in each of these areas and, more specifically, who had successfully introduced or vetoed key police initiatives.

Dahl's most general conclusion was that New Haven did not have the pyramidal power structure that Hunter described for Regional City; rather, power in New Haven was diffused among a variety of groups and individuals. Thus, Dahl determined that leaders in any one of the three issue areas were unlikely to be influential in either of the other two. Among fifty major leaders, only three exerted influence in more than one area. All of them were public officials: the mayor, his predecessor, and the chief of the redevelopment agency (Dahl 1961: 181). Leaders, in other words, are specialized. Moreover, the sources of leadership are diverse. In none of the issue areas were leaders predominantly drawn from a "single homogeneous stratum of the community" (Dahl 1961: 182).

According to Dahl, things had not always been this way in New Haven. In its early years, the city was ruled by professional men drawn from a small circle of patrician families. But by the twentieth century, the patricians were largely irrelevant. They were replaced by a diversified cast of political characters, including popular politicians who often had roots in New Haven's ethnic communities, and professional public administrators. Relevant political resources, once concentrated in the hands of the patricians, were now dispersed. No one group was able to monopolize wealth, prestige, knowledge, electoral attractiveness, and control of mass media.

Two "strata" were the focus of particular attention in Dahl's study. Shying away from social class or terms suggestive of class, he labeled them "economic notables" and "social notables." The notables were New Haven's top wealth and prestige classes. By Dahl's account, the two were discrete groupings with little overlap: only 5 percent of the individuals on his combined list of notables fell into both categories (Dahl 1961: 68). (This conclusion, to which we will return below, is clearly at odds with the findings of the community prestige studies we examined in Chapter 2).

Dahl was interested in the notables because of the significance that had been attributed to them by the elite theorists. One test of their importance for New Haven's power structure was representation among the fifty leaders influential in the three issue areas. Dahl counted only seven economic and social notables on this list. To these few might be added an additional four "leaders," which were actually corporations that had influenced specific decisions.

Dahl's research convinced him that the social notables had largely withdrawn from the public arena that had in earlier epochs been dominated by their patrician ancestors. The economic notables appeared to be more active, but their participation was largely limited to the one issue area that was of particular interest to them: urban renewal. Even in this context, their influence did not seem overwhelming. A careful examination of key urban renewal decisions led Dahl to conclude that the economic notables were simply one of several groups attempting, with varying degrees of success, to influence the city's policies.

Thus, Dahl did not find a concentration of community power in the upper class, but neither did he see an even distribution of influence throughout the class structure. He attributed little or no direct influence over important government decisions to those below the line "dividing white-collar from blue-collar occupations" (Dahl 1961: 230). None of the fifty key leaders, for example, had a blue-collar occupation. Influential individuals, Dahl reported, were typically drawn from "middling social levels" (Dahl 1961: 229). But from Dahl's point of view, the class origins of the powerful are less important than the fact that power is diffused, specialized, and based on diverse political resources, so that no social group can regularly impose its will on the community.

THE DECISIONAL METHOD AND ITS CRITICS

The response to Dahl's work in some ways paralleled the earlier reaction to Hunter's study. The book especially appealed to those who saw in it a sophisticated vindication of American democracy. If Dahl's New Haven did not quite fit the idyllic world of the traditional high school civics text, it did maintain that power was not wholly concentrated in a small, upper-class elite and that most citizens of whatever class had access to some significant political resource. Moreover, *Who Governs?*, like *Community Power Structure*, popularized a method of inquiry. The decisional method was widely imitated, but like the reputational technique, it did not escape criticism. Among the principal objections raised were the following:

1. *The conclusions of a decisional community power study are necessarily dependent on the specific decisions selected for investigation.* Yet there are no generally accepted criteria for specifying a list of community issues that is in some sense "representative" of local power arrangements. Dahl, for instance, has been criticized for selecting education as an issue area, since, as he concedes, the city school system is of little interest to the notables, who generally live in the suburbs. They are therefore prohibited from holding school board positions, and they typically send their children to private or suburban public schools (Dahl 1961: 70). The party nomination process is likewise a questionable test of community power, since it involves a matter of institutional procedure rather than conflict over public policy alternatives.

2. *The decisional method is biased toward attributing power to the government officials and civic leaders who are most directly and publicly associated with each issue area.* It is least likely to uncover a small elite of the sort described by Hunter, which is quietly involved in a broad array of issues, setting the basic goals that a larger group of more specialized leaders publicly pursue.

3. *By focusing specifically on decisions*, Dahl's method misses the significance of latent concerns that are not allowed to become public issues requiring decisions. For example, when Hunter studied community power in Atlanta, most questions involving equity for black citizens were in this category of "nondecisions." By manipulating public opinion or institutional procedures (such as the operation of legislative committees), the powerful are frequently able to suppress consideration of matters they prefer to ignore. (For criticism of decisional method, see Aiken and Mott 1970; Bonjean and Grimes 1974; Domhoff 1978; Bachrach and Baratz 1974).

DOMHOFF: NEW HAVEN RESTUDIED

One of Dahl's critics among class theorists, G. William Domhoff, reexamined New Haven politics in the 1950s. Domhoff's book, *Who Really Rules?* (1978), refuted several of Dahl's principal claims, including the notion of a sharp distinction between economic and social notables and the idea that the economic notables had relatively little influence over New Haven's urban renewal program.

Domhoff challenged the criteria that Dahl used to define his social and economic notables. Domhoff's top prestige class, delineated by the membership of city's prestigious private clubs, was broader than Dahl's, though still limited to approximately 1 percent of the population. His top economic class is defined in terms of a network of interconnected business enterprises, banks, and corporate law firms at the center of New Haven's private sector. The leaders of these organizations were Domhoff's economic notables. Defined this way, Domhoff's top economic class was larger than Dahl's, but included fewer small-business owners and fewer managers of big corporations headquartered outside New Haven. It consisted essentially of the leaders of the largest local enterprises—about 300 people.

By Domhoff's criteria, 60 percent of New Haven's economic notables were also social notables (Domhoff 1978: 30). Dahl, of course, had contended that the two groups were almost mutually exclusive—which he took as evidence that New Haven was not dominated by a single elite. Domhoff had knocked down part of the support for Dahl's basic argument.

Domhoff also chose to reexamine Dahl's conclusions about decision making in urban renewal because he regarded it as the most significant of the three issue areas considered by Dahl. Although Domhoff did not point it out, it is also the issue that would be most likely to concern the pocketbooks of the rich and thus the most likely one to get them active in decision making. (See also a book on the same subject by one of Dahl's associates: Wolfinger 1973).

Dahl portrayed the program as the work of a newly elected Democratic Mayor, Richard C. Lee, and his redevelopment chief. The role of the economic notables, by this account, was essentially passive. With the help of documentary material unavailable to Dahl, Domhoff told a very different story. He showed that the urban renewal effort in New Haven was actually shaped by the prior initiatives of certain upper-class interests, both national and local. Nationally, Domhoff traced the story back to the late 1930s. At that time, a national debate over urban renewal policy set those who would emphasize building adequate housing for the poor against those who were concerned with preserving downtown real estate values and anxious to replace central city slums with commercial and other nonresidential structures. The first alternative was backed by labor and liberal groups, the second by major business interests. By the early 1950s, it was clear that the second alternative had emerged victorious.

Domhoff stressed the influence of several national, business-dominated organizations, such as the Urban Land Institute and the U.S. Chamber of Commerce, which played key roles in defining and promoting the conservative urban renewal alternative. The Institute carried on research and advised cities. The U.S. Chamber lobbied on a national level and showed its local affiliates how to design renewal programs beneficial to downtown business interests.

Locally, Domhoff emphasized activities of the New Haven Chamber of Commerce, which, he concluded, was dominated by members of his network of economic notables. In the early 1940s, the chamber persuaded a reluctant mayor to appoint a special commission to develop a comprehensive renewal plan. The commission was headed by a business executive active in the chamber; it produced a plan that, without essential modification, became the basis for the renewal program of the 1950s. After the war, an independent redevelopment agency was set up by the city. The agency's board was dominated by business figures. Its minutes suggest that the Chamber of Commerce was the only group that took an active interest in its proceedings. Immediately after Lee's election in 1953, the chamber began to press the mayor-elect for action on the urban renewal program. Lee was already familiar with the chamber's views on the matter, since he had been employed by the organization in the early 1940s when its interest in urban renewal first emerged. Domhoff did not deny Mayor Lee's leadership role in the 1950s, but he showed that there was much more to the story.

In sum, Dahl's *Who Governs?* substantially underestimated the influence of business interests on urban renewal in New Haven.[1] Dahl was apparently misled by the unjustifiably restrictive way he defined his economic notables, by his lack of access to relevant documentary evidence, and by his failure to consider national developments affecting local events. The last concern points to a significant problem of theoretical perspective. Dahl appears to have proceeded on the assumption that the power structure of New Haven was self-contained. Domhoff reminded us that communities are integrated into a larger society and therefore potentially subject to the decision making of national elites.

COMPARATIVE COMMUNITY POWER STUDIES

Ultimately, the power structures of Atlanta or New Haven are of minimal importance to people outside Atlanta and New Haven. They are of wider

---

[1]Domhoff also argued that Dahl had underestimated the influence of Yale University (located in New Haven) on the redevelopment program, though his evidence in this regard seems less compelling.

interest only if studying them can tell us something about community power generally. For this reason, scholarly interest in this field has turned to comparative analyses of the dozens of individual community power studies that have been done since the publication of Hunter's book (Walton 1970; Clark 1971; Bonjean and Grimes 1974). The object of such studies is to search out systematic patterns in the variations among communities. Researchers have been especially interested in determining the community characteristics that are associated with pluralistic versus pyramidal (elite-dominated) power structures.

One of the best comparative studies is Walton's (1970) analysis of research covering sixty-one separate communities. Walton tested a long list of hypotheses about community power that had been suggested by earlier work. Among his principal conclusions was one that confirmed the suspicions of many who were familiar with the literature: Investigators who employed the reputational technique (typically sociologists) were most likely to uncover pyramidal power structures, and those who used the decisional method (generally political scientists) were most likely to find pluralistic configurations. It would seem that many investigators were predisposed toward one or another image of the power structure before research had even begun and, perhaps unwittingly, chose the method that tended to verify their assumptions. In order to avoid the spurious results that might be introduced by this tendency, Walton based his conclusions on a multivariate analysis that controlled for the method employed by the researcher.

One important finding relates directly to social class. Walton found a general agreement in the studies he analyzed that decision makers are "overwhelmingly" recruited from the upper-middle class and include a particularly high proportion of business leaders. Another key finding concerns the influence of local affiliates of large corporations. Walton determined that communities in which a high proportion of business firms are absentee-owned are the most likely to have pluralistic power structures. Local managers of national firms tend to stay out of town politics, since their future careers are within the corporation not the community. Local business leaders, on the other hand, have deep roots in the community, along with knowledge and connections that make them effective participants in politics. In fact, they are likely to have inherited a family tradition of involvement in community affairs.

The perspective of the affiliate manager is reinforced by the parent corporation's own attitude toward local politics. Within the community, the corporation's concerns may include taxation, water supply, labor relations, and police protection. But these matters are typically settled before the local plant is established, and the corporation may simply shift its operations elsewhere if it becomes dissatisfied with conditions. Firms engaged in extractive operations (for example, mining) are an important category of exceptions to this generalization. Such firms cannot easily pull up stakes and leave, and they are often deeply involved in local politics (Goldstein 1962).

Absentee firms appear to encourage pluralistic power arrangements by reducing community dependence on local business interests, without becoming directly involved in local politics. However, Walton's finding should not obscure the extent to which community affairs are indirectly but powerfully shaped by national corporate interests. A corporation's implicit threat to rechannel investments away from a community is a potent influence on local decision making, even when corporate mangers are not among community decision makers. Moreover, corporate interests may collectively influence local communities by shaping national policies that affect them. Domhoff's portrayal of the development of urban renewal policy is a relevant example. Thus, a community may be inclined toward pluralistic decision making and yet have few critical decisions to make. We must shift our focus to a national level of power if we are to understand fully the forces that shape local communities.

MILLS: THE NATIONAL POWER ELITE

C. Wright Mills' book, *The Power Elite* (1956), was published shortly after Hunter's community study and provoked a parallel controversy on national power arrangements. Mills, whose work on the middle class we referred to in Chapter 3, was a major critic of American civilization during the 1950s, a self-satisfied era in our national life characterized by Mills as "a material boom, a nationalist celebration, a political vacuum" (Mills 1956: 326). Mills' conception of the national power structure centered on the growing significance of three major interlocking institutions: the modern corporation, the executive branch of the federal government, and the military establishment. He saw each of these institutions becoming enlarged and centralized. A few hundred major corporations were taking the place of thousands of smaller competing firms, which once typified the economy. The federal executive had gathered enormous powers and resources previously nonexistent or scattered among other units of government. The military, once small and decentralized, had developed into a colossal bureaucracy, commanding a war machine of unprecedented scale and destructive power. (For evidence supporting Mills with regard to these trends, see Dye 1990.)

As the corporations, the federal government, and the military grew, they eclipsed and subordinated other institutions (Mills 1956: 6):

> No family is as directly powerful in national affairs as any major corporation; no church is as directly powerful in the external biographies of young men in America today as the military establishments; no college is as powerful in the shaping of momentous events as the National Security Council. Religious, educational, and family institutions are not autonomous centers of national power; on the contrary, these decentralized areas are increasingly shaped by the big three in which developments of decisive and immediate consequence now occur.

The implication Mills drew from these trends was that the basis of national power had become control over the three key institutions. Those who sit at the "commanding heights" of the corporate, political, and military hierarchies make the critical national decisions. Who are they? In the corporations, an amalgam of very rich families with corporate-based fortunes, plus the ranking executives of the top national firms, whom Mills collectively labeled the "corporate rich"; in the federal government, the president, vice president, cabinet, heads of major agencies, and members of the White House staff; among the military, the generals and admirals. These three institutional elites together constitute the power elite.

Mills argued that the emergence of a national elite undercut important traditional bases of power. As we saw in the preceding section, community elites decline in importance as the power of national institutions grows. Investment decisions that are crucial for a community may be made in a distant corporate boardroom. Even the significance of wealth as a power resource is reduced. Very sizable personal fortunes are of trivial significance relative to the assets of any major national corporation. However, the largest family fortunes, such as those of the Rockefellers, Mellons, or Fords, are typically invested in national corporations, and in this form, wealth retains its significance as a basis of power.

Mills conceived of the political system as consisting of three tiers. The top tier is, of course, the power elite. The bottom tier, which Mills labels "mass society," encompasses the great majority of the population. Subject to large-scale national institutions beyond their control or comprehension, and misinformed by media that are dominated by national elites, the members of the mass are passive participants in the political system. Between the mass society and the power elite are "the middle levels of power," comprising a multitude of competing interest groups from labor unions to the gun lobby, whose most typical arena of conflict is the Congress. The middle levels of power are the source of most political news, but they are not the locus of the most important political decisions. In the welter of competing interests, none can impose itself. This "semi-organized stalemate," as Mills characterized it, only serves to reinforce the dominance of the power elite.

The key decisions are, of course, reserved for the power elite. But which are the key decisions? Mills is unambiguous on this point. Two issue areas are of sweeping importance: national economic policy and military affairs–foreign policy. These matters unmistakably, often brutally, intrude into the lives of ordinary men and women as boom or bust in the economy and peace or war abroad. Mills saw decision making in both areas shifting from the Congress to the executive branch of government, where the viewpoints of corporate and military elites are richly represented. He was keenly aware of the terrible fact that a small circle of men had assumed the right to launch a cataclysmic nuclear war. And, while most economic decisions under capitalism had always been private decisions, Mills felt that the leaders of major national corporations were much freer to determine their own course than were their smaller predecessors, subject to the restraints of a competitive market. We might add that since the publication of Mills' book, the emergence of the multinational corporation places many important economic decisions even further beyond the control of national politics (Barnet and Muller 1974).

Mills' conclusions were clearly influenced by the period during which he was writing. The country had recently come out of World War II and was entering a period of protracted cold war. General Eisenhower, a military man, was in the White House, and other officer-heroes were serving in civilian posts. These transitory developments probably led Mills to overestimate the significance of the military sector of his power elite. In the postwar world, the country faced a series of key decisions in foreign policy and international economic policy, which could be dominated by a handful of men if the major corporate, political, and military institutions they represented were in agreement. But even as Mills wrote, new issues were emerging, such as the civil rights question, which could not be easily contained at the elite level of power but would have to be fought out at the middle level, particularly in Congress, and even in the streets.

MILLS, HIS CRITICS, AND THE PROBLEM OF ELITE COHESION

Much of the criticism directed at elite theory in the community-power literature is also relevant to Mills' work. For example, pluralists have argued that Mills' methodology, like Hunter's, is flawed by a failure to examine actual decisions as a test of the power of the elite. In this section, we will focus on the substantive issue that has provoked the most debate among Mills' readers: the problem of elite cohesion—that is, the extent to which the members of a hypothesized elite hang together in pursuit of common objectives and in opposition to other groups. (For other lines of criticism, see the community-power discussion above and Domhoff and Ballard 1968.)

The cohesion issue was effectively posed in a classic essay by Dahl, directed at both Hunter and Mills, which accused the elite theorists of "confusing a ruling elite with a group that has a high *potential for control*" (Dahl 1967: 28). To be politically effective, potential for control must be coupled with "potential for unity." Thus, Dahl contended, the American military has the potential to impose a dictatorship on the nation, but that potential means nothing unless military leaders are in agreement on that objective. The implication is that Mills defined an elite in terms of key positions in organizations that control important resources (potential for control) without ever demonstrating a political consensus among the members of the elite (potential for unity).

Mills' pluralist critics are clearly predisposed to the belief that unity among power contenders is difficult to achieve, especially in contemporary America. This inclination was strikingly illustrated in the best-known pluralist alternative, David Riesman's version of the American political system, which was developed in a more general book on American society and culture, *The Lonely Crowd* (1953). In the course of his treatment of national power, Riesman asked two questions: "Is there a ruling class left?" and "Who has the power?" His answers were, respectively, no and no one. As the first question implies, Riesman believed that the country had had a ruling class in the past. Early in the history of the republic, it consisted of the landed gentry and mercantile interests that made up the Federalist leadership and, later, of captains of industry. But by the 1950s, the ruling class had been supplanted by an amorphous constellation of "veto groups," organized representatives of specialized interests that included "business groups, large and small, the movie-censoring groups, the farm groups and the labor and professional groups, the major ethnic and major regional groups" (Riesman 1953: 246). The veto groups are distinguished from the powerful of previous eras by their inability to take positive initiatives to impose their own will. Feeling themselves powerless, chary of offending other groups, their function is largely defensive, "to neutralize those who might attack them" (Riesman 1953: 247).

How can any decision be taken in such a political context? Is anyone in charge? Riesman's reply was that leadership may be needed to initiate something new or halt something in progress, but little leadership is needed to maintain the status quo. To the extent that anyone exercises power, it is in terms of very specific and narrow issues. Power that might be effective over a broad range of issues or in the face of big questions that affect the nation as a whole is smothered by the action of the veto groups.

Mills characterized Riesman's amorphous power structure as "a recognizable although a confused statement of the middle levels of power, especially as revealed in congressional districts and in Congress itself" (Mills 1956: 244). Power at that level is indeed a semi-organized stalemate. In Mills' view, Riesman was guilty of a mindless empiricism that equated all interest groups and all issues. To describe the corporate elite and the Polish-American lobby as two veto groups misses the point. Clearly, certain groups are more powerful than others; moreover, some groups may recognize a community of interest with others, thus creating a definite structure of power. As we have seen, Mills was only interested in the big economic and foreign policy questions and regarded most of the issues that consume the attention of the middle levels as trivial.

Cohesion presented a particular problem for Mills. He had to demonstrate both that the three distinct elites are internally cohesive and that they are drawn together into a single power elite. Mills did present a series of mechanisms through which elite unity might be achieved. They fall into essentially two categories: social-psychological and structural. The social-psychological mechanisms include similarities in origins, education, career, and lifestyles, which produce "a similar social type" and contribute to ease in informal association (Mills 1956: 19). Mills presented evidence on these topics for each elite. He noted that elite men tend to be drawn from upper-class or upper-middle-class, urban, white, Protestant families and that they are likely to be educated in Ivy League schools. There is a significant overlap between the world of the power elite and upper-class "society," with its elaborate links among "proper" families, select prep schools, class values, and notions of style. (These commonalities, as Mills conceded, apply more to the civilian elites than to the military.)

In addition, members of the three elites have similar career experiences, even if they do not move through the same institutions. The experience of managing a large organization is shared by the corporate rich, the "political directorate," and the chiefs of the Pentagon. The character of modern bureaucratic life has tended to blur the distinction between leadership in a large corporation, a civilian department of government, and an army. Mills contended that these shared elements of background and careers and the considerable material rewards attached to elite position tend to make members of the power elite conscious of the differences between themselves and the great mass of the population and to draw them together; they develop a form of upper-class consciousness, which leads them to view the world from a similar perspective.

The structural mechanisms of cohesion examined by Mills concern the more-or-less-formal connections between institutions. One critical link is the interchange of personnel among the three institutions, especially the movement of representatives of the corporate world into and out of top political positions, a question we explore at greater length below. Another tie between these two is the dependence of political candidates on financing from the corporate rich, which we will also examine further. The military is closely allied with the corporations, who are its suppliers, while militaristic foreign policy pursued by the political directorate strengthens its ties to the generals and admirals. All three elites are, of course, compelled to consider each other by virtue of the inevitable interdependence of institutions operating on such a scale.

Pluralists remain unconvinced by Mills' treatment of the problem of cohesion, but a Marxist critic appears to suggest that Mills has been all too successful in demonstrating elite unity. In an essay entitled "Power Elite or Ruling Class?" Paul Sweezy observed quite accurately that there is an unresolved tension in Mills' book between two views of the elite. The first is based on social class: Mills provided evidence that "those who occupy the command posts do so as representatives or agents of a national ruling class which trains them, shapes their thought patterns, and selects them for their positions of high responsibility" (Sweezy 1968: 123). Much of the evidence Mills presented for elite unity seems to point in this direction, especially the material on social background and recruitment patterns. The second view emphasized three "major institutional orders"; here Mills approached the corporate, military, and political realms as distinctly separate domains with autonomous leadership, which comes together to form the power elite. Sweezy was highly skeptical of the second view, particularly given the evidence that Mills presented for the former. The American military, Sweezy contended, is firmly under civilian control, and the political elite is dependent on the class that rules the corporation; thus, the justification for thinking in terms of institutional elites collapses. Mills never dealt at length with this line of criticism, although in a breezy reply to critics on the left, he commented, "They want to believe that the corporation and the state are identical. . . . I don't believe it's quite that simple" (Mills 1968: 224).

THE NATIONAL CAPITALIST CLASS: ECONOMIC BASIS

Sweezy's criticism suggests an alternative to both the Millsian and the pluralist views of national power: the identification of power with the class that controls income-producing wealth, the capitalist class. This conception, which has its roots in Marx, is similar to Mills' economic elite but assumes that the "corporate rich" have largely subordinated competing elites to their will. We have chosen the term *capitalist class* in preference to Sweezy's *ruling class* so as to avoid the implicit suggestion about political arrangements.

Tentatively, we can define the capitalist class as consisting of people who receive more than half of their income from property sources. Although such people surely exist at all income levels, only at very high levels—on the order of several hundred thousand dollars a year—does dependence on property income become the predominant pattern (see Table 4–1). A basic division within this class is that between local and national capitalists. The locals are led by the community's most affluent business owners, such as bankers, real estate investors, and department store owners. The national figures are those who own and/or manage the major corporations—those Mills dubbed "the corporate rich."[2] In the remaining pages of this chapter, we will explore the economic and social bases of the capitalist class and examine the mechanisms through which it makes its influence felt in national politics. Since we have touched on the local capitalist class in the community studies we reviewed earlier, and since our interest here is national power, we will restrict most of the discussion that follows to the national sector of the capitalist class.

In Chapter 4 we found that households with annual incomes over $1 million receive most of their income from accumulated wealth, in such capitalist forms as interest, dividends, and rents (see Table 4–1). We discovered that wealth, especially corporate wealth, is highly concentrated in the United States. For example, the richest 10 percent of households own 84 percent of publicly traded stock, most of it in the hands of the top 1 percent of households (Table 4–4).

---

[2]For some purposes, the national sector of the capitalist class can be conceived as divided into two subgroupings: (1) those who are tied to the smaller corporations oriented toward domestic markets and have few investments abroad; they tend to be more conservative and less internationalist in political perspective, and (2) those who are linked to the larger multinational corporations with extensive investments abroad; they are more internationalist and more liberal on certain political issues.

Here we take a closer look at the wealth of the very rich. The asset portfolios of the top 2.5 million wealth holders in the United States, about 1 percent of the population, are analyzed in Table 8–1. All have net worths in excess of $500,000. Twenty-five thousand are worth more than $10 million. The table reveals a significant difference between the merely comfortable and the truly wealthy. Those with the largest fortunes have the highest proportions of their total holdings in the form of corporate stock and other business assets. For people worth between $500,000 and $1 million, real estate (including the individual's own residence) is the largest asset category. For those worth over $10 million, stock is the largest category, making up about half the value of total assets. The average wealth holder at this level owns over $11 million in corporate stock.

Even these data tell us relatively little about the largest American fortunes. For that purpose, we turn to the list compiled annually by *Forbes* magazine of the 400 wealthiest individuals in the United States. The net worths of the 400 individuals listed by *Forbes* in 1991 ranged from $275 million to $5.9 billion. Table 8–2 contains a sampling of the *Forbes* 400, with information on the size and source of their fortunes. The second part of the table samples *Forbes'* intriguing (though less systematic) list of the 100 largest *family* fortunes.

The 400 personal fortunes are concentrated in four fields: manufacturing, real estate, media, and petroleum. *Forbes* (which touts the 400 list as evidence of America's open class structure) considers barely a quarter of these fortunes to be "inherited." But the accompanying mini-biographies suggest that an additional 20 percent of the 400, perhaps more, build their fortunes on existing family wealth. For example, Amory Houghton, Jr., holds a $420 million stake in Corning Glass, a company founded by his great-great-grandfather in 1851. The Newhouse brothers inherited the "largest privately held newspaper chain in the country," in addition to Condé Nast, publisher of *Vogue* and other magazines. (Father Samuel Newhouse bought Condé Nast for $5 million in 1959 as an anniversary present for his wife: "She asked for a fashion magazine," he later claimed, "and I went out and bought her *Vogue*") (*Forbes* 1991: 152). Oddly, neither the Houghton nor the Newhouse fortunes are considered by *Forbes* to be inherited.

By focusing on individual net worth, the *Forbes* 400 and similar listings underestimate the concentration of wealth at the top. Many, if not most, of the largest fortunes are held in common by members of extended families. These clans are based on descent from a founding ancestor, who typically accumulated the initial fortune during the late-nineteenth-century expansion of American capitalism. Some are represented among the 400. For example, the list includes three descendants of John D. Rockefeller; ten descendents of Thomas Mellon, a banker contemporary of Rockefeller; and eleven Du Pont heirs.

But in many cases, shares in these established family fortunes are sufficiently dispersed after a generation or more that no individual controls

TABLE 8–1 Assets of Top Wealth Holders by Net Worth, 1986

Size of Net Worth[a]	Number of Persons	Mean Value of Total Assets	Value as a Percentage of Total Assets								
			Corporate Stock	Real Estate	Cash	Noncorporate Business Assets[b]	Bonds	Mortgages and Notes	Life Insurance Equity	Other Assets[c]	Total
$500,000 to $1 million	1,548,300	$763,946	21.9	33.9	13.7	4.6	8.4	3.9	1.5	11.9	100
$1 to 2.5 million	710,000	$1,618,095	27.8	29.2	9.6	7.6	10.3	3.9	1.0	10.6	100
$2.5 to 5 million	150,300	$3,820,885	36.6	23.9	7.0	8.6	10.2	3.4	0.8	9.4	100
$5 to 10 million	55,500	$7,426,306	41.3	21.6	5.7	9.6	11.8	2.5	0.4	7.0	100
$10 million or more	25,000	$23,102,440	49.5	13.5	5.1	11.0	9.5	4.0	0.2	7.1	100
Total	2,489,100	$1,565,090	32.0	26.7	9.4	7.6	9.7	3.7	1.0	9.9	100

[a]Reflects debt not shown separately.

[b]Net value of sole proprietorships, farms, and share of partnerships.

[c]Includes intangible and depletable assets, annuities, pensions, and personal property.

SOURCE: Schwartz and Johnson 1990 (based on estate tax data).

the minimum $275 million required for inclusion on the *Forbes* 400 list. The Kennedys, for example, are not represented, although *Forbes* estimates their fortune at $350 million. The Kennedy clan is included on *Forbes'* (probably incomplete) list of family fortunes.

The economic power of families like the Rockefellers rests on the fact that their estates are jointly administered (although individual members may also maintain separate investments). Family holdings may center on a single enterprise (Ford Motor Company) or encompass a network of linked interests, typically revolving about a bank. For example, the Mellons

TABLE 8–2 Large Fortunes

Personal Fortunes	Estimated Net Worth (millions of dollars)	Primary Source of Wealth
Sam Moore Walton family	22,000	Wal-Mart Stores
John Warner Klug	5,900	Metromedia
Samuel and Donald Newhouse	5,600	Publishing
William Henry Gates	4,800	Microsoft
Ann Cox Chambers	4,000	Inheritance
Edward F. Mars, Sr. family	2,000	Candy
Walter E. Annenberg	1,600	Publishing
Stephen D. Bechtel family	1,300	Engineering
Edward Gaylord	1,100	Media, real estate
Joan Beverly Kroc	1,000	Inheritance (McDonald's)
Edward P. Bass	940	Oil
Joseph A. Albertson	930	Albertson's Inc.
Laurance S. Rockefeller	900	Inheritance
William Randolph Hearst, Jr. family	875	Inheritance (newspapers)
James Cargill	835	Inheritance (Cargill)
Paul Mellon	800	Inheritance
Dorrance Hamilton	790	Inheritance (Campbell Soup)
Michael Milken	750	Junk bonds
Nelson Peltz	600	Leveraged buyouts
William Clay Ford family	600	Inheritance (Ford)
John J. Louis	545	Inheritance
James H. Marshall III	525	Oil
Roy Park	515	Park Communications
Jeremy Jacobs	500	Sports concessions
Amos B. Hostetter	500	Cable television
William Russell Kelly	500	Kelly Services
Roy Edward Disney	470	Inheritance
Charles F. Dolan	440	Cable television
Samuel Zell	400	Real estate, investments
Roy M. Huffington	400	Oil
James E. Davis	400	Winn-Dixie Stores
Joseph Mandel	385	Premier Industrial
Irenéé Du Pont, Jr.	375	Inheritance
Gerrish Milliken	325	Textiles
Robert J. Congel	310	Shopping malls
David T. Chase	300	Real estate, media
Kenneth Behring	300	Developer
Michael Dell	300	Computers

Continued

TABLE 8–2 *Continued*

Extended Family Fortunes	Estimated Net Worth (millions of dollars)	Primary Source of Wealth
Du Pont	8,600	Inheritance (Du Pont Co.)
Rockefeller	5,000	Inheritance
Mellon	4,300	Inheritance
Busch	1,300	Anheuser-Busch
Upjohn	1,260	Inheritance (Upjohn Drug)
Hillenbrand	1,100	Caskets
Lilly	1,000	Pharmaceuticals
Johnson	940	Inheritance
Andersen	750	Windows
McClatchy	500	Newspapers
Sammons	500	Cable TV
Houghton	420	Corning Glass
Kennedy	350	Inheritance

SOURCE: *Forbes*, October 21, 1991.

control the Mellon Bank of Pittsburgh and a series of associated enterprises (Lundberg 1968: 150–154; Herman 1981: 217–221).

The concentration of personal wealth is paralleled by the concentration of wealth in the corporations themselves. The dominant position of the largest corporations, which Mills observed, seems to have been further consolidated since he wrote. In 1950, shortly before the publication of *The Power Elite*, the 100 largest—among nearly 200,000 industrial corporations—already controlled approximately 40 percent of all industrial assets; by 1986, their share had grown to 61 percent. Finance capital was even more concentrated. The 25 largest banks controlled over 50 percent of all bank assets in 1988 (Dye 1990: 20–21).

However, as concentration of corporate assets has grown, stockholding in individual companies has become dispersed. The largest corporations have hundreds of thousands of stockholders, and diffusion of stock ownership is so great that individual holdings of even 5 percent of total stock are uncommon. One result of this tendency is a change in the relationship between the ownership and the management of large corporations. Firms that once would have been controlled by families or small groups of investors are now typically dominated by their top officers, who are seldom owners of proportionally large blocks of company stock but rather people who have worked their way up through the ranks. A major study conducted recently by Edward Herman of the University of Pennsylvania's Wharton School of Business and Finance indicated that sixty-four of the 100 largest industrial corporations were controlled by hired managers; an additional fourteen were controlled by some combination of managers and "outside" members of the board of directors (who are most typically officers of other corporations); and only twenty-two of the firms were controlled by owners (Herman 1981: 60).

Some writers detect in these developments a "managerial revolution," which they conceive in terms of a sharp split between the interests of stockholders and corporate managers (Berle and Means 1932; Burnham 1941; Galbraith 1967). From this viewpoint, a manager who is neither an owner nor subject to the dictates of owners is free to substitute personal goals for those of the stockholders. According to these theorists, the traditional owner-entrepreneur operated a firm with one consideration in mind: profit. The professional manager is not quite free to ignore profit but, with little direct stake in the corporation's earnings, may pursue a broader array of objectives, including corporate growth, technological progress, the welfare of employees and customers, and public esteem for self and company. One well-known writer was so moved by his image of corporate high-mindedness that he dedicated an essay to celebrating "The Soulful Corporation" (Kaysen 1957).

Mills' conception of the corporate rich consisting of both top executives and major stockholders is a direct response to these ideas, which pose yet another problem of elite cohesion. What has taken place, he insists, is not a "managerial revolution" but a "reorganization of the propertied class . . . into a new corporate world of privilege and prerogative" (Mills 1956: 147). There has been a diffusion of stockholding in individual corporations, but (as we have demonstrated) the ownership of corporate stock remains concentrated among the top few percent of wealth holders. As a result, "the narrow industrial and profit interests of specific firms and industries and families have been translated into the broader economic and political interests of a more genuinely class type" (Mills 1956: 147). The professional manager is included in the new propertied class. He shares the privileges, interests, and outlook of the established rich.

Contemporary evidence bears Mills out on this point. In 1990, the average chief executive officer of a major American corporation held $49 million of his company's stock and received $1,592,000 in total annual compensation. These figures are from *Forbes'* annual survey of executive compensation in the 800 largest U.S. corporations (*Forbes* 1991: 236–237).

In addition to the direct incentives offered to executives, there are other forces upholding the interests of owners in the management of corporations. For one, major stockholders have not been wholly eliminated as an influence in the corporate world. As indicated above, Herman's study of the top 100 nonfinancial corporations showed that 22 percent were controlled by owners (Herman 1981: 60). An earlier *Fortune* survey of the top 500 firms concluded that 30 percent were family run. It is probable that proprietary interests are even stronger below this level of firm size. Even in the large firms that are management dominated, there are frequently small blocks of stock substantial enough to allow their owners influence over some key decisions (Herman 1981: 60, 101–102; see also Zeitlin 1974).

In addition, managerial behavior is constrained by forces outside the corporation that indirectly represent capitalist-class interests. Poor economic performance by a publicly held firm may limit its access to credit and may subject managers to the risk of removal through a corporate takeover. The influence of those whom Herman labels "the wealthy investor community" is subtle and pervasive:

> Several score institutions and several thousand wealthy investors account for a great deal of stock ownership, and this group is informed and powerful. Many top officers and outside directors are members of this wealthy investor community, and its interests filter into corporate attitudes, ideology, and standards in a variety of ways—directly through substantial representation on boards and in some cases by outright control; indirectly through transactions with financial interests and continuous inquiries and suggestions from owners and creditors and their representatives (brokers, security analysts) (Herman 1981: 101).

If this community is unhappy with the economic performance of a firm, its managers see the price of their stock falling, encounter increasing difficulties in obtaining credit, and face the risk of displacement through a corporate takeover. But the ideological influence to which Herman refers operates on a more subtle plane. The attitudes of the investor community pervade the boardrooms and executive offices so as to focus all discussion on profit-oriented performance criteria. All other considerations are simply "ruled off the agenda" (Herman 1981: 102). This orientation has been thoroughly inculcated in a manager by the time he or she has risen to the top ranks of the corporation. If anything, the professional manager appears to be under greater pressure to conform to orthodox business standards than the founding entrepreneurs or their heirs (Herman 1981: 103).

There is, then, little support for the idea of a "managerial revolution," at least in the sweeping form that it has been put forward by some writers. Corporations are typically run by their managers, but this does not appear to alter corporate goals. Since managers share the objectives and rewards of wealthy stockholders, it seems reasonable to accept Mills' conception of a capitalist class consisting of owners and managers tied to a corporate system.

An elaborate web of association ties the members of this class to one another and reinforces their common perspective. Interlocking directorates connect the largest firms to a vast national network of banks and corporations (Dooley 1969; Pennings 1980; Koenig et al. 1979). The leaders of the corporate world encounter each other through industry organizations, business lobby groups, participation in government advisory groups, service on the boards of foundations and related nonprofit organizations, and connections to the institutions of upper-class society, which we describe in the next section (Herman 1981: 187–242; Dye 1979).

THE NATIONAL CAPITALIST CLASS: SOCIAL BASIS

Parallel to the economic basis for a national capitalist class in the corporate economy, there is a social basis in an upper-class social world built on prestige and exclusive patterns of association. Among the institutions identified with this world are the select prep school, the *Social Register*, and the proper metropolitan men's club. These three have, in fact, been widely used by researchers as formal indicators of upper-class membership (Domhoff 1970: 9–23).

We have already described the appearance of the *Social Register* in early industrial America, when the new rich were being socially merged with the established upper classes. Although occasionally capricious in its inclusions and exclusions, a century after its creation, the *Social Register* remains, in Mills' words, "the only list of registered families . . . the nearest thing to an official status center that this country, with no aristocratic past, no court society, no truly capital city, possesses . . ." (1956: 57).

A small circle of prestigious prep schools, such as St. Paul's (New Hampshire), Hotchkiss (Connecticut), Foxcroft (Virginia), and Chapin (New York), draw their students from upper-class families (Domhoff 1970: 9–32). These day schools and boarding schools tend to be concentrated in the Northeast, but they draw many students from throughout the country. They are secular or nominally Episcopalian (the religious affiliation most common in the upper class) and traditionally single-sex institutions, although many have become coeducational in recent years. A few are older than the republic, but most were founded or experienced their major expansion around the time the *Social Register* appeared. They have, moreover, served a similar function: the integration of old prestige and new money (Baltzell 1958: 292–319).

Prep school graduates are informally referred to as "preppies," a term with mixed undertones of admiration and contempt, which is used more loosely to refer to various aspects of an upper-class lifestyle (Birnbach 1980). The extension of the term is not inappropriate. Admission to one of these select institutions is typically indicative of acceptance to upper-class status. The style and values the prep schools inculcate in their students equip them for participation in an upper-class social community. The network of personal ties that develop among prep school students and their families will serve them well in their subsequent careers and social lives.

The prep schools contribute to a pattern of upper-class endogamy by bringing students and their siblings into contact with potential marriage partners, both directly and through debutante parties and other upper-class social functions to which prep school students are likely to be invited. A study of weddings covered on the *New York Times* society page, a traditional monitor of upper-class social life, found that 68 percent of the brides and grooms were either listed in the *Social Register* or had attended an exclusive prep school. Twenty-four percent of the marriages were between individuals both listed in the *Social Register* (Blumberg and Paul 1975). This is a high figure, given that the *Register* covers less than 0.5 percent of the national population, but it misses the full extent of upper-class endogamy. Since 68 percent of the new spouses can be considered upper class by the two criteria mentioned, the level of endogamy should be more than 36 percent. (This figure is based on the extreme assumption that all non-upper-class individuals in the study are marrying upper-class individuals, thus diluting endogamy.)

Most major metropolitan areas have one or two men's clubs, such as New York's Knickerbocker, San Francisco's Pacific Union, or Philadelphia's Philadelphia Club, with a distinctly upper-class membership. The major clubs perform much the same function as the *Social Register* and the prep schools: They indicate who can be considered part of "proper society." They also provide an informal setting where upper-class associations can be developed and maintained and, on occasion, important business or political matters can be discussed free from outside scrutiny (Baltzell 1958: 336–354; Domhoff 1970, 1974).

The prep schools, top universities, and metropolitan men's clubs draw not only from their own regions but from a national upper-class population. For example, over 30 percent of the members of San Francisco's Bohemian Club reside outside the San Francisco area (Domhoff 1974: 30). The national scope of the upper class is also evident in marriage patterns; the society page study referred to earlier found that 54 percent of the marriages reported were intercity (Blumberg and Paul 1975: 74). The *Social Register*, perhaps in recognition of the national character of the upper class, has recently abandoned its city-by-city editions and begun publishing a single national edition.

To what extent is the national social class we have been describing identical with the national wealth class examined in the previous section? We would be surprised if the two did not overlap substantially, since our studies of wealth and prestige on a local level suggest an intimate connection. In Chapter 3, we noted that virtually all of the founders of great fortunes in the late nineteenth or early twentieth centuries had traceable descendants listed in the *Social Register* by 1940. More recently, Blumberg and Paul (1975) tabulated occupational data on the fathers of brides and grooms in their society page study; nearly 60 percent of the men whose occupations were identified were corporate executives. Using *Social Register* listings, education at top prep schools, and membership in the exclusive metropolitan clubs as criteria, Domhoff (1967: 51) found that 53 percent of the directors of the top manufacturing and financial corporations could be classified as upper class. Using slightly broader criteria, Dye (1979: 170) concluded that only 30 percent of the presidents and directors of major corporations were of upper class *origins*. What these studies demonstrate at the very least is substantial overlap between the top wealth and prestige classes at the highest levels of American society.

For the student of national power arrangements, the significance of the link between upper-class society and corporate wealth is related to the problem of cohesion raised earlier. The achievement of consensus on specific policy issues and more generally the maintenance of class solidarity is made easier because those who own and control the major concentrations of national wealth encounter each other in a private sphere of informal relations. The schools and clubs are merely the outer manifestations of this realm whose deeper meaning resides in shared experience, mutual intimacy, and the bonds of friendship and kinship that produce a consciousness of common identity and shared values. Domhoff points to group dynamics research in social psychology, which has established that physical proximity among the members of a group, frequent association, a group reputation of high prestige, and an informal atmosphere all contribute to group solidarity. These are characteristic features of the upper-class world we have been describing and prepare us for another basic conclusion of the same research: "Members of socially cohesive groups are more open to the opinions of other members and more likely to change their views to those of other members" (Domhoff 1974: 89–90, 96). Baltzell, in his earlier study of upper-class Philadelphia, made a related point, which he phrased in terms of social control: An upper-class community inculcates and sustains "a mutually understood code of conduct" in its members. Upper-class people are especially subject to the "norms and sanctions of their peers. A man caught in an act of dishonesty or disloyalty fears, above all, the criticism of his class of lifelong friends" (Baltzell 1958: 61).

MECHANISMS OF CAPITALIST-CLASS POWER: PARTICIPATION IN GOVERNMENT

In the next few pages, we will discuss the institutional mechanisms that extend the power of the capitalist class well beyond the economic sphere that is its immediate concern. Among the most potent of these mechanisms is direct participation in national government, especially at the top levels of the executive branch. Although few recent presidents have been drawn from the upper class, they have typically placed upper-class men in key cabinet positions and leaned on them for advice. Journalist David Halberstam captured the essence of this dependence in the opening pages of a perceptive book on the Kennedy-Johnson years. He records a meeting between John F. Kennedy, the month before he took office, and Robert Lovett, a Wall Street investment banker with impeccable social credentials and elaborate corporate connections—in Halberstam's words, "the very embodiment of the Establishment." Kennedy tried to persuade Lovett, a Republican who had voted for Kennedy's opponent, to accept a major cabinet post:

> Lovett declined regretfully . . . , explaining that he had been ill. . . . Again Kennedy complained about his lack of knowledge of the right people, but Lovett told him not to worry, he and his friends would supply him with lists. Take Treasury, for instance—there Kennedy would want a man of national reputation, a skilled professional, well known and respected by the banking houses. There were Henry Alexander at Morgan, and Jack McCloy at Chase [Manhattan Bank], and Gene Black at the World Bank. Doug Dillon too. Lovett said he didn't know their politics. Well, he reconsidered, he knew McCloy was an independent Republican, and Dillon had served in a Republican administration, but, he added, he did not know the politics of Black

and Alexander at all (their real politics of course being business). At State, Kennedy wanted someone who would reassure European governments: They discussed names, and Lovett pushed, as would Dean Acheson, the name of someone little known to the voters, a young fellow who had been a particular favorite of General Marshall's—Dean Rusk over at [the] Rockefeller [Foundation]. He handled himself very well, said Lovett. The atmosphere was not unlike a college faculty, but Rusk had stayed above it, handled the various cliques very well. A very sound man (Halberstam 1972: 16–17).

Ironically, the three top cabinet appointments made by Kennedy, a liberal Democrat, reflected the advice of Lovett and others like him. As secretary of the treasury, he chose C. Douglas Dillon, an investment banker connected with Dillon, Read, and Company, a major Wall Street firm started by Dillon's father; as secretary of defense, Robert McNamara, who had just been made president of Ford Motor Company; and as secretary of state, Dean Rusk, then president of the Rockefeller Foundation, a man with many admirers in corporate circles (Burch 1980: 175–177).

Cabinet recruitment studies show that Kennedy's appointments followed a pattern established by his predecessors and maintained by his successors. Cabinet officers are overwhelmingly drawn from the top of the class structure, most notably from the national capitalist class and the national prestige class we have been describing. Mintz (1975) examined the social and occupational backgrounds of people who served in the cabinet between 1897 and 1973. She found that nearly two thirds of the fathers of cabinet officers had had professional or managerial occupations. Seventy-eight percent of the cabinet officers themselves had been corporate officials or partners in corporate law firms; at least 60 percent could be classified as members of a national upper class on the basis of *Social Register* listings, exclusive prep school attendance, or membership in any one of a long list of selective social clubs; and over half of the cabinet members fit both categories.

Using slightly different criteria, Burch concluded that nearly two thirds of cabinet officers and major diplomatic appointees from 1933 to 1980 were linked to the corporate world or sizable family fortunes (Burch 1980: 378–384). The occupational backgrounds of the men covered by Burch's study were as follows:

Lawyers	29%
Big businessmen	27
Small businessmen	4
Government service	21
Other	19

Clearly, most ranking federal officials were lawyers, big businessmen, or government career people. Burch noted a substantial representation of corporate lawyers and investment bankers in the first two categories and suggested, as have Mills and others, that such men play a crucial mediating role between business and government, akin to the coordinating roles they play within the corporate world. Our own, more focused survey of the men who held the top cabinet posts (State, Defense, and Treasury) from Kennedy's inaugural cabinet in 1961 through Reagan's in 1981 reveals that 60 percent were drawn from major industrial corporations, financial institutions, or corporate law firms (*Who's Who in American Politics 1979; Who's Who in America 1980*).

President George Bush did not break the mold. He selected a corporate lawyer (James Baker) as Secretary of State and the former head of a major Wall Street securities firm (Nicholas Brady) as Secretary of the Treasury. A majority of the members of his inaugural cabinet had backgrounds in corporate or family businesses or corporate law (*Current Biography* 1982–1988). But if business is well represented among top federal decision makers, there is no parallel representation of labor. Since 1913, only six men have served in the cabinet who were in any way connected with the labor movement. Most served for short terms and were Secretaries of Labor (Burch 1980: 377; *Statesman's Year-Book* 1979–1990).

What can be said of the class background of Congress? Approximately 75 percent of the senators and representatives who served in the 102nd Congress (1991–1992) came to elective office from positions in business and the professions, principally law (data from LTV 1990). Another 15 to 20 percent had spent their entire careers in politics. The representation of blue-collar workers or white-collar employees in the 102nd Congress was scanty indeed. Only two of the 535 members of the Congress had been union leaders—a background that is common for legislators in some Western democracies. Evidence going back to 1906 shows that this pattern of recruitment to Congress has been consistent through the twentieth century (Nagle 1977).

Income and wealth data on members of Congress can be gleaned from the financial disclosure forms they are required to file each year. According to the disclosures covering 1988, approximately 70 percent of senators and 33 percent of members of the House of Representatives had incomes in excess of $150,000 the previous year. Fewer than 2 percent of American families had incomes at that level (*USA Today* 1990; U.S. I.R.S. 1990).

According to Senator Daniel Patrick Moynihan, "At least half of the members of the Senate today are millionaires. That has changed the body. We've become a plutocracy. . . . The Senate was meant to represent the states; instead it represents the interests of a class" (*New York Times*, Nov. 25, 1984). In fact, Moynihan's Senate colleagues include some of the wealthiest men in America—among them, Jay Rockefeller, John Danforth (of the Ralston Purina clan), Edward Kennedy, and, until his accidental death in 1991, H. J. Heinz III, heir to the "57 varieties" fortune.

Unfortunately, Congress has been less forthcoming about its members' wealth than about their incomes. The financial disclosure forms provide only sketchy information on the asset holdings and systematically conceal the magnitude of large fortunes. For example, Senator Heinz's disclosure indicated that he held assets worth at least $13 million; *Forbes* (1990: 321) listed the senator among its 400, estimating his net worth at $500 million. The financial disclosure forms do reveal that at least 40 percent of both House and Senate members hold assets worth over $500,000. And at least 25 percent of senators and 10 percent of House members hold assets worth over $1 million.[3] In contrast, only 1.6 percent of Americans own assets worth over $500,000, and just 0.5 percent hold assets in excess of $1 million.[4]

Congress, then, is recruited from the upper levels of the class structure, although the typical senator or representative appears to be drawn from a stratum a notch below that of the typical cabinet officer. He or she is also more likely to be a career politician rather than a top executive or corporate lawyer on temporary assignment. The law and business backgrounds of many members of Congress suggest a smaller-scale, more localized version of the corporate world so richly represented in the cabinet. If the cabinet is recruited from the national capitalist class, the Congress draws on local upper-middle and capitalist classes.

These generalizations apply with less force to the Senate, which is wealthier on average than the House and includes some men who are heirs to national capitalist fortunes. A study of the 92nd Congress (1971–1972) classified 20 percent of senators as upper class on the basis of social credentials, including *Social Register* listings, prep school attendance, or membership in exclusive clubs. The corresponding figure for members of the House was 5 percent (Zweigenhaft 1975: 124).

What do cabinet and congressional recruitment patterns tell us about national politics? It certainly cannot be argued that class background allows us to predict the political behavior of individual decision makers. For example, Edward Kennedy and John Warner, two of the wealthier men in the Senate, stand at opposite ideological poles of mainstream American politics. But we have already seen that the behavior and opinions of people in the aggregate are shaped in important ways by class position; in the next chapter, we will see that this generalization is applicable to politics. Moreover, we have noted how informal association shapes receptivity to opinions. The congressman whose personal associations are largely upper class or upper-middle class is likely to be more open to viewpoints common at the top of the class structure than to those prevalent toward the bottom.

The precise effect of these tendencies is not easy to gauge, but in Chapter 9 we will point out certain issues, such as health care and unemployment, that are of much greater concern to working-class people than to people of higher class rank. It is probable that the failure of the United

---

[3] *USA Today,* May 31, 1990. In the financial disclosure statements, assets are listed in broad categories. Assets worth $300,000 or $30 million would be thrown into the same category, "over $250,000." The value of a primary residence, the principal asset of the majority of American families, is not included.

[4] Schwartz and Johnson 1990. These figures refer to 1986.

States to develop a national health care system or adopt policies designed to counter high unemployment rates are both related to the recruitment patterns of cabinets and congresses.

MONEY AND POLITICS

The most obvious resource available to the rich is money, used to finance legitimate political activities, or on occasion, to purchase illegitimate influence over officeholders. In some periods of American history, the application of political cash has produced blatant distortions of public life. For example, in 1913, a lobbyist for the National Association of Manufacturers (NAM) publicly acknowledged that he had bought legislative favors with bribes and influenced House leaders to appoint congressmen favorable to NAM to House committees and subcommittees (*Congressional Quarterly* 1976: 654, 662). More recently, the Watergate revelations, which drove Richard Nixon from office, included evidence that a decision to raise milk price supports had been exchanged for huge campaign donations from milk producers, that the Nixon reelection committee had accepted money from foreign interests in violation of federal law, and that a long list of major corporations had made illegal campaign donations (Alexander 1980: 73–80). Nixon also had followed the traditional practice of using high-level jobs to reward major contributors. Ruth Farkas, a sociologist and member of the board of a New York department store, made one of the largest donations to the campaign and was subsequently appointed ambassador to Luxembourg. According to congressional testimony, Ms. Farkas had turned down an earlier offer of the ambassadorship to Costa Rica with the objection, "'Isn't $250,000 an awful lot of money for Costa Rica?'" (Alexander 1980: 58).

The publicity generated by the abuses of the Nixon campaign and the accelerating cost of running for office produced pressure for campaign financing reform, and significant legislation was passed in the 1970s. The new laws limit individual campaign donations to $1,000 per candidate in each election and $25,000 annually to all federal candidates; restrict group donations, which must be made through registered Political Action Committees (PACs), to $5,000 per candidate per campaign; provide for federal financing of presidential campaigns on a matching basis; and require public disclosure of campaign donations and expenditures.

But, as interpreted by the courts, the legislation does not limit *independent* expenditures made by individuals or groups on behalf of a candidate, as long as they do not coordinate their spending with the candidate's campaign organization. The law does not restrict so-called "soft money" donations that escape federal regulation by being channeled through state and local party organizations. Nor does it limit the amounts that candidates or their families can spend on their own campaigns, except in the case of presidential contenders who accept federal financing.

Self-financing is especially important for nonincumbent candidates. In 1988, for example, new members of the House of Representatives spent an average of $80,000 of their own money—more cash than the average citizen has on hand—in their successful campaigns (Makinson 1990: 3). Some candidates appear to have bought their way into the United States Senate. In 1978, the future Senator Heinz loaned his own campaign $2.5 million (Alexander 1980: 22). In the 1980s, Herb Kohl put $7 million into his own campaign, not quite matching Jay Rockefeller's $10 million. Both won Senate seats by narrow margins (Makinson 1990: 3; Alexander and Haggerty 1987: 92, 95).

Soft money is the biggest loophole in the campaign legislation. National parties can collect unlimited sums on behalf of state and local affiliates. Although this money does not go directly to federal candidates, it can easily be spent in ways that benefit their campaigns. In 1988, some 249 individuals gave $100,000 or more to Republican soft-money accounts. A single individual gave $500,000. Two years earlier, Joan Kroc, heir to the McDonald's fast-food fortune, gave the Democrats a cool $1 million in soft money (Makinson 1990: 11).

The major effect of the legislation of the 1970s is to constrain the influences of very rich individuals on candidates while increasing the political weight of certain organized groups, including corporate groups. In 1972, some individuals gave hundreds of thousands of dollars to political candidates. Direct contributions to candidates on that scale are no longer legal. As a result, the wealthy contributor has become less important than the corporate PAC.

The weight of political action committees in federal campaigns has grown steadily since the early 1970s, as has the proportion of PAC money from corporate sources. The law does not permit corporations or labor unions to contribute money from their treasuries to political candidates. But they can use their resources to run PACs that solicit donations from union members or corporate executives and stockholders. Unions pioneered this form of fundraising, but by 1988, corporate PACs were contributing nearly three times as much as labor PACs to federal candidates (Makinson 1990: 19).

Clearly, campaign finance reform has not severed the traditional connection between wealth and politics, although it has certainly altered the character of the relationship between them. The limit on individual campaign donations—still relatively high from the viewpoint of middle-income people—compels political fund raisers to shift their focus slightly downward in the class structure. Politicians, who still depend on individual contributors for the greater part of their campaign financing, must spend more of their time at upper-middle-class cocktail parties (unless they depend on direct-mail solicitation, a technique most likely to produce results for candidates at the extremes of the ideological spectrum) (Malbin 1979: 26).

Koenig (1980) argued convincingly that the corporate core of the national capitalist class is the principal beneficiary of recent legislation. Although the very rich individuals among them have had the political power of their personal fortunes reduced, the class as a whole has probably enhanced its influence. The various corporate committees do, in fact, tend to contribute to the same sympathetic candidates (*Congressional Quarterly* 1980: 3408). Thus, for the upper class, the reorganization of political giving is a phenomenon parallel to the earlier reorganization of corporate ownership: Both contribute to upper-class cohesion by making class interests and power more collective and less individual. At the same time, campaign financing reform contributes to the legitimacy of the political system by eliminating some gross forms of corruption without undermining the ability of the corporations to use campaign money to assure business-oriented government.

BUSINESS LOBBIES

What corporate interests do not achieve through direct participation in government or by financing political campaigns, they may attain by persuading public officials to adopt (or abandon) particular politics. Especially since the emergence of the modern corporate economy, business representatives have devoted elaborate efforts to shaping the decisions of Congress and the federal departments and regulatory agencies that carry legislation into practice. The terms *lobby* and *lobbyist* are often employed to cover these efforts, although in their strictest sense, they refer only to organizations or individuals professionally engaged in influencing legislators. At the turn of the century, as the case of the NAM lobbyist cited earlier suggests, business lobbies—backed by abundant flows of cash—moved Congress with dependable ease. In recent decades, major business lobbies have generally employed more subtle methods, in part because competing interests are better organized, and the possibilities of unfavorable publicity are greater. (On contemporary lobbying techniques, see Green 1979: 27–59 and *Congressional Quarterly* 1976: 653–667.)

Currently, the principal business lobby organizations in Washington are the U.S. Chamber of Commerce and the Business Roundtable. Also important are the National Federation of Independent Business, representing small business, and the National Association of Manufacturers, which speaks for smaller industrial corporations. The chamber gains its special strength from its ability to mobilize pressure on members of Congress through its 2,500 local affiliates. The local capitalist class, which dominates the affiliates, is likely to include important elements of the power structure in a legislator's home district, as well as people who belong to the same social networks as the legislator. When the Washington office wants to pressure senators or representatives on a vote, it can systematically mobilize letters, phone calls, and personal visits from a local business owner, a legislator's former law partner, or a fellow member of the local country club.

The Business Roundtable consists of the chief executive officers (CEOs) of approximately 200 of the largest corporations. Its power is based on the formidable resources controlled by these corporations and the prestige of those who lead them. The roundtable typically operates more quietly than the chamber. Its stock-in-trade is the personal visit from a CEO and the carefully crafted economic study or legal brief supporting its position. The leaders of the roundtable have access to members of the House and Senate and even to the President—entree that no ordinary lobbyist could hope to duplicate. A congressional aide commented, "A visit from a CEO has an unbelievable impact, as perhaps it should. It shows a commitment" (Green 1979: 29).

The power of these two groups was demonstrated in 1978 when they collaborated with other business lobbies to fight a bill creating a consumer protection agency. The legislation had already passed the House or Senate on five separate occasions; it was backed by the President, the Speaker of the House, and 150 consumer, labor, and other groups; and it had been supported two to one by the public in surveys conducted by the Harris organization. As part of its campaign, the Roundtable had retained Leon Jaworski, a highly paid, Washington-wise lawyer known for his role as a special prosecutor in the Watergate investigation, to lobby against the legislation, and employed the North American Precis Syndicate to disseminate prepared editorials and cartoons to some 4,000 newspapers across the country, never acknowledging the business group as the source. According to the syndicate, the material was used approximately 2,000 times. Two undecided congressmen told lobbyists for the bill that they were for the first time receiving sizable contributions from small-business groups encouraged by the U.S. Chamber of Commerce. Another representative was informed by his campaign finance chairman that his biggest contributors were against the legislation. A Florida congressman expressed fear that "the Chamber will run a candidate against me in the primary." The bill failed in the House 227 to 189 (Green 1979: 53–56).

There are, as pluralist writers would remind us, a multitude of lobby groups operating in Washington, many of them directly opposed to business lobbies on key issues. But their mere existence is not confirmation of the pluralist image of the political order as an athletic league in which all major social interests are represented with more-or-less-equal strength. Some interests are not represented at all. Others cannot compete in the same league as the representatives of business (as the fate of the consumer agency bill suggests). Business lobby efforts are certainly better financed and probably better organized than those of any other sector of the society. The Chamber of Commerce alone had an annual budget of $20 million in 1978; the American Petroleum Institute, which represents the major oil companies, spent $30 million (Green 1979: 28, 34). Eighty percent of the 1,000 largest corporations maintain their own representatives in Washington.

The superior organization of business interests is in part a reflection of a more general phenomenon referred to earlier: People toward the top of the class structure have higher rates of social participation. Only a minority of blue-collar workers are members of labor unions, but virtually every business belongs to some representative organization.

The authoritative Washington journal *Congressional Quarterly,* in its *Guide to Congress,* noted that two broad lobby coalitions confront each other over most critical legislation: a conservative alliance, led by the major business groups, and a liberal alliance, generally led by labor unions and civil rights groups. In the next chapter, we will have an opportunity to assess the relative strengths of these two camps, when we examine some recent legislative conflicts over legislation affecting labor unions.

POLICY-PLANNING GROUPS

A step removed from the conflictual world of political campaigns and legislative battles is a quieter and less visible realm of organizations dedicated to formulating and disseminating broad proposals for national policy. We have already noted the influence of one of them, the Urban Land Institute, which was set up to study urban renewal prospects. Like that institute, the most influential policy organizations—groups such as the Council on Foreign Relations, the Council for Economic Development, and the Business Council—have been created and financed by the corporate elite, which plays a prominent role in their activities. A social-network analysis of the boards of major corporations, memberships of exclusive men's clubs, and participants in key policy-planning organizations reveals an elaborate pattern of overlap (Domhoff 1975).

Similar to the policy groups, both in terms of their functions and their links to the national upper class, are the major charitable foundations and the policy research "think tanks." Foundations such as Rockefeller, Ford, Lilly, and Kellogg (all named for the wealthy families that endowed them) fund research and pilot projects to test policy ideas. Many of the best-known think tanks are clustered in Washington, where they are able to feed their research findings and policy recommendations to sympathetic politicians, lobbyists, and journalists.

One of the key functions of the policy groups, foundations, and think tanks is to back the careers of appropriately conservative public policy intellectuals, many of whom will be channeled into government positions. The career of former Secretary of State Henry Kissinger is illustrative: Kissinger moved from a university position to Washington via an active role at the Council on Foreign Relations and close ties to the Rockefeller family (Lundberg 1975; Collier and Horowitz 1976). His predecessor from the Kennedy-Johnson years, Dean Rusk, moved to State from the presidency of the Rockefeller Foundation. As we have seen, Rusk was urged on Kennedy by a representative of the corporate establishment. The policy organizations also provide a private, informal setting in which corporate leaders can seek a consensus on significant policy questions before they become public issues.

In the 1980s, two Washington think tanks, the Heritage Foundation and the American Enterprise Institute (AEI), played especially prominent roles. They promoted conservative proposals that the Reagan administration turned into policy, helped staff the new administration, and worked through the media to popularize right-wing ideas. In his first term, Reagan filled 34 positions with people from AEI and 36 with people from Heritage. Walter Wriston, AEI board member and chairman of Citicorp in the early 1980s, described AEI's influence with a line from a Barry Manilow song: "I write the songs the [whole] world sings" (Blumenthal 1986: 54).

INDIRECT MECHANISMS OF CAPITALIST-CLASS INFLUENCE

Capitalist-class influence over government is not limited to the direct means we have been describing (recruitment to decision-making positions, campaign financing, lobbying, and domination of policy-planning institutions). The capitalist class can also affect government policy indirectly, through its control of the economy and the mass media. A defining characteristic of a capitalist society is the existence of a relatively small class that controls most productive wealth and therefore independently makes investment decisions that can decisively affect the welfare of other classes. Although governments in capitalist societies have limited control over what business leaders do, their political fortunes are closely linked to business decisions. The connection is typically phrased in terms of "business confidence." If business leaders lack confidence in a government or its policies, they are not likely to risk their capital in new investments. The resulting decline in aggregate level of investment will soon be reflected in a rising level of unemployment, which, in turn, will subject the government to pressure from an electorate dissatisfied with the state of the economy. If the government wants to rectify the situation without making basic changes in the capitalist economic order, it must find a way to regain the confidence of investors. What sorts of government policies are likely to alienate business confidence? Basically, any which threaten business profits, from "excessive" taxation of corporate income to the imposition of expensive regulations designed to reduce pollution or guarantee worker safety. The precise factors that lead to a loss of confidence are less important than the essential fact that this mechanism gives the capitalist class an indirect veto over government policy.

Two aspects of the business-confidence veto are particularly worth noting. One is that it can influence a government without actually curtailing investment. The mere risk of such action is enough to persuade decision makers to reconsider a proposed policy. In fact, the possible effect of government action on investor behavior is frequently raised as an issue in public policy debates. The other is that the veto mechanism does not require conscious, concerted action by members of the capitalist class to be effective. Isolated investment decisions made on the basis of an objective assessment of potential risk and profitability may collectively produce a downturn in business activity and subject a government to popular pressure (Bloch 1977).

Governments are also subject to the limits imposed by private control of the mass media, through which people receive information about public affairs. In the United States, virtually all significant media are owned by the local and national capitalist classes (though they might just as well be organized as cooperatives, like the respected French paper *Le Monde*, as semiautonomous public bodies, like the British Broadcasting Company, or as organs of political parties, like a number of European papers). Control of the media has become highly concentrated, and the principal media companies are now counted among the largest corporations.

Four TV networks provide Americans with most of the news they hear. Two, ABC and CBS, are represented on the *Forbes* 500 listing of the largest U.S. corporations. NBC is owned by one of the top firms on the list, General Electric. Newcomer CNN is owned by media magnate Ted Turner, who is worth an estimated $1.3 billion (*Forbes* 1990). The corporations that own the two most influential daily newspapers, the *New York Times* and the *Washington Post*, and the two major newsmagazines, *Time* and *Newsweek* (owned by the *Post*), are also represented on the *Forbes* 500 list. Fifteen newspaper chains account for more than half of the country's daily circulation. Most American newspapers get their national and international news from a single source, Associated Press (AP), a cooperative owned by the media it serves (Dye 1990: 121–123).

Capitalist-class influence over the media is not limited to the power of ownership. Since the media are operated for private profit and most of their income comes from corporate advertising, media managers are sensitive to pressure from advertisers. The networks are additionally subject to the influence of the affiliated stations that broadcast their programs to local audiences. Thus, Edward R. Murrow's brilliant and controversial current affairs program, "See It Now," on CBS television, was forced off the air after its corporate sponsor withdrew and no regular replacement could be found. In the 1950s, programs that dealt with the issue of racial discrimination or employed black actors could not appear on network television because corporate sponsors refused to be associated with them and some affiliates (particularly in the South) refused to carry them. There are, however, relatively few confrontations between advertisers and networks over such matters. As an executive of a major advertising agency explained, direct interference by the advertiser is seldom necessary, "because the producers involved and the writers involved are normally pretty well aware of what might not be acceptable" (Barnouw 1978: 54; see also Tuchman 1974).

As the ad man's comment suggests, the main power that the capitalist class exercises over the media is the power to impose implicit limits on what is "acceptable." The media, in turn, operate on their audiences not by imposing specific ideas, but by defining the subjects that are appropriate for consideration and delineating the range of reasonable opinion. In other words, they define the public agenda. (Their ability to do so is increased by the concentration of media control.) Thus, until the late 1950s, racial inequity was not a national issue, though it was most certainly a serious national problem. During the Cold War, the question that so troubled Mills, the possibility that our leaders were pursuing policies that could carry us toward nuclear annihilation, were never raised in a serious fashion by the major media.

THE CAPITALIST-CLASS RESURGENCE

Who has the power? This is the question that pluralists, elitists, and class theorists are trying to answer in the debate we examined at the beginning of this chapter. The question assumes that the distribution of power is stable—it does not vary over time. But this is probably not a safe assumption. It seems clear, for example, that the capitalist class and business interests in the U.S. have gained at the expense of other competitors for power since 1970 (Blumenthal and Edsall 1988; Blumenthal 1986; Edsall 1984; Ginsberg and Shefter 1990).

In the early 1970s, there was a growing sense of vulnerability and declining power in capitalist circles. The economic system was changing in ways that seemed threatening and unpredictable. Wages had been rising; profits had been stagnating; productivity had been declining. The international economy, dominated by the United States since the end of World War II, was becoming much more competitive. The U.S. economy was twice shaken in the 1970s by abrupt leaps in the world price of oil. Increasing government regulation in areas from environmental practices to consumer protection and workplace safety seemed to be raising the cost of doing business. Many business leaders felt politically isolated.

John Harper, then chairman of Alcoa, recalled the period from the happier perspective of the 1980s:

> We [corporate leaders] were not effective. We were not involved. What we were doing wasn't working. All the polls showed business was in disfavor. We didn't think that people understood how the economic system works. We were getting short shrift from Congress. I thought we were powerless in spite of the stories of how we could manipulate everything (Blumenthal 1986: 77).

Beginning in the early 1970s, business leaders took a more direct and aggressive role in national politics. Corporations, as we have seen, seized the opportunity presented by the 1974 PAC legislation. Business lobbyists perfected the upscale "grassroots" campaign—mobilizing local business leaders, stockholders, depositors, suppliers, or dealer networks to influence Congress. In 1972, Alcoa's Harper joined other top corporate leaders to found the Business Roundtable. About the same time, the Heritage Foundation was started, with the help of a $250,000 donation from Colorado brewer Joseph Coors, and the American Enterprise Institute began its transformation from an inconsequential research center into a key player in the world of public policy. Both received financial backing from major corporations and foundations endowed by wealthy families such as the Mellons, the Pews, and the Olins (Blumenthal 1986; Edsall 1984: chapter 3).

These efforts to reassert the power of the privileged have paid off. In the late 1970s, business won a series of key legislative battles—for example, defeating both the consumer protection agency bill and labor reform legislation that would have made it easier for unions to organize workers. In earlier chapters, we saw that federal taxes on high incomes, inherited wealth, and corporate earnings have all been reduced since the early 1970s. The real incomes of workers have declined, and the distribution of income has become more concentrated. In the next chapter, we will see that the power of organized labor declined as the power of business grew. The power shift that began in the 1970s contributed, in the 1980s, to three successive victories of Republican presidential candidates distinctly sympathetic to the needs of the wealthy and the corporations.

The last two decades have seen an unmistakable transformation of the structure of power. However, there is no reason to assume that the new power structure is unalterable. The changes that have taken place may themselves set off further change, in directions we cannot now imagine.

CONCLUSIONS

We began this chapter by reviewing studies of community power structures. Although we reached no definitive resolution of the debate between elite or class and pluralist views, we did learn that local decision makers are overwhelmingly drawn from the upper and upper-middle classes and that local affairs are strongly influenced by the decisions made by national elites, especially corporate elites.

The latter conclusion forced us to think about national power structures. We examined the work of C. Wright Mills, which alerted us to institutional developments that have tended to concentrate power in the leadership of huge organizations. Within Mills' power elite, the corporate sector appeared to be dominant over the political and military sectors. This led us to focus our attention on the national capitalist class, and we learned the following. A small class controls most corporate stock, and although major corporations are typically run by their top executives, there is no division of interest between these hired managers and big stockholders who do not hold such positions; we may regard the two groups, as did Mills, as members of a single class. There is considerable overlap between the national capitalist class and the national upper class represented by such institutions as the *Social Register* and exclusive men's clubs and prep schools; the upper class provides some of the social glue that binds the members of the capitalist class together. Finally, the national capitalist class has powerful means to shape national politics; these include placement of its members in top decision-making positions, campaign financing, lobbying, creation of policy-planning organizations, exercise of the business-confidence veto, and control of the public agenda through the mass media. Looking back over the last two decades, we concluded that the power of the capitalist class has grown.

However, a pluralist would be quick to point out that neither the formidable array of political resources available to the capitalist class nor recent indications of growing capitalist-class power are definitive proof of domination by that class. Any such broad conclusion would have to be based on a more comprehensive analysis of the political system than is possible within the bounds of this book. We are, however, committed to examining how social classes participate in the political system and how they interact in the political arena. With those goals in mind, we will amplify the picture we have painted here in the next chapter, which deals with class consciousness and conflict between classes in electoral and industrial contexts.

SUGGESTED READINGS

Bottomore, Tom. 1966. *Elites in Modern Society.* New York: Pantheon Books.
  Short, lucid survey of elite theory.

Domhoff, G. William, and Hoyt B. Ballard, eds. 1968. *C. Wright Mills and the Power Elite.* Boston: Beacon Press.
  Excellent set of critical essays on Mills' power elite thesis.

Dye, Thomas R. 1990. *Who's Running America: The Bush Era.* Englewood Cliffs, N.J.: Prentice-Hall.
  The national elite, sector by sector.

Edsall, Thomas B. 1984. *The New Politics of Inequality.* New York: Norton.
  The growing power of the affluent in American politics and the resulting shift in national policies.

Judis, John B. 1991. "Twilight of the Gods." *Wilson Quarterly* 5 (Autumn): 43–57.
  Intriguing account of the rise and fall of the "American establishment" of bankers, corporate lawyers, and scholars who once made U.S. foreign policy. Helpful annotated bibliography.

Makinson, Larry. 1990. *Open Secrets: The Dollar Power of PACs in Congress.* Washington, D.C.: Congressional Quarterly.
  A mother lode of data on campaign financing. Want to know where your member of Congress gets campaign money? Look here.

Schwartz, Michael, ed. 1987. *The Structure of Power in America: The Corporate Elite as a Ruling Class.* New York: Holmes & Meier.
  Business unity, politics, and policy.

9 Class Consciousness and Class Conflict

I believe that leaders of the business community, with few exceptions, have chosen to wage a one-sided class war today in this country.

Douglas Fraser, president of the United Auto Workers (1978)

We owe the concept of class consciousness to Karl Marx. Its role within his theory was pivotal, joining individual experience to broad social structures and transforming alienated individual resentment of the capitalist present into decisive striving for the socialist future.

Class consciousness implies an *awareness* of membership in a group defined by a relationship to production, a sense that this shared identity creates *common interests* and a common fate, and, finally, a disposition to take collective *action* in pursuit of pursuit of class interests. At some points in his work, Marx implied that only a group whose members experience such a consciousness can be defined as a class. Elsewhere he carefully distinguished between a *class-in-itself* and a *class-for-itself*. The first is a class in a formal, definitional sense; its members share a social position (defined by the analyst) but are unaware of their common situation. The latter is a class in an active, historical sense: Its members are aware of common interests, they engage in militant action focused on goals that they conceive as being in direct opposition to those of other classes—indeed, are defined by that opposition. Thus, embodied in Marx's conception of class consciousness—especially in the notion of a class-for-itself—is the expectation of class conflict.

In its fullest sense, class consciousness is not just an aspect of public opinion ("What percentage of blue-collar workers supported Roosevelt?"), but an intense, collective involvement in the events of a critical historical juncture. It develops out of a long series of strikes against bosses who exploit workers and riots against authority that brutalizes the masses. It culminates in urban mobs roaming the streets and burning the buildings that symbolize upper-class domination and in peasants seizing the land they work, and it ends with a revolutionary seizure of power in the name of the oppressed: Paris in 1871, Mexico in 1910, Moscow in 1917, Peking in 1949, Havana in 1959.

Revolution is rare, but simmering class struggle is endemic. Slave revolts, violent strikes, local mobs on a rampage—these occur regularly in many societies. And more institutionalized and controlled forms of class struggle, such as union organizing campaigns and political movements that seek legislative power to help the underprivileged, are considered a normal and healthy part of a democratic society. Historians study past revolutions; sociologists usually focus on the early stages in the development of class consciousness, which could, under very specific historical circumstances, lead toward revolutionary consciousness but is more likely to result in peaceful change or historical stagnation. In this chapter, we will examine the extent to which people are aware of sharing a class identity and class interests, the social factors that advance or retard the development of this consciousness, and its relationship to political opinion and behavior. The final sections of the chapter will focus on class conflict as reflected in two arenas: electoral politics and labor relations.

MARX AND THE ORIGINS OF CLASS CONSCIOUSNESS

One of Marx's major objectives was to isolate the social forces that could produce the transformation of a class-in-itself into a class-for-itself; he hoped that by understanding the process, he could determine how to intervene and speed it up. Specifically, he asked, What inherent tendencies of capitalist society are likely to produce a class-conscious proletariat? Here are the factors that are especially stressed in his work:

1. *Concentration and communication.* The process of industrialization in capitalist society concentrates the proletariat in big cities, working-class neighborhoods, and large factories. This process promotes communication among workers, leading to a recognition of common problems and facilitating efforts at political organization.
2. *Deprivation.* Marx expected a progressive impoverishment of the proletariat, if not in an absolute sense, at least relative to the rising productive capacity of the industrial economy and the wealth of the bourgeoisie.
3. *Economic insecurity.* Marx was convinced that the proletariat's sense of deprivation would be exacerbated by the periodic experience of unemployment during the downturns in the capitalist economy, which, he observed, is quite subject to boom-and-bust cycles.
4. *Alienation at work.* Marx identified the mindless, repetitive, unsatisfying quality of factory-type labor with capitalism. (For a modern rendition of this argument, see Braverman 1974). Such labor is fundamentally at variance with human nature as Marx understood it and is therefore a spur to the development of class consciousness.

5. *Polarization.* The swings of the capitalist economy drive smaller enterprises out of business; their owners are forced into the proletariat, while control of the economy becomes further concentrated at the top. The result is the steady depletion of the middle ranks and the corresponding development of a society polarized between a miniscule, affluent bourgeois minority and an impoverished proletarian majority.

6. *Homogenization.* Within the proletariat, Marx observed a lowering of skill levels and therefore an equalization of wage levels produced by adaptation to the simple requirements of machine tending in the modern factory. This tendency leads to a less stratified, more homogeneous proletariat, which, because of its shared condition, is all the more disposed to unified political action.

7. *Organization and struggle.* In order to defend itself, the proletariat would be drawn increasingly into working-class parties and labor organizations. Participation in such organizations and the experience of struggle against capitalist employers and the bourgeois state and its police and armies would promote the development of hostile class consciousness and revolutionary struggle.

The revolutions that Marx's theory anticipated in the advanced industrial countries never came. However, class-based revolutions in industrializing agrarian states (Mexico, Russia, and China, for example) have been a characteristic feature of twentieth-century history. In the industrial nations, working-class parties and labor movements reshaped political systems and economic life. In both cases, the factors listed above have significantly changed events. Understood as variables that can depend upon specific circumstances, they can even help us understand the failure of revolution in the advanced countries. For example, the homogenization of the labor force that Marx predicted did not occur, as we saw in Chapter 3. Although property relations became polarized in the sense that most productive property was concentrated in the hands of a small minority, the relative differentiation of the bureaucratically organized occupational structure and the substantial range in incomes even among manual workers inhibited the emergence of a sense of common identity and the shared experience that are the basis of class consciousness. Furthermore, the profits from world trade helped avoid the impoverishment of the industrial workers. We might say that Marx was sociologically correct in identifying the key processes, even if he was historically wrong in predicting their outcome.

LEGGETT: WORKING-CLASS CONSCIOUSNESS IN DETROIT

We begin our examination of the processes that generate class consciousness with John Leggett's classic study of blue-collar workers in Detroit. Leggett (1968) employed survey research to measure the individual differences that make some workers more class conscious and politically militant than others. He reported on interviews with 400 manual workers—a random sample of seven ethnically varied Detroit neighborhoods. The sample was designed to allow Leggett to make comparisons among ethnic groups.

Leggett attempted to measure class consciousness with questions like these:

> What political party do you prefer and why?
> When business booms in Detroit, who profits?
> Would you join other working-class people in anti-landlord protests over high rents and poor housing?
> Should the wealth of the country be divided up more equally?[1]

He was looking for answers that could be considered class conscious. For example, a class-conscious answer to the second item would be "rich people" or "big business." He assumed that a higher number of such answers, especially in response to the strongest questionnaire items, was indicative of a higher level of class consciousness. Following this logic, Leggett developed what sociologists call a Guttman scale to measure an individual's level of class consciousness. Those who gave consistent or nearly consistent class-conscious responses Leggett labeled *class militants*. They were about one third of the sample. All but 10 percent of the people interviewed gave at least one class-conscious response.

Leggett was particularly interested in two sources of class consciousness, economic insecurity and working-class organization, both suggested by Marx. He considered "uprootedness"—the experience of being torn out of an agrarian region and thrust into an urban industrial milieu—to be one source of economic insecurity. His reasoning was that "the prepared," workers from industrialized regions, would not be plagued by the problems of adaptation to economic life that would plague "the uprooted," who should therefore exhibit higher levels of class consciousness. Two subgroups within the sample could be classified as uprooted: southern-born blacks and European-born Poles and Ukrainians. When uprooted members of these two groups were compared with prepared members of the same groups, it became clear that uprootedness did in fact contribute to class consciousness. Fifty-two percent of the uprooted were class militants (fell into the top two categories), as against 22 percent of the prepared. Further analysis of the data established two significant points: (1) Workers who were not only from agrarian regions but actually had farm experience were more class conscious than the uprooted generally, and (2) that tendency was increased among those who joined labor unions in the city. Sixty-four percent of the men who combined these characteristics were class militants (calculated from Leggett 1968: figure 4–4, p. 64).

---

[1] Most of these questions have been paraphrased.

Leggett also found higher class consciousness among unemployed workers and among those who were old enough to have experienced the economic dislocations of the 1930s Depression era. These results again confirmed the economic insecurity thesis.

Black workers in Detroit revealed much higher levels of class consciousness than whites: 57 percent of the blacks were militants, compared to 22 percent of the whites. Leggett found more class militants among blacks than among any other group or subgroup, and the difference between blacks and whites exceeded every other comparison he recorded in the book. Leggett used blacks as an example of a *marginal minority,* which he defined as a "sub-community of workers who belong to a subordinate ethnic or racial group which is usually proletarianized and highly segregated" (Leggett 1968: 14). Such groups are subject to greater economic insecurity than the general population. They are more or less culturally homogeneous and relatively isolated from middle-class contacts. Their members typically live in ethnic neighborhoods and belong to ethnically identified formal organizations—churches, social clubs, political groups. These social features might be expected to amplify working-class consciousness. They were partially shared by communities of Polish and Ukrainian workers, whom Leggett classified as *semimarginal* and who were, in fact, much more class conscious than workers of British and German descent, though less so than blacks.

In an effort to sort out the relative importance of different sources of class consciousness, Leggett performed a correlational analysis of eight different predictors of class consciousness: race-ethnicity, union membership, uprootedness, downward mobility, generation, skill level, employment status, and personal income. Leggett's analysis showed that these eight factors collectively explained a respectable 40 percent of the variance in the scale of class consciousness. Most of the result could be attributed to three key factors: race-ethnicity, union membership, and uprootedness.

Ultimately, the reason for studying class consciousness is the assumption that it influences political behavior. Leggett measured the political effect of class consciousness by asking respondents whether they had backed Governor G. Mennen Williams (1948–1960), a progressive who was identified as prolabor and supportive of liberal economic measures and civil rights initiatives. He found that Williams was supported by 76 percent of the class militants in the Detroit sample but only 50 percent of the other workers he interviewed (Leggett 1968: 122).

The relatively mild items on the Detroit questionnaire certainly do not tap the intense common commitment and revolutionary drive that Marx had in mind when he wrote of class consciousness. Nor can the opinion snapshot we get from this kind of survey research tell us how workers will respond to an unfolding crisis, such as a strike or political upheaval. Nonetheless, Leggett did show that Detroit workers were aware of class and class interests and that this consciousness was reflected in their political behavior. He also demonstrated that consciousness was advanced, as Marx led us to expect, by working-class organization and economic insecurity. However, the Detroit study uncovered sources of economic insecurity whose significance was underestimated by Marx. Current unemployment and memories of the Depression fit neatly with Marx's focus on the boom-and-bust dynamics of the capitalist economy, but the same cannot be said of uprootedness and marginal minority status. In this chapter, we will frequently return to the theme of ethnic subordination suggested by the last factor.

THE ORIGINS OF WORKING-CLASS CONSCIOUSNESS IN COMPARATIVE PERSPECTIVE

Empirical studies of class consciousness such as Leggett's are relatively rare, in part because they typically depend on data gathered through costly surveys. But there is another way of getting at the origins of class consciousness: Instead of looking at the social correlates of class-oriented opinion, we can look at the correlates of class-oriented political behavior represented in voting patterns. This admittedly back-door approach to the topic is justified by the evidence we have from Leggett and other authors that class consciousness does in fact predict voting behavior; besides, this focus has the distinct advantage of turning dozens of voting studies from all the world's democracies into grist for our mill. Here we will limit ourselves to looking at influences on working-class voters. Later in the chapter, we will look at voting from a slightly different perspective involving all classes.

Our task is simplified by the fact that two sociologists, Lipset (1960) and Szymanski (1978), have synthesized the relevant research. Nearly all the studies they examined present data on working-class support for leftist or liberal parties—in the United States, the Democratic party; and in Europe, communist and socialist parties. We can summarize much of what Lipset and Szymanski found with the following list of social correlates of leftist/liberal voting among workers:

> Lower skill level.
> Unemployment.
> Union membership.
> Larger city.
> Larger workplace.
> Economically advanced region.
> Minority ethnic or religious group.
> Specific occupations: miner, fisherman, sailor, longshoreman,
>   forestry worker.
> Male.
> Working-class origin.

Many of these variables are already familiar from Leggett's research. They are also, by and large, consistent with Marx's theorizing about class consciousness. For example, Marx's emphasis on "concentration and communication" anticipates the significance of larger workplaces, big cities, and occupations such as miner and longshoreman that tend to isolate workers in separate communities (Kerr and Siegal 1954: 190–191). Marx stressed the sense of economic insecurity produced by unemployment and the influence of organization, represented here by union membership. His emphasis on alienation at work is reflected in the significance of lower skill levels and larger workplaces. On the other hand, Marx did not attach great significance to gender (which has not, in fact, played an important role in U.S. elections) or ethnicity.

RICHARD CENTERS AND CLASS IDENTIFICATION

Our concern with class consciousness is based on the idea that it provides a link between objective class position (measured in terms of occupation, income, or wealth) and political behavior. That is, we assume that people who recognize and articulate their class position are more likely to promote their class interests. The most systematic and sustained effort to investigate this linkage on a national scale grows out of the work of Richard Centers (1949), who focused on one aspect of class consciousness, *class identity*, or the sense of belonging to a particular social class.

Centers began by noting that in previous public opinion surveys (such as the famous one conducted by *Fortune* magazine in 1940), about 80 percent of Americans consistently called themselves middle class. Some popular writers seized upon these figures to proclaim that America was almost completely a middle-class country—that if we had any class consciousness at all, it simply meant that we mostly thought of ourselves as belonging to the same big group. But Centers noticed that the figure quoted came from the following survey item, which offered only three alternatives:

What social class do you consider that you belong to?
1. Upper class
2. Middle class
3. Lower class

He also noticed that when respondents were asked the question in open-ended form (without a specific list of answers from which to choose), many called themselves working class. Centers made a reputation for himself by changing the wording of the fixed answers, adding working class as one of the possible replies. When he asked a nationally representative cross section of 1,097 adult white men in 1945 which class they belonged to, offering them the four alternative answers, he got the following replies:

Upper class	3%
Middle class	43

Working class	51
Lower class	1
Don't know	1
"Don't believe in classes"	1
Total	100%

In subsequent samples, he got almost the same distribution. Centers rightly concluded that Americans do not like the term *lower class* and that this dislike was the main conclusion to be drawn from the *Fortune* survey, not that they actually thought of themselves as overwhelmingly middle class. His confidence in the results of the surveys was strengthened by the fact that only a tiny minority of respondents refused to accept one of the labels suggested in the question. Even fewer were inclined to deny the existence of social classes. Centers was moved to say, "The authenticity of these class identifications seems unquestionable" (Centers 1949: 78).

Since 1945, numerous national surveys have employed questions modeled on Centers' forced-choice item and obtained similar results. For example, since 1956, the SRC national election surveys have included the following question, offering a more restricted choice than Centers:

> There's been talk these days about different social classes. Most people say they belong to the middle class or the working class. Do you think of yourself as belonging to one of these classes? (If yes) Which one? (If no) If you had to make a choice, would you call yourself middle class or working class?

The results of the SRC surveys, recorded in Table 9–1, are similar to those obtained by Centers. But note that the percentage of working-class identifiers has eroded about 8 percent since 1956. It is not clear why this is happening. In a study covering the 1960 to 1968 surveys, Schreiber and Nygreen (1970) discarded the tempting hypothesis that the trend reflects the white-collarization of the labor force, after finding that the decline in working-class identification was no different for manual workers than for the sample as a whole. Apparently, occupational reshuffling is not the explanation.

TABLE 9–1 Class Identification in SRC Surveys, 1956–1988 (percent)

	Survey Year			
	1956	*1964*	*1984*	*1988*
Middle class	39	43	47	44
Working class	58	53	50	50
Don't know, other	1	1	1	4
Reject idea of class	2	2	1	1
Total	100	100	100	100

Sample sizes ranged from 483 to 2,040.

SOURCES: Schreiber and Nygreen 1970: 351; and Center for Political Studies, computed for this table.

A few years after Centers published his findings, Kahl and Davis (1955) examined class identification as part of a Cambridge, Massachusetts, study. Before posing Centers' question, the investigators asked their respondents to describe the class system "in this part of the country," in their own words. They followed with an open-ended class-identification question that avoided giving answers to interviewees: "What social class do you think you are in?" The response to Centers' item was just about what he had obtained nationally, but the preliminary questions elicited a very different pattern of replies. Responding to the open-ended items, over 20 percent of the sample denied the existence of class or answered using occupational or other categories unrelated to Centers' class categories. Among those who spontaneously selected labels that were on Centers' list, middle class was the most popular choice. However, many of these middle-class identifiers shifted to working class when presented with the forced-choice question, as did many who initially avoided class labels.

These findings throw cold water in Centers' confident assertion quoted previously. We can assume that the forced-choice question is in some sense a measure of class consciousness, since, as we will see, responses to it are related to both class position and political attitudes. However, the Cambridge research tells us that we cannot use the answers as literal descriptions of the way people freely conceive of their own class positions.

CORRELATES OF CLASS IDENTIFICATION

What types of persons chose the particular labels offered in class-identification surveys? Centers regarded occupation as the principal basis of class identification. When he sorted respondents by occupation, he found that 70 percent or more of professionals and businessmen considered themselves middle class, while over 70 percent of manual workers considered themselves working class. A series of election-year surveys conducted from 1956 to 1968 by the University of Michigan's Survey Research Center (SRC) obtained similar results (Schreiber and Nygreen 1970). But as even these figures indicate, there was not complete consensus. Centers' data conform to a pattern that is by now familiar: The results were fairly clear-cut at the extremes of the class structure but somewhat ambiguous in the middle. In particular, over a third of sales and office workers among Centers' respondents and about half of that group among the SRC respondents labeled themselves working class (Centers 1949: 86, Hamilton 1975).

Centers' view that occupation is the main determinant of identification is supported by Hodge and Treiman's (1968) analysis of the 1964 SRC survey. They found that occupation (of the family's main earner) was a stronger predictor of class identification than either family income or respondent's education, but the three variables considered together left much of the variance in class identification unaccounted for. Objective class position seemed to be a crucial but not decisive determinant of this aspect of class consciousness.

Hodge and Treiman experimented with several additional variables in an effort to isolate other influences on class identification, but they only found one additional factor that was significantly and independently related to class identification: patterns of association. The class positions of friends, neighbors, and kin proved to be strong influences on the formation of class identification. The data available to measure this variable were somewhat crude but serviceable; the survey simply determined whether respondents had *any* friends, *any* neighbors, or *any* relatives in several broad occupational categories. For example, 57 percent of those who had high-status friends, neighbors, and kin, but no low-status contacts in any of these categories, identified themselves as upper or upper-middle class. Only 7 percent of those in the opposite situation did so. Of course, the contacts of a respondent can partly be explained by his or her status, but not entirely so. Hodge and Treiman found association to have an independent influence on class identification. Put differently, your class identification depends partly on your objective class position and partly on whom you know. Together, occupation, income, education, and association explain only about one fifth of the variance in class identification.

MARRIED WOMEN AND CLASS IDENTIFICATION

Centers' original class-identification surveys referred only to men, but the SRC data cover both men and women. If we are interested in the relationship between objective class position and identification, the inclusion of women raises an intriguing problem: Does a working wife base her class identification on her husband's occupation, her own occupation, or some combination of the two? Hodge and Treiman implicitly allowed for the influence of the wife's job on the self-placement of both spouses when they measured earnings in terms of *family* income. But by defining occupation for all respondents as "main earner's occupation"—the husband's in most households—they ducked the questions posed above.

Table 9–2, drawn from a paper by Ritter and Hargens (1975), considers the effect of both spouses' occupations on the class identification of wives. Their data are from four SRC surveys, conducted between 1960 and 1970, which asked respondents to choose between working-class and middle-class identification. The table indicates that both a woman's own occupation and her husband's influence her choice. For example, reading down the leftmost column shows that the proportion of wives of professionals who identify themselves as middle class falls off appreciably as the woman's own occupational rank declines. A comparison of the polar cases— professional wives with husbands in the lowest blue-collar positions and professional husbands with wives in such occupations—suggests that the effect of husband's position is slightly greater. After further analysis, Ritter

TABLE 9–2 Percentage of Women Reporting Middle-Class Identification, by Own and Husband's Occupation

Occupation of Wife	N	Occupation of Husband				
		Professional, Technical, and Managerial	Clerical and Sales	Craftsmen and Foremen	Operatives, Service, and Laborers	All Occupations
Professional, technical, and managerial	127	85	67	56	44	72
Clerical and sales	213	69	48	32	27	45
Operatives, service, and laborers	213	35	50	24	15	22
All occupations	553	73	52	35	21	43

SOURCE: Ritter and Hargens 1975: 939.

and Hargens concluded that the influence of wife's occupation and husband's occupation on her self-placement are probably close to equal. They also found a significant, though less powerful, influence of a woman's father's occupation on her own class identification.

A recent analysis of national survey data from 1974 to 1986 refines Ritter and Hargens' conclusions about wives' class identification. Beeghley and Cochran (1988) wondered why many working wives (a majority in some studies) still fit the pattern that has long been assumed in stratification research—that is, they base their class identification largely on their husbands' class characteristics. The key, Beeghley and Cochran found, was the women's attitude toward gender roles. The surveys asked respondents whether they supported the Equal Rights Amendment (ERA) and whether they believed that a woman should hold a job if her husband was able to support her. Using answers to these items, respondents' attitudes toward gender roles were classified as traditional or egalitarian. As expected, working wives with traditional gender views base their class identity on their husband's occupation. Those with egalitarian views considered their own and their husband's jobs, along with other class factors pertaining to both spouses.

Neither of these studies estimates the effect of wives' occupation on the overall pattern of class identification for men and women. The effect is limited by the fact that wives' and husbands' occupations are correlated. That is, men with better jobs tend to be married to women with better jobs. As we would expect, the data reveal few respondents in the high-low polar categories alluded to above. Only about 17 percent of the wives of white-collar men hold blue-collar jobs. Although the proportion of wives of working-class men holding white-collar positions is significantly larger, it is likely that such women are concentrated in the lowest sorts of clerical or sales positions. The data entry operator and the sales clerk at Woolworth's come to mind.

We may speculate that women and men answering the class identification question are not simply weighing two occupations but are thinking in terms of a standard and style of living, a set of associations, and particular values and attitudes that are shared by the members of a household. Traditionally, all of these things were largely dependent on the husband's job, and it was reasonable to assume that the class positions of most individuals were socially fixed by the characteristics of the (male) heads of their families. But rising female participation in the labor force, growing family dependence on dual incomes, and corresponding changes in gender norms all conspire to undermine this conventional wisdom. These studies are consistent with the idea that family continues to be the relevant basis of class position, but the standing of families is no longer solely dependent on the activities of men.

CLASS IDENTIFICATION, POLITICAL OPINION, AND VOTING

Our interest in class identification, like our concern with class consciousness generally, stems from the idea that consciousness is a link between class position and political attitudes and behavior. After Centers had satisfied himself that class identification was closely correlated with occupation, he went on to explore its relationship to ideological differences. For this purpose, he included in the survey a "conservatism-radicalism battery" of six questions like the following:

> In strikes and disputes between working people and employers, do you usually side with the workers or the employers?

> Which of these statements do you most agree with? (1) The most important job for the government is to make certain that there are good opportunities for each person to get ahead on his own. (2) The most important job for government is to guarantee every person a decent and steady job and standard of living (Centers 1949).

Centers placed respondents on a rough ideological scale, based on the number of their answers that went in a conservative or radical direction. At one end were the "ultraradicals," who consistently sided with workers and supported government intervention. At the other end were the "ultraconservatives," who consistently backed employers and favored individual initiative.

How do conservatism and radicalism relate to class identification? *Perhaps the main finding of Centers was that answers to the identification question were predictive of answers to the ideological battery.* Let us contrast the percentages of self-labeled middle-class and working-class persons who fell into the various ideological categories, as shown in the table at the top of the next page (Centers 1949: 20). The conservative persons were more likely to come from the middle class and the radicals from the working class (all five differences within rows were statistically significant). Note, however, that

	Middle Class	Working Class
Ultraconservative	35%	12%
Conservative	33	23
Indeterminate	21	33
Radical	7	19
Ultraradical	4	13
Totals	100%	100%

the differences are due entirely to the middle-class persons, who were conservative and consistently avoided radical answers. But the working class gave as many conservative as radical answers (and a great many working-class persons were "indeterminate," which means they gave inconsistent answers).

Centers further discovered that if he held occupation constant and varied class identification, he got substantial variations in ideology. Thus business, professional, and white-collar people were generally conservative, *but those who called themselves middle class were much more so* than the minority who called themselves working class. Similarly, the minority of manual workers who called themselves middle class were more conservative than the rest of the manual workers, who called themselves working class (though this difference was less marked than the preceding one). The details are shown in Table 9–3. Comparisons such as these led Centers to say that his data supported the "interest-group" theory of social class behavior, for it seemed that appropriately class-conscious members of a stratum had attitudes more typical of that stratum than did the minority whose objective occupational status and subjective identification were at variance.

Centers succeeded in demonstrating that class identification, which he used as an indicator of class consciousness, is a strong predictor of political orientation. His data do not quite support his contention that class identification is the critical link between class position and political orientation— that is, that occupation produces class identification, which in turn shapes

TABLE 9–3 Occupational Stratum, Class Identification, and Conservatism-Radicalism

Stratum and Class Identification	N	Conservatism-Radicalism (Percent)			
		Conservative	Indeterminate	Radical	Total
Urban business, professional, and white collar:					
Middle class	298	74	20	6	100
Working class	100	47	30	23	100
Urban manual workers:					
Middle class	83	37	30	33	100
Working class	318	25	34	41	100

SOURCE: Centers 1949: 126.

political orientation. It may well be that the class identity item used in surveys, which simply asks respondents to select one from a list of class labels, is too loose a measure to be a real stand-in for class consciousness. Campbell and his associates (1954, 1960) got better results when they looked separately at respondents who indicated that they thought of themselves as middle or working class before they heard the question. For these people, class identity did indeed function as a link between objective class position and subjective opinion, as Centers (and Marx) would expect.

FRAMES OF REFERENCE

The various sets of data concerning self-identification or class consciousness in America, and the connections between identification and ideology, do not automatically fit together, because the researchers used different techniques. However, if we borrow some general ideas from social psychology, and also attempt to apply to American data some interpretations that were first suggested by British researchers exploring very similar data, we can arrive at a tentative synthesis.

Elizabeth Bott has pointed out that "people do not experience their objective class position as a single clearly defined status." We might add that such clear definitions are the result of calculated decisions on the part of academic researchers; *they* are the ones who create concepts such as "upper-middle class" or "bourgeoisie" and through hard thinking attach some specific empirical criteria for membership. (Granted, they may start with words in popular usage, but by the time they have finished their ratiocinations, the original words have taken on new meanings.) Naturally, they endow their concepts with connotations that derive from the researchers' own general philosophy; thus, Warner thought of prestige strata and the Marxists of actual or latent conflict groups. Then the researchers go into the field and try to discover the degree to which the populace thinks as the concepts suggest they should, and the investigators feel a growing sense of triumph the more closely they can fit the data to the concepts.

But, said Bott:

> When an individual talks about class, he is trying to say something, in a symbolic form, about his experiences of power and prestige in his actual membership groups both past and present. These membership groups— place of work, friends, neighbours, family, etc.—have little intrinsic connection with one another, especially in a large city, and each of the groups has its own pattern of organization. The psychological situation for the individual, therefore, is one of belonging to a number of segregated, unconnected groups, each with its own system of prestige and power. When he is comparing himself with other people or placing himself in the widest social context, he manufactures a notion of his general social position out of these segregated group memberships. . . . The group memberships are not differentiated and related to one another; they are telescoped and condensed into one general notion (1954: 262).

The man—or woman—in the street is aided in conceptualizing by ideas and terms that have diffused into popular culture from intellectual debate. Thus, especially in Europe, there are many factory workers who have long been subjected to propaganda that stems from Marxist ideology. Naturally, they not only use class-conflict terminology; they actually perceive their own position and interpret their everyday experiences in terms of conflict. Similarly, the American middle classes have been bombarded with propaganda about our equality and absence of classes. Consequently, they tend to perceive as individual differences experiences that a European would see as common class experiences.

Therefore, any individual's self-perception in a stratification order is a combination of (1) actual experiences in a wide variety of contexts in many membership groups and (2) verbal theories about society, which are usually vague and somewhat contradictory common-sense notions that have filtered down from the theorizing of intellectuals and propagandists. Consequently, the social reality of identification that we are studying is complex rather than simple, and when we simplify it (as we must for certain purposes) into categories such as middle class or working class, we do violence to the original facts. The simple and neater the scheme, the further it is from reality.

Let us complicate the picture even more. Bott went on to point out that

> the individual performs a telescoping procedure on other people as well as on himself. If they are people who have the same, or similar, group memberships as himself, he is likely to feel that they have the same general position and belong to his own class. If they are outsiders, his knowledge of them will be indirect and incomplete so that there is plenty of room for projection and distortion. . . .
>
> The suggestion advanced here is that there are three steps in an individual's creation of a class reference group: First, he internalizes the norms of his primary membership groups—place of work, colleagues, friends, neighborhood, family—together with some more hazy notions about the wider society; secondly, he performs an act of conceptualization in reducing these segregated norms to a common denominator; thirdly, he projects his conceptualization back on to society at large. . . . The main point is that the individual himself is an active agent. He does not simply internalize the norms of class which have an independent external existence. He takes in the norms of certain actual groups, works them over, and constructs class reference groups out of them (1954: 263–265).

Finally, Bott reminded us that because the conceptualizations were both hazy and tied to a variety of actual experiences in the life history of an individual, they could shift in the course of an interview. Sometimes respondents would think in terms of the people they knew as children in their hometown; sometimes of their workmates; sometimes of their dreams for their own children. These shifts, plus those of the forms of the questions being asked, would lead them to shift frames of reference.

The interviews that Bott conducted in London paralleled the American data and showed at least four different models of the class structure that could be used by different respondents or by the same respondent at different points in an interview:

1. Two-valued power models.
2. Three-valued prestige models.
3. Many-valued prestige models.
4. Mixed power and prestige models.

The power model tends to be reduced to a two-level system, since it takes two to make a fight, and respondents tend to think primarily of themselves (and their colleagues) versus all the rest. Respondents whose own experiences had involved considerable conflict would be likely to think of a two-level system of conflict. *But they would almost all be members of the working class.* Middle-class persons—even those with a personality predisposition to conflict—would have certain experiences and interests that would make it difficult for them to accept the Marxist conception. For one thing, most of their personal conflicts would have been with other members of the middle classes; they compete as individuals with other individuals. They would want to feel that their success was a result of personal qualities, such as intelligence and ambition, so that they could claim full credit for their positions. If they feel something of a failure, they might blame it on bad luck or even the unfairness of the higher-ups—but they still must acknowledge that some people (manual workers) are even worse off; so they resist the notion of basic class conflict between haves and have-nots. They do not want to have to identify with the workers in order to blame the bourgeoisie and the system for their troubles.

People who think in terms of prestige, and who recognize the possibility of rising and falling, almost inevitably use a model with at least three levels. Prestige implies someone above you and someone below you. Only the really "down and out" will admit that they are on the bottom; self-respecting workers will always look down on bums beneath them and gain psychic satisfaction from their own superiority. Thus, contented workers who do not stress conflict and who feel some personal success will think in terms of at least a three-valued prestige hierarchy with themselves in the middle, bums on the bottom, and business and professional men on top. Middle-class persons also use the three-valued model, but they shift the dividing lines between the levels and lump all workers together as the ones below them and recognize the upper class as the group on top. You can almost always get a person who uses a three-valued system to make finer distinctions just by pushing the questioning; as you narrow the focus of attention to any part of the system, the respondent begins to think of subtler differentiations among persons. *The prestige concept, unlike the power idea, is infinitely divisible.*

Bott suggested that the three-valued prestige model is most common among the people who think of themselves as belonging in the middle. She found that the many-valued model was used by those who placed themselves in the working class but felt some incompatibility in their position.

Thus, those who were somewhat better educated than their colleagues or who had aspirations to a more cultured outlook would admit that they were occupationally rather near the bottom but saw themselves as intellectually higher up. They would resolve the difficulty by seeing a society of many layers, with themselves second from the bottom and more like the level above them on some characteristics than on others.

The mixed prestige-power models were used by intellectuals who tried to reconcile Marxist theorists with the more complex facts of contemporary life (see also Bott 1964 and Ossowski 1963).

We can now hypothesize that the men in the Cambridge sample who shifted from "middle" on the open-ended questions to "working" on the closed probably used the many-valued prestige model in their thinking. They knew that they were occupationally in the working class, but when they had the chance to think in other terms (such as education or values), they could see themselves as toward the middle of a complex system. Those who denied class at first but then called themselves workers when forced to make a choice also probably thought of a many-layered system without clear lines of demarcation. Those who stuck to "middle class" through both forms of the questions were most likely to use the three-layered model. Those who remained with "working class" through both forms probably came closest to the power model, which simplifies and consolidates shades of difference.

ELECTIONS AND THE DEMOCRATIC CLASS STRUGGLE

Although no advanced industrial country has experienced the convulsive class revolution envisioned in the *Communist Manifesto*, most have passed through periods of bitter class confrontation and continue to experience less dramatic, institutionalized struggles over conflicting class interests. Class conflict is especially evident in two realms: electoral politics and labor relations. Most of the remainder of this chapter will be devoted to these areas.

Elections in modern democracies have been characterized as manifestations of "democratic class struggle" in recognition of the representative role of political parties (Anderson and Davidson 1943; Lipset 1960: chapter 7). Synthesizing the available evidence in 1960, Seymour Martin Lipset wrote:

> Even though many parties renounce the principle of class conflict or loyalty, an analysis of their appeals and their support suggests that they do represent the interests of different classes. On a world scale, the principal generalization which can be made is that parties are primarily based on either lower classes or the middle and upper classes (p. 230).

In most parliamentary systems, parties can be arrayed on a spectrum from right to left, with the former upholding the interests of the privileged classes and the latter attacking them on behalf of the less fortunate.

The constituencies of major parties in four European democracies can be read from the election data in Table 9–4. In the French multiparty system, business owners, executives, and professionals tend toward the right-wing parties. Clerical workers are spread out across the political spectrum.

TABLE 9–4 Political Preference by Occupational Class in Five Countries

France: Voting in 1981 Presidential Election

	Party					
Occupational Class	Communist (Marchais)	Socialist (Mitterand)	UDF (Liberal) (Giscard)	Gaullist (Chirac)	Other	Total
Professional/ management	5	16	29	35	15	100
Self-employed	8	15	32	31	14	100
Clerical workers	15	29	21	16	19	100
Manual workers	28	30	20	12	10	100
Farmers	4	22	31	34	9	100
Retired, etc.	10	28	41	13	8	100

Netherlands: Party Preference of the Electorate in 1986

	Party					
Self-Reported Social Class	PvdA (Socialist Labor)	VVD (Liberal)	D'66 (Liberal)	CDA (Christian Democrat)	Other	Total
Upper class/ upper-middle class	19	39	12	26	5	100
Middle class	29	18	6	37	9	100
Upper working class	47	11	5	28	9	100
Working class	53	3	5	30	9	100

Great Britain: Voting in 1983

	Party			
Occupational Class	Labor	Alliance	Conservative	Total
Professional and managerial	12	26	61	100
Office and clerical	21	24	55	100
Skilled manual	35	27	39	100
Semiskilled and unskilled	44	28	29	100
Trade unionists	39	28	32	100
Unemployed	45	26	30	100

Germany: Voting in 1976

	Party			
Occupational Class	Social Democrat	Free Democrat	Christian Democrat	Other	Total
Self-employed	30	8	62	—	100
White-collar	46	12	41	1	100
Workers	52	5	42	1	100

Continued

TABLE 9-4 *Continued*

United States: Presidential Vote in 1988

Occupational Class	Party	
	Democratic (Dukakis)	Republican (Bush)
Professionals or managers	40	59
White-collar workers	42	57
Blue-collar workers	50	49

United States: Vote for U.S. House of Representatives in 1990

Occupational Class	Democrats	Republicans	Total
Professionals or managers	49	51	100
White-collar workers	52	48	100
Blue-collar workers	55	45	100

SOURCES: France—Frears 1988; Netherlands—Daalder and Koole 1988; Great Britain—Charlot 1985; Germany—Kolinsky 1984; United States, 1988—Pomper 1989; United States, 1990—*New York Times*, November 8, 1990. (Figures in Pomper 1989 leave 1 percent of the U.S. vote unspecified.)

Manual workers lean toward the Socialists and Communists. In the Netherlands, the upper classes vote for the liberal parties or the Christian Democrats; the majority of workers support the Socialists. The British Labor Party has traditionally drawn working-class support, while the Conservatives have run strong among the privileged.

The relationship between classes and parties has traditionally been looser in the United States than in most Western democracies. Democratic political systems typically have at least one major party that identifies itself as socialist and presents itself as a partisan of the working class (e.g., the Communist, Socialist, Social Democratic, and Labor parties of Western Europe). The Democratic Party in the U.S. has traditionally been regarded as the party of the "common man," but it has never called itself socialist, and it has become increasingly coy about appealing directly to working-class interests. Nevertheless, the Democrats have done better among working-class voters than middle-class voters for as long as anyone has bothered to keep track (Lipset 1960: chapter 7; Abramson et al. 1991: 139)

A definite class gradient is evident in the preferences of American voters in 1988 and 1990. But the percentage gap between the professional/managerial and blue-collar strata was modest (6 to 10 percent) and smaller than the corresponding differences in the other countries. The income data suggest greater polarization, but only because the top and bottom income classes represent fairly narrow extremes (respectively, 5 and 12 percent of voters in 1988).

Explanations for the "peculiarism" of the American party system often revolve around the role of ethnicity in our politics. The American party system is rooted in the relationships among class, ethnicity, and politics as they were worked out in the period 1928–1948. Samuel Lubell (1956) has interpreted this period in a classic study of electoral behavior. Lubell described what might be called the emergence of the urban ethnic working class. The generation preceding 1925 witnessed the greatest mass immigration that this country has ever experienced, and most of the newcomers went to work in the factories of the big cities. Simultaneously, many farmers were streaming into the cities, where they could make more money in the new industries; during and after the First World War, this internal migration included hundreds of thousands of blacks. By the 1920s, these people and their children were becoming voters in great numbers. For the first time, the urban workers approached dominance in national politics, for it was not until then that more people lived in cities than on farms. And these urban workers were primarily Democrats; by 1928, the Democrats outvoted the Republicans in most of the big cities. It is Lubell's central thesis that although economic issues have always been important in our politics, in earlier years they were more closely connected with regional interests and conflicts, but by 1928 the parties had taken on the color of class parties—with the Republicans representing business (and successful farmers) and the Democrats representing workers (and unsuccessful farmers). This one split tended to override regional differences and divisions based on noneconomic issues (though of course they were still significant):

> Never having known anything but city life, this new generation was bound to develop a different attitude toward the role of government from that of Americans born on farms or in small towns. To Herbert Hoover, "rugged individualism" evoked nostalgic memories of a rural self-sufficiency in which a thrifty, toiling farmer had to look to the marketplace for only the last fifth of his needs. The Iowa homestead on which Hoover grew up produced all of its own vegetables, its own soap, its own bread. . . .
>
> In the city, though, the issue has always been man against man. What bowed the back of the factory worker prematurely were not hardships inflicted by Mother Nature but by human nature. He was completely dependent on a money wage. . . . A philosophy that called for "leaving things alone" to work themselves out seemed either unreal or hypocritical in the cities, where nearly every condition of living groaned for reform. . . . If only God could make a tree, only the government could make a park (Lubell 1956: 33–34).

Franklin Delano Roosevelt, who came to office in 1933 in the depths of the Depression, was the first president to take advantage of these developments. He built an electoral coalition of Catholic and Jewish immigrants and their sons and daughters, blacks (in the urban North where they had the vote), and working-class people generally. As we will see later in this chapter, the labor movement grew rapidly under Roosevelt, and unionized blue-collar workers were among his strongest supporters. Roosevelt also managed to hang on to white Southerners, who had traditionally supported the Democratic Party.

The Democratic Party that emerged from the New Deal period was a party with strong working-class support but not a working-class party. Ethnicity blurred class lines. Most white Protestant workers supported the

Democrats, but some of them (more than among Jewish or Catholic workers) favored the Republicans. The Democratic party also had important upper-class backers, drawn from the "ethnic rich"—men such as John F. Kennedy's multimillionaire father Joseph, a Catholic, or Jewish financier Bernard Baruch. The ethnic rich bankrolled the party, and although they were probably more ideologically flexible than their Protestant counterparts in the Republican party, they were nonetheless a conservatizing influence on their own party.

The New Deal coalition that Roosevelt built in the 1930s enabled the Democrats to dominate American politics for decades. But beginning in 1968, the Republicans aggressively challenged the political status quo. At the same time, the New Deal coalition was losing internal cohesion. The result, apparent by the 1980s, was not a decisive realignment in national politics such as occurred under Roosevelt, but rather an inconclusive standoff. The Republicans took more or less permanent charge of the White House, and the Democrats retained control of both houses of Congress (Blumenthal and Edsall 1988; Ginsberg and Shefter 1990; Abramson et al. 1991).

This odd remaking of national politics was connected with important changes in American society. In the prosperous years after World War II, many of the children and grandchildren of immigrants moved into the middle and upper-middle classes. Even if they retained their Democratic affiliation, these people often became more conservative and more open to the political message of the Republicans. They were increasingly likely, for example, to see themselves as beleaguered taxpayers rather than beneficiaries of government programs. The class differentiation of the descendants of Catholic and Jewish immigrants convinced Democratic leaders to dilute their party's already weak working-class identity. As second and third generation Americans moved to the suburbs, the urban Democratic political machines they had supported went into decline, undercutting the party's ability to mobilize voters.

At the same time, Democratic positions on issues such as Vietnam, affirmative action, welfare, and abortion were straining the loyalty of working-class supporters. These issues, which were ably exploited by the Republicans, divided a large sector of blue-collar Democrats from the party's upper-middle-class activists. Affirmative action issues in the workplace split black and white working-class Democrats, especially under the unstable economic conditions that prevailed in the 1970s and 1980s. Of course, the Democrats lost much of their traditional white support in the South because of the party's identification with the cause of civil rights.

Finally, the relative weights of business and labor in national politics began to change in the 1970s. In response to an increasingly competitive economic environment, organized business interests began to lobby more aggressively, give more money to politicians, and confront the labor movement more directly. In a shifting economy, the proportion of unionized workers in the labor force was declining—a trend accelerated by the policies of Republican administrations. Organized labor had been one of the mainstays of the New Deal coalition. Now its ability to mobilize blue-collar voters for the Democrats, its capacity to finance political campaigns, and the strength of its voice in Washington were all waning. Under these circumstances, Democratic officeholders inevitably became less attentive to the needs of organized labor.

The class differentiation of the descendants of immigrants, the emergence of issues that divided traditional Democratic constituencies, the decline of labor, and the gains of organized business interests in national politics all contributed to the decline of the New Deal coalition. They also seem to have further weakened the class basis of the American party system.

CLASS, ETHNICITY, AND POLITICS

The most thorough class analysis of the American electorate to date is Richard Hamilton's *Class and Politics in the United States* (1972). Hamilton's work is especially valuable because it consistently takes both class and ethnicity into account. Unfortunately, the book is twenty years old and largely based on a study of the 1964 election. In this section, we will try to make up for that by replicating part of his study with data from the 1984 election. The two contests make a neat comparison: In 1964, liberal Democrat Lyndon Johnson crushed conservative Republican Barry Goldwater; in 1984, conservative Republican Ronald Reagan (the man who nominated Goldwater in 1964) subjected liberal Democrat Walter Mondale to an equally humiliating defeat. In retrospect, it appears that 1964 and 1984 marked the respective high points of liberalism and conservatism in late twentieth-century American politics.

Table 9–5 compares our 1984 analysis with Hamilton's results from 1964. Data in both cases were drawn from the National Election Survey series conducted by the University of Michigan's Survey Research Center. Following Hamilton, we divided respondents into four classes by distinguishing working from middle class (manual vs. nonmanual occupations) and then splitting both into upper and lower groupings on the basis of income.[2] We also employed Hamilton's two ethnic categories, which we retitled for brevity's sake as WASPS and Ethnics. The first refers to white Protestants and the second to all others—notably, African-Americans, Catholics (including most Hispanics), and Jews. (The 1984 figures in the

---

[2]Respondents constituting a national sample of adults were placed in occupational categories (derived from the 1980 census) based on the most recent job held by the head of the respondent's household. Managerial, professional, technical, sales, and clerical jobs were considered middle class. All others were considered working class, including farm occupations. The middle class was divided into two categories at $30,000 in 1983 income. The working class was divided at $20,000.

TABLE 9–5 Presidential Preference, by Class and Ethnicity: Percent Voting for Democratic Presidential Candidate

	1964		1984	
Class	WASPS	Ethnics	WASPS	Ethnics
Upper middle	37	74	18	40
Lower middle	52	84	37	49
Upper working	68	76	27	43
Lower working	81	92	35	48

SOURCES: 1964 from Hamilton 1972: 437; 1984 computed for this table from Center from Political Studies 1984.

table may not be perfectly compatible with Hamilton's from 1964, but we can use the table to compare the general pattern of Presidential preference in the two elections.)

The overall pattern in the table is easy to anticipate. In both contests, Ethnics and people toward the bottom of the class hierarchy were more likely to vote for a liberal Democrat. But note that preference is much more decisively structured by class among WASPS than among Ethnics. In fact, among Ethnic voters, class virtually washes out as an influence.

An apparent anomaly in the 1984 data is the relatively high Democratic vote among lower-middle-class WASPS. In both the Presidential vote and the U.S. House vote (Table 9–4), this grouping was more likely to favor Democrats than any another class among WASPS. This phenomenon may reflect the proliferation of routinized, low-wage office jobs, which probably do not merit the label lower-middle class. (Our class model, introduced in Chapter 1, classifies such jobs as working class.)

The 1984 data reinforce one of Hamilton's basic points: that upper-middle-class WASPS are an extraordinarily conservative segment of the electorate. In 1964, they were the *only class* to give Goldwater majority support. By 1984, the entire electorate had moved to the right, but upper-middle-class WASPS still stood out in their support for a conservative Republican, separated from both upper-middle-class Ethnics and lower-middle-class WASPS by approximately 20 points. Hamilton demonstrated, using national survey data going back to the 1950s, that they were likely to take very conservative stands on national issues.

On the basis of the same data, Hamilton argued that the electorate was, on the whole, quite liberal on social and economic issues—certainly more liberal that anyone would judge from national policy. How did his claim hold up in the conservative 1980s? Table 9–6, based on the 1984 survey, shows considerable support for spending on domestic social programs. Of course, from 1964 to the mid-1970s, there was an enormous expansion of federal spending on such programs. But by 1984, the national political debate focused on which programs to cut and by how much. Ironically, there was more support for liberal programs than anyone could guess from politicians' speeches or media coverage.

In the 1984 survey, people were asked whether they favored cutting,

TABLE 9–6 Support for Increased Spending on Social Programs, by Occupation (1984)

Occupation	Percent
Managerial	30
Professional	32
Sales	27
Technical	40
Clerical	46
Crafts	40
Operatives	52
Service	46
Laborers	48
Farm	28
All occupations	39

SOURCE: Center for Political Studies 1984 (computed for this table).

maintaining, or increasing federal spending on social security, food stamps, public assistance (welfare), medical care, and job programs. The table indicates, by occupation, the percentage of respondents who favored expanding at least three of these five programs. About 40 percent of those interviewed ("all occupations") favored such increases, while fewer than 20 percent wanted reductions in three or more programs.[3] There is an obvious class difference in opinions on this issue, with a gap of approximately 20 percent separating upper-middle-class occupations and the lower-working-class categories.

There is a paradox, of which Hamilton is well aware in his analysis of American political orientations. If, with the exception of an upper-middle-class minority, Americans tend to take liberal positions on bread-and-butter economic issues and identify themselves as Democrats, how is it that liberal Democrats are not consistently elected to office and liberal measures are not regularly enacted? The flood of liberal legislation in the 1960s and 1970s, even under relatively conservative Republican administrations, appeared to lay this question to rest, but in the political world of the early 1990s, it gains renewed relevance.

Especially significant is the fact that the people toward the top of the class structure are better informed politically and more likely to participate in political activities than those toward the bottom. Surveys conducted by the U.S. Census Bureau (1989b, 1991e) show that voter participation accelerates as income rises. In the November 1990 balloting, for example, people from families with incomes over $50,000 were nearly twice as likely to vote as those with incomes under $10,000 (see Table 9–7). Members of families with incomes over $35,000 constituted 43 percent of the voting-age population but 51 percent of actual voters. Thus, middle- to upper-middle-income households are overrepresented. The result is a middle-class electorate. Of course, astute politicians can read the numbers, and they tend

---

[3]The second figure, from the same source, is not included in the table.

TABLE 9–7   Percent Voting, by Family Income, 1990

Family Income	Percent Voting
Under $5,000	32
$5,000–9,999	31
$10,000–14,999	38
$15,000–19,999	39
$20,000–24,999	41
$25,000–34,999	46
$35,000–49,999	51
Over $50,000	59

Note: Self-reported voting as proportion of voting-age population. (Election statistics show that self-reporting overstates actual turnout.) Restricted to members of families.

SOURCE: U.S. Census 1991e.

to craft their message for the middle class. A logical alternative, suggested by sometime presidential candidate Jesse Jackson among others, is to match a liberal platform with a massive effort to increase the electoral participation of lower-income people, both black and white. But many Democratic leaders are reluctant to take this path, fearing that it will send more middle-class voters to the Republicans and exacerbate the party's problems with white voters generally (Ginsberg and Shefter 1990: 15).

Hamilton also argued that voters are not presented with clear issue choices in elections. And voter concerns are not always reflected in public discussion. Using the SRC surveys from the 1950s, Hamilton showed that a majority of the electorate favored national health care legislation, which nonetheless failed to become a political issue (Hamilton 1972: chapter 3). Four decades later, public preoccupation with the issue of health care seems even more intense than it was in the 1950s. For example, a poll conducted by *The Wall Street Journal* in 1991 found that health care and the state of the economy were the two national issues that most concerned Americans.[4] The poll also showed that 69 percent of respondents favored the creation of a universal health care system like Canada's. Despite this level of public concern, health care was not a significant issue in the 1988 or 1990 elections, and there was no expectation that the current Congress would act on national health care.

TRENDS IN VOTING AND CLASS PARTISANSHIP

The percentage of Americans who bother to vote on election day is low by international standards and has declined sharply since the 1960s. In Western European countries, election turnouts are typically in the 75 to 95 per-

---

[4]July 28, 1991. Respondents were presented with a list of six key issues. Forty percent chose health care as the first or second most important issue. The economy was chosen by an equal proportion.

cent range. But only 51 percent of the age-eligible population participated in the 1988 U.S. Presidential election, and only 38 percent voted in the 1990 Congressional elections (U.S. Census 1989b, 1991e; U.S. General Accounting Office 1990). As we have seen, there are significant class differences in U.S. voter turnout.

The influence of social class on voter preferences, historically less in the United States than in comparable countries, appears to have declined in recent decades. However, the trend is erratic from one election to the next. In 1948, the populist message of Democratic presidential candidate Harry S. Truman produced an enormous class difference: About 78 percent of manual workers supported Truman, but only 23 percent of the managerial/professional class voted for him. The gap between these working-class and upper-middle-class voters was 55 percent. But in the next election, the Republican candidate, war hero General Dwight D. Eisenhower, proved popular among all classes. The gap between working class and upper-middle class dropped dramatically, to 19 percent. Since then, the class gap in Presidential elections has varied between 10 and 18 percent, with a perceptible long-term downward trend, especially among whites (Campbell et al. 1954; Pomper 1989).

The gap between races in voting choice is much greater than that between classes, and it shows no clear pattern of change over time. For example, in 1988 there was a gap of 51 percent by race between the parties, with blacks favoring the Democratic candidate: 92 percent of blacks voted for Dukakis, but only 41 percent of whites did so. With the exception of 1976, that gap remained above 50 percent from 1968 through 1988 (Abramson et al. 1991: 134, table 5–2). Black voters, because they are overwhelmingly Democratic, because they are disproportionately working class, and because their electoral participation has grown considerably since the civil rights revolution of the 1960s, have been able to make up for part of the decline in the class gap in voting among whites.

Why is the class gap shrinking in American elections? Certainly not because objective class inequalities are diminishing. We have seen, for example, that income inequality is actually increasing and that the occupational structure is not becoming more egalitarian. But it may be that the way that people perceive their place in the class structure is changing. We noted a modest downward trend in the proportion of people who identify themselves as working class in national surveys. Institutions that encourage working-class people to see themselves as a separate group with distinct interests, such as labor unions and urban political machines, have declined since the Roosevelt era.

Finally, the messages of the political parties and their candidates matter. As the New Deal coalition came unglued, the Democrats attempted to regain lost ground by moving toward the political center. In the 1970s and 1980s, Democratic campaigns typically failed to emphasize issues such as progressive taxation or health care, which might have special working-class appeal, for fear of alienating more privileged voters. At the same time, the Republicans sought out issues (such as abortion and affirmative action) that are likely to separate working-class Democrats from their traditional loyalties.

Despite these trends, the class gap has persisted in U.S. elections. When the gap widens, the Democrats tend to win. For that reason, class continues to enter into the calculations of successful politicians and political strategists.

CLASS CONFLICT AND THE LABOR MOVEMENT

The preceding sections focused on electoral politics as an arena of "democratic class struggle." Here we shift our attention to industrial conflict, the confrontation between capitalists and laborers in the workplace. American labor history has been distinguished by an ironic combination of violent struggle and limited class consciousness. Violence has grown out of tenacious capitalist resistance—not so much to specific economic demands as to the very right of workers to organize labor unions. The class consciousness of American workers has been limited, in the sense that they and their leaders have typically sought circumscribed goals—basically union recognition, economic security, and decent working conditions. Labor unions, as we will see, have played a key role in supporting the passage of liberal welfare measures whose main benefits have flowed not simply to their own members but to working-class and lower-class people generally. But they have rarely sought a fundamental reordering of economic and political relationships designed to benefit the working class, which is to say that they have not developed a socialist class consciousness. In both ways, the American experience differs from that of industrial democracies generally. Elsewhere, employer resistance has been less evident, and the labor movement has been more committed (in rhetoric at least) to a socialist transformation. In regard to the first point, a leading historian observed on the eve of the New Deal:

> Employers in no other country, with the possible exception of those in the metal and machine trades of France, have so persistently, so vigorously, at such costs, and with such a conviction of serving a cause opposed and fought trade unions as the American employing class. In no other Western country have employers been so much aided in their opposition to unions by the civil authorities, the armed forces of government, and their courts (Lorwin 1933: 355, cited in Greenstone 1977: 19).

Nearly a half century later, an American labor economist writing on industrial conflict in Europe noted, "The resistance of U.S. employers to unionizing efforts has no serious counterpart in Europe today" (Kassalow 1978: 97, cited in Brody 1980: 248).

In Chapter 3, we sketched the history of the labor movement up to World War I. A brief review of the decades that followed will place contemporary developments in a meaningful framework. World War I was a period of rapid expansion for the unions, especially those belonging to the American Federation of Labor (AFL), which had formed a wartime political alliance with the Wilson administration. Free to organize workers without government or employer harassment, the AFL unions grew from 2 million to 4 million members in two years (Brooks 1971: 134).

Coasting on its wartime momentum, the AFL continued to grow and gained some important strike victories in the immediate postwar period. But a reaction soon set in, fed by a national "red scare." In 1919, 300,000 steelworkers nationwide went out on strike, protesting twelve-hour work-days and bare-subsistence wages. The response of employers and the federal and local governments that supported them was decisive and often brutal (Brooks 1971: 139–144; Boyer and Morais 1975: 202–209).

The steel strike collapsed after three bitter months. Its fate was indicative of what was to come in the 1920s. In particular, it revealed many of the weaknesses of working-class consciousness and organizations, as well as the effectiveness of the techniques used by the capitalists to resist unionization. The strike showed a working class divided by differences in skill level, race, and ethnicity. Blacks were used as strikebreaking workers. Language diversity kept many immigrant workers from communicating with one another or with native-born members of their class. All these divisions were systematically exploited by employers.

Clearly, violence played a major role in the suppression of strikes. Violence in turn depended on the fact that local and national authorities sided with the capitalists. In effect, civil liberties were routinely suspended in strike situations. Without the normal protection of the laws, workers could be physically intimidated (often by thugs hired for this purpose), union organizers harassed, and leaders jailed. Much of this activity was coordinated by special firms that sold "union-busting" services (Litwack 1962: 95–115).

Resistance to the labor movement was so effective in the 1920s that the proportion of the nonagricultural labor force organized in unions declined from approximately 20 to only 10 percent (Brooks 1971: 148). Most of the core that remained by 1930 was in the relatively conservative AFL craft unions, which showed little interest in organizing the masses of less-skilled mass-production workers employed in large-scale industry. At this late date, even the legal right of workers to bargain with their employers through labor unions remained in doubt. In short, there was little to indicate that the American labor movement was on the verge of an era of militant action, dynamic growth, and expanding national power.

The change, of course, came with the Great Depression and the concomitant shift in national politics, particularly during the years 1933 to 1937. A labor historian has described this period as "the highwater mark of class struggle in modern American history" (Davis 1980: 47). Indeed, some historians believe that without the legal reforms that were produced by the struggles of the period, all-out revolution would have ensued. The transformation of labor relations that emerged from this period was the product of a clash between intense worker militancy and dogged capitalist resistance. When the United Textile Workers announced an industrywide strike in 1934, *Fibre and Fabric,* the New England trade journal, declared, "A few hundred funerals will have a quieting influence" (*Fortune* 1937: 122). Before this bitter, violent strike had completed its three-week run, thousands of National Guard troops had been mobilized in seven states, and twelve strikers and one deputy had been killed. The union lost.

But during these years, there were more labor victories than defeats. Some of the most significant were gained through a new tactic, the sitdown strike; workers forced concessions from employers by taking physical control of the workplace. First used in the rubber industry in 1936, the innovation, which appealed to the militant mood of workers at the time, spread rapidly. One labor official remembers 1937 as the year he received calls daily saying, "My name is Mary Jones; I'm a soda jerk at Liggett's; we've thrown the manager out, and we've got the keys. What do we do now?" (Brooks 1971: 180).

By 1938, the right to union representation had been written into law through the National Labor Relations Act, known as the Wagner Act, and unions had successfully established themselves in the mass-production industries, such as steel, automobiles, rubber, and electrical goods, which stood at the center of the American economy. A conjunction of social and political developments made these accomplishments possible. Of critical importance were the developments in overcoming the division within the working class and the labor movement that had plagued earlier unionization drives. The significance of ethnic differences declined as the sons and daughters of immigrants joined the labor force. No longer cut off from one another by language barriers, more confident of their place in American society than their parents had been, and more demanding of their rights, these second-generation Americans helped recast labor relations just as they contributed to the revamping of national electoral politics.

Differences between skilled and unskilled or semiskilled workers in manufacturing did not disappear, but a barrier to unionization was removed when the CIO (Committee for Industrial Organization) was formed within the AFL in 1935, with the explicit purpose of organizing workers on an industry-by-industry basis rather than on the craft basis that was typical of the AFL unions. The following year, the more aggressive CIO broke with the tradition-bound AFL, retitling itself the Congress of Industrial Organizations. The CIO strove to remove another source of weakness by organizing both black and white workers.

As CIO organizers set about their task, the American working class was in an extraordinarily militant mood. The story of the drugstore soda jerk may be apocryphal, but it suggests the atmosphere of the times. The rank and file frequently ran ahead of union organizers, who found themselves forced to restrain premature action they were not in a position to support. Working-class solidarity grew to the extent that big strikes attracted workers from other industries and localities, who came to offer moral and even physical support (Greenstone 1977: 44).

A key to the worker militancy of the 1930s was the experience of the Depression. We have seen that economic insecurity feeds class consciousness. The insecurity that workers experienced during this period was connected to the breakdown of the entire economic system. The confidence that workers had had in their employers was shattered by the recognition that even such powerful companies as U.S. Steel and General Motors were subject to the vagaries of the marketplace. When capitalists responded to the Depression by laying off workers, reducing benefits, and speeding up work, they lost the loyalty of many workers. An elderly Ford worker complained that he had devoted his working life to "helping to create a millionaire," and an unemployed machinist about to lose his home concluded, "The bankers and industrialists who have been running our country have proved their utter inability, or indifference, to put the country in a better condition" (Brody 1980: 77).

But neither the reduced factionalization nor the growing militancy of the working class could have accomplished the transformation of industrial relations of the 1930s without the changed political context represented by the New Deal. As the experience of the 1920s made clear, if government was hostile to labor, or at least so indifferent as to ignore patently illegal forms of employer resistance, unionization was impossible. The new state of affairs was clear from the passage of the Wagner Act in 1935. The act guaranteed the right of workers to form labor unions, prohibited employers from interfering with the exercise of that right, and set up a National Labor Relations Board (NLRB) thereafter, to ensure that these provisions were carried out in practice. Shortly thereafter, the LaFollette Civil Liberties Investigating Committee began hearings "which, through exposure, largely neutralized the repressive weapons heretofore used in the fight against trade unionism" (Brody 1980: 139).

Both the labor act and the LaFollette investigation would have been unthinkable under earlier administrations. Ironically, neither was encouraged by Roosevelt, whose own attitude toward labor was ambivalent (Brody 1980: 138–146). He did not initiate many of these reforms but was so dependent on working-class political support that he was forced to go along with them.

THE POSTWAR ARMISTICE

The organizational base that the labor movement had built in the 1930s was strengthened during World War II, through close cooperation between the government and the unions in support of the war effort. By 1945, approximately 36 percent of the nonagricultural labor force was organized (Dubofsky 1980: 9). A wave of strikes (which had been prohibited during the war) in 1945 and 1946 gained substantial wage increases for workers, in sharp contrast to the repression of unions in the wake of World War I.

Yet the labor movement was not, as some thought at the time, invulnerable. In 1947, capitalist-class interests gained passage of an important piece of antilabor legislation, the Taft-Hartley Act. In the next couple of years, as a new "red scare" developed, an upheaval within the labor movement resulted in the removal of Communists and other leftists from the positions of leadership they had held in some unions and the expulsion of a number of left-leaning unions from the CIO. The Taft-Hartley Act made it easier for employers to resist union organizing, and the loss of the left deprived the union movement of its most effective organizers, people who had played a major role in the union drives of the 1930s. Both contributed to the collapse of a major organizing effort in the South and severely weakened other efforts to extend union coverage.

Despite these defeats, the labor movement had for the first time established a secure place for itself. In the process, it had significantly reformulated the relationship between the capitalist and working classes in key areas of the economy. Unionization not only provided mechanisms to press for better wages and hours but also afforded some control over working conditions and protection from the traditional petty tyranny of the foreman. Though the majority of workers were never represented by unions, the unionization of some firms and industries posed an implicit threat that tended to constrain the behavior of capitalists in other sectors.

If the 1930s represented a period of explicit class conflict during which these changes were forced on the capitalist class, the decade after the war was the time when the details of a class armistice were worked out. When leaders of the major labor organizations met in Washington with key business representatives in 1945, their purpose was to "lay the basis for peace with justice on the home front." The conference was convened by President Truman at the suggestion of conservative Senator Arthur Vandenberg, who told the president, "Responsible management knows that free collective bargaining is here to stay . . . and that it must be wholeheartedly accepted" (Brody 1980: 175).

Although many business leaders were willing to accept the existence of unions, they were determined to preserve for the capitalist class what they termed the "right to manage." At stake was participation in decisions regarding such matters as investment (including plant openings and closings), product design and production methods, and the pricing of final products. Had labor gained a share in these decisions, as did some contemporary European unionists, the labor movement could have had meaningful influence on employment and other basic economic questions. Unions would thus have been able to represent, not just their own members, but the working class more broadly. That "property rights" in fact remained the dominant influence was demonstrated over the next decade; business did not so much as concede to unions the right to examine the books of the enterprises with which they bargained. The right to manage gradually ceased to be an issue (Brody 1980: 173–213).

By the late 1950s, the shape of the industrial peace was unmistakable. Unions were firmly established among blue-collar workers at the core of the economy in heavy industry. Here they could gain substantial benefits for their members as long as they did not interfere with management pre-rogatives—benefits that would allow a large segment of the working class a life of relative affluence. "The labor movement," concluded auto union leader Walter Reuther, "is developing a whole new middle class" (Brody 1980: 192). Serious industrial conflict was banished to the periphery of the economy—to the smaller firms, weaker economic sectors, and backward regions (especially the South). If such conflict was relatively infrequent, that was in good measure because the labor movement had grown satisfied and unaggressive. By the end of the decade, the giant AFL-CIO (the two had remerged in 1955) was behaving, in the words of labor economist Richard Lester like a "sleepy monopoly" (Dubofsky 1980: 8).

Oddly enough, labor came to play a more dynamic role in national electoral and legislative politics than it did in the workplace. During the 1930s, the labor movement began to cast off its traditional "volunteerism," the determination to steer clear of politics except when its own direct inter-ests were involved. In the postwar period, labor emerged as a major sup-porter of the Democratic party and a broad array of liberal social and eco-nomic programs. Writing in the late 1960s, David Greenstone (1977) concluded that the relationship between labor and the Democratic party, although unofficial, had come close to that between Social Democratic par-ties and labor unions in some Western European countries. The political goal of the activists within the labor movement was "to complete the trans-formation begun by the New Deal and make the Democrats the genuine party of the common man in America" (Brody 1980: 229). Having assumed this burden, labor came to represent not just union members but the work-ing and lower classes generally.

Labor was politically active in two broad arenas: electoral and legisla-tive. In the former, the labor movement promoted liberal candidates, both within and on behalf of the party, raised a substantial part of the party's campaign money, and fielded thousands of campaign workers. In areas such as Detroit, where unions were especially strong, the party and the union's political organization became virtually indistinguishable (Green-stone 1977: 119–140). In Washington, labor maintained a formidable lobby-ing apparatus. During the 1960s and early 1970s, the labor lobby played a major role in obtaining passage of liberal legislation in such areas as civil rights, health care, minimum-wage protection, public employment pro-grams, nutrition programs for the poor, and occupational health and safety (Greenstone 1977: xvii–xxiii, 319–360). Clearly, the benefits of such legisla-tion flowed toward the lower portions of the class structure. Ironically, union lobbyists could not match their successes in social legislation with victories in labor legislation, such as the long-sought repeal of bothersome portions of the Taft-Hartley Act.

As we indicated in the preceding chapter, union lobbyists became the leaders of broad liberal coalitions that confronted business lobbies and other conservative groups over critical pieces of legislation. In effect, a class cleavage ran through the center of national legislative politics. Although that cleavage disappeared from view over many issues (such as the Vietnam war), it was still apparent in the struggles over budget and tax policies in the 1980s.

LABOR IN DECLINE

The 1980s were years of devastating decline for the American labor movement. The clearest measure of labor's fate was the drop in the proportion of workers who belonged to unions. Membership rates, which had been slowly eroding since the mid-1950s, plunged in the 1980s. By 1990, only 16 percent of the nonagricultural labor force was unionized—the lowest figure since the 1930s (U.S. Labor 1980: 412; 1991).

Labor's problems stemmed to some extent from basic shifts in the economy and occupational structure, which we discussed in Chapter 3. Manufacturing, the bastion of union power since the 1930s, was itself in decline. Increasing foreign competition and distorted monetary exchange rates bolstered the sales of imports in the U.S. market and priced American goods out of markets abroad. American manufacturers responded by closing older U.S. plants and moving operations to low-wage havens abroad or to the anti-union states of the Sunbelt. In the course of the decade, one in every three jobs in heavy industry disappeared (Berman 1991: 7).

Employment growth in the 1970s and 1980s was strongest among those categories of workers who are the most difficult to organize: white-collar employees, service workers, employees of small establishments, and female workers. (Most of these groups are among those we have found to be the least class conscious.) The only area in which the unions have made significant gains in recent years is in the public sector, whose future employment growth is uncertain at best. In short, labor's natural economic base has been crumbling.

Labor's organizational decline was soon reflected in political weakness. The 1978 battle over the Labor Reform Bill in Congress gave an early warning of what was to come. Both labor and business regarded the measure, which sought to restore the effectiveness of the 1935 Wagner Act in protecting workers' rights, as a critical test of strength. The legislation was all-important to union hopes of regaining lost ground. Over the years, employers had discovered that they could stave off union organizing efforts by legal maneuvering and other tactics designed to discourage union activists and postpone representation elections. The longer the delays stretched out, the greater the probability that a union drive would collapse. The basic provisions of the bill were designed to guarantee speedy elections to determine whether workers wanted union representation, to ensure prompt decisions from the NLRB in unfair-labor-practice cases, and to stiffen penalties for violation of existing labor laws, such as the firing of union sympathizers.

The bill evoked one of the most extensive and expensive lobby campaigns in the history of the republic. The AFL-CIO spent $3 million on its campaign in support of the bill. A coalition of business groups, including the Business Roundtable, National Association of Manufacturers, U.S. Chamber of Commerce, and National Federation of Independent Businessmen, spent approximately $5 million to secure its defeat. Some 20 million pieces of mail were generated by the two campaigns (Cameron 1978: 80). Both made extensive use of state and local affiliates to pressure members of Congress, who found themselves deluged by communications from businessmen and labor leaders in their states or districts (Cameron 1978; Green 1979: 51–53).

The reform bill glided through the House and found strong majority support in the Senate. However, senators sympathetic to business mounted a determined filibuster to keep the matter from coming to vote (an ironic parallel with the problems the bill was intended to remedy), and business lobbies applied strong pressures to sustain the filibuster. Labor supporters failed by two votes to gain the 60 percent of the Senate needed to close off debate and force a vote on the bill.

Unionists were stunned by both the undemocratic character of their defeat and by the composition of the coalition that had confronted them. They were used to the idea that the lesser capitalists represented by the Independent Businessmen, the Chamber of Commerce, and the NAM harbored strong antiunion sentiments. But the fact that the Business Roundtable—representative of the largest American corporations, the firms with which the unions had made their peace in the 1950s—would support initiatives designed to thwart union organizing came as a terrible revelation. It opened old wounds. In the midst of the conflict over the bill, Lane Kirkland, a ranking AFL-CIO official who later became president of the federation, had charged the corporations with embarking on "a campaign to kill the hopes of the most oppressed and deserving workers in this country. It is class warfare. . . ." (Zeitlin 1980: 32). Shortly after the Senate defeat, Douglas Fraser (1978), president of the United Auto Workers, resigned from the semiofficial Labor–Management Group in Washington with a direct attack on his Business Roundtable colleagues in the group: "I believe leaders of the business community, with few exceptions, have chosen to wage a one-sided class war today in this country. . . . The leaders of industry, commerce, and finance in the United States have broken and discarded the fragile, unwritten compact previously existing during a past period of growth and progress."

Ronald Reagan, elected in 1980, proved to be the most anti-union president in years. Reagan signaled his intentions early in his presidency by firing thousands of striking air traffic controllers and putting their union out of business. Under his administration, the protections of labor law were further diluted by pro-management appointments to the NLRB and the Labor Department. One experienced union lawyer commented that the labor board had come to operate "like a bloodless bureaucratic death squad" (Geoghegan 1991, quoted in Berman 1991: 7).

In the 1978 elections, for the first time, business PACs outspent labor PACs (*New York Times*, Aug. 4, 1981). In Congress, the ability of labor lobbyists to move (or block) legislation was waning.[5] The AFL-CIO's attempt to recoup political losses by throwing the confederation behind a pro-union presidential candidate in 1984 ended in embarrassing failure.

Business, encouraged by the new atmosphere in Washington and often threatened by competition, assumed a more aggressive stance in labor relations. Many firms forced unions to yield "givebacks" of favorable wage rates and working conditions won in earlier negotiations. Others sought to destroy their unions by forcing decertification elections or provoking a strike and hiring nonunion replacement workers. In the Sunbelt, corporations found a legal and political atmosphere in which union organizing was extremely difficult. Even large, established corporations were taking advantage of the services of a militant new breed of labor consulting firms.

The new consultants pose an intriguing contrast with the union-busting industry of the 1920s and 1930s. The firms that once supplied tear gas, small weapons, and strikebreakers have been supplanted by practitioners of manipulation, sophisticated in their application of social scientific knowledge. For example, the consultants show management how to convey to employees an artificial sense of participation in company decisions and how to use systems of subtle rewards and punishments to influence employee attitudes or create anxieties that work against the union. Employers are advised to avoid certain categories of workers, because they are more likely to be receptive to union appeals. Consultants also instruct corporations in the advantages they can glean from calculated violations of labor law (Langerfeld 1981).

Organized labor contributed to its own problems. After the purges of leftist activists in the late 1940s, the well-paid officers of many unions had grown complacent, in a few cases corrupt, and distant from the problems of ordinary workers. Such leaders may have been qualified to conduct the daily business of established unions in quiet times, but they were less effective against the rising tide of political and economic change.

Labor's defeats in recent years have real implications for the dynamics of the American class system. The unions always spoke for a much wider

---

[5]For example, for an entire inflation-ravaged decade, labor was unable to obtain an increase in the minimum wage, whose purchasing power steadily declined. Also, in 1991, the unions lost a major legislative battle with business over the U.S.–Mexico free-trade agreement. The agreement posed a serious threat to wage rates and union organization in the United States.

constituency than their own members. In the workplace and in national politics, however indirectly and imperfectly, organized labor has represented the interests of working-class Americans against those of the capitalist class. At work, the unions set a standard that even the employers of nonunion workers had to acknowledge. In national politics, the labor movement was an integral part of a liberal coalition that was able to usher through Congress an increase in the minimum wage, a progressive tax measure, or worker safety legislation. In both arenas, labor's fall has altered the balance of power among classes in ways that will effect this country for years to come.

CONCLUSIONS

This chapter has examined one central theme: class consciousness. Marx first raised the issue, since he believed that subjective awareness of one's objective position in the property and occupational system would lead to a sense of belonging with others of similar position and promote conflict with those above or below. Those experiences in turn would generate political beliefs or ideologies that interpreted the world and suggested appropriate forms of organized action to advance class interests.

Contemporary sociologists have used Marx's views to create specific hypotheses that could be tested in empirical research. They wanted to explain the origins of class consciousness as indicated in various forms of verbal response, including use of class terms of self-identification, and they wanted to pinpoint the consequences of identification for political belief and action. They have consistently found, in the United States and other countries, that objective position in the occupational system is an important influence on identity, with manual workers usually considering themselves to be working class and nonmanual workers calling themselves middle class. These terms carry connotations much broader than just a classification of jobs: They imply a way of life, a set of friends, and a system of values. And they cover all members of a family, not just the principal breadwinner, whose job is usually used to define class membership. In the United States, very small percentages of people consider themselves either upper class or lower class (or poor); almost half put themselves in the working class, and a slightly smaller proportion consider themselves middle class.

But identification flows from other causes besides occupation, so it reflects more than one social fact of life. Particularly important is racial or ethnic membership. When people belong to a minority group suffering from discrimination that leads to economic insecurity and social stigma, they respond by identifying with other underdogs in the system. And underdogs in general develop ideologies of liberal or leftish tone that demand changes in society to promote more equality and justice. These ideologies are even more pronounced among individuals and groups who have had these additional experiences: uprootedness, such as movement from one country to another or from farm to city; a high rate of interaction within their own group and isolation from other groups; two or three generations of similar social status; and membership in organizations, especially unions, that promote class conflict. All of these specific indicators, except perhaps for aspects of ethnicity, can be deduced from Marx.

In the Western democratic countries, class consciousness has not expressed itself in the sort of revolutionary upheaval that Marx anticipated. Instead, class conflict has been channeled into electoral competition and labor politics. The democracies typically have a right–left spectrum of parties, with the left parties tending to draw their members from the bottom of the class structure and the right parties from among the privileged.

The United States fits this general pattern, but the class identities of our major parties have always been somewhat blurred by ethnicity and regional traditions, and the class gap in party support has never been as great as the corresponding difference in European systems. From the 1930s until the 1970s, the Democratic Party dominated national politics. The party was built on a working-class base, reinforced by traditional ethnic and regional loyalties. In recent years, the strength of this so-called New Deal (or Democratic) coalition has eroded, producing a stalemate between the major parties in national politics.

In labor politics, the American labor movement consolidated its strength at the same time that the New Deal coalition emerged under Roosevelt, in the early 1930s. It was to become, in fact, one of the pillars of the Democratic Party. Through the 1930s, American labor history was marked by violence and management resistance to the right of workers to unionize. The post–World War II period produced an "armistice" in labor relations, as the major corporations accepted unions as a fact of life. But that has come undone in recent years. Business, supported by conservative Republican administrations in Washington, has become much more aggressive in its confrontations with labor. Union strength has declined sharply, contributing in turn to the decline of the Democratic Party.

These developments have shifted the class balance in American politics in favor of the capitalist class.

SUGGESTED READINGS

Aronowitz, Stanley. 1973. *False Promises: The Shaping of American Working Class Consciousness.* New York: McGraw-Hill.
> *The development of working-class consciousness in historical perspective, emphasizing the effect of consumer culture on working-class views.*

Brody, David. 1980. *Workers in Industrial America: Essays on the 20th Century Struggle.* New York: Oxford University Press.
> *A lively introduction to the history and historiographic literature of the labor movement in the twentieth century.*

Edsall, Thomas, and Mary Edsall. 1991. *Chain Reaction: The Impact of Race, Rights and Taxes in American Politics.* New York: Norton.

   The "wedge" issues that splintered the Democratic class-ethnic coalition.

Ehrenreich, Barbara. 1989. *Fear of Falling: The Inner Life of the Middle Class.* New York: Pantheon Books.

   The evolving consciousness of the upper-middle class and its consequences for American politics and culture. Sources range from Hollywood to Dr. Spock.

Geoghegan, Thomas. 1991. *Which Side Are You On: Trying to Be for Labor When It's Flat on Its Back.* New York: Farrar, Strauss & Giroux.

   A labor lawyer's personal account of the 1970s and 1980s.

Rieder, Jonathan. 1985. *Canarsie: The Jews and Italians of Brooklyn against Liberalism.* Cambridge, Mass.: Harvard University Press.

   Political ethnography. The disintegration of the Democratic coalition from the bottom up.

Thompson, E. P. 1963. *The Making of the English Working Class.* New York: Vintage Press.

   Classic historical portrayal of the development of working-class consciousness.

10 The Poor, the Underclass, and Public Policy

We stand at the edge of the greatest era in the life of any nation. . . . Even the greatest of all past civilizations existed on the exploitation of the misery of the many. This nation, this people, this generation, has man's first chance to create a Great Society; a society of success without squalor, beauty without barrenness, works of genius without the wretchedness of poverty.

Lyndon B. Johnson, 1964

We have decided to abjure a glitzy, splashy, high-profile announcement of new [poverty] programs and a grand new strategy. We concluded that there were no obvious things we should be doing that we weren't doing that would work.

Keep playing with the same toys. But let's paint them a little shinier.

Bush administration officials, 1991

Americans added a new word to their political lexicon in the 1980s: *homelessness*. The term was not a recent addition to the dictionary, but it was being used in a new way—to name a social problem. And while homelessness had once referred to people without fixed residence, who drifted from place to place, it was now being applied to people who literally had no access to conventional housing, people who slept in doorways, packing crates, bus stations, and shelters. No one knew exactly how many there were. The best national estimates from the mid-1980s suggested that there might be 500,000 homeless on an average night and that some 1.5 million people would find themselves homeless sometime in the course of a year (Rossi 1989: 20–70; Wright 1989: 19–32).

Although 500,000 people could populate a city the size of New Orleans, they constitute only a fraction of a percent of the nation's population. Yet the homeless attracted intense national attention. Homelessness had been little known since the Great Depression of the 1930s. In the 1980s, it suddenly became visible on city streets across the country. There was general recognition that the homeless population was growing and a broader sense that the homeless were simply the ominous tip of the iceberg of poverty.

In fact, poverty rates *were* rising. By the government's official standard, 13.5 percent of Americans were poor in 1990, more than at any time since the late 1960s (U.S. Census 1991b). And the poor were becoming poorer. In a book on the homeless, sociologist Peter Rossi calculated that there were as many as 7 million unmarried, working-age adults in 1987 who could be described as "extremely poor"—people with incomes so low that they were one step from homelessness. This was, Rossi found, more than twice as many as there had been in 1970 (Rossi 1989: 72–81). Against a background of rising childhood poverty rates, the nonpartisan National Commission on Children (1991: 124) offered the following observation on hunger in its 1991 report to the President and the Congress:

> While there is debate over the prevalence of childhood hunger in America, there is no doubt that the problem has increased over the past decade. Recent estimates of the number of children who experience hunger range from 2 million to 5.5 million. The increase is closely related to the high rates of childhood poverty and may become even more severe in the 1990s, if poverty among families with children is not reduced.

As the decade of the 1990s began, most Americans and their leaders in Washington were aware of poverty but little inclined to do anything about it. People were more concerned about the shaky performance of the national economy and their own (in many cases) shrinking or stagnant incomes. They were not as confident as they had once been in the capacity of government to tackle big social problems. But as we will see in this chapter, the national will to fight poverty has waxed and waned in our history.

In this chapter, we will deal with several key questions. How is poverty defined and measured? How many Americans are poor? Who are the poor? What are the long-term trends in poverty? What are the causes of poverty? How has government policy responded to poverty? We will begin by going back to the 1930s, when poverty was first recognized as a problem requiring federal action, and the modern system of social programs was born under Franklin Roosevelt's New Deal.

THE BEGINNINGS OF WELFARE: ROOSEVELT

The Great Depression hit the nation with devastating effect. The unemployment rate was above 20 percent from 1932 through 1935 and did not go below 15 percent until the eve of war in 1940—and simultaneously, the wages of those who had jobs were pushed down. Per capita income was cut by a third. Private charities were overwhelmed, and the attempts of local governments to provide some form of assistance were totally inadequate. With the election of Franklin D. Roosevelt in 1932, the federal government moved quickly and devised entirely new approaches to the problems of unemployment and poverty (Piven and Cloward 1971: chapters 2 and 3). The first step was direct "relief" in the form of cash payments to any family in desperate need, financed by "emergency" grants from the federal to the state governments, which made up their own rules of distribution within broad federal guidelines. In 1934, one sixth of the population was receiving assistance. In addition, the Civilian Conservation Corps provided some jobs for unemployed young men in reforestation and similar projects.

By 1935, it was decided to expand the job programs and phase out direct relief. The Works Progress Administration was set up to finance all sorts of construction projects and other activities, mainly organized by local governments. From 2 to 3 million people (mostly men) were employed in the later years of the decade by the WPA and another million people in other projects (U.S. Census 1975: I, 339). The responsibility for direct relief was turned back to the states and localities, but with some help in the form of "categorical" programs, in which the federal government would share the costs of support for impoverished people classified as unemployable for specific reasons: the blind, the physically or mentally disabled, the elderly, and mothers who had small children but no husbands to provide for them. (The small program for mothers eventually grew into AFDC, Aid to Families with Dependent Children.) It was assumed that much of the aid dispensed under these programs would eventually fade away as the new programs of the federal social security system took hold and as the country worked its way out of depression and unemployment.

The core of that new system of social security was OASDI—Old Age, Survivors and Disability Insurance. Enrolled individuals (especially in the earlier years, not everyone was included) would receive full coverage after ten years of contributions made by employees and employers to the trust funds. Retired persons would get permanent pensions, as would those who were disabled to the point of not being able to work. Widows and orphans (survivors) of insured workers would get benefits. There was also established a system of unemployment insurance, administered by the states under federal guidelines, that would give temporary payments to insured workers during periods of layoff, usually up to twenty-six weeks. All insurance-type benefits were automatic "entitlements" to those enrolled; they did not require the "means test" to demonstrate poverty that was required for direct relief.

Each of these new programs dealt with a piece of the general problem of distress—that is, with a special category of people. Some programs emphasized care for the elderly; others gave help to young mothers without husbands; others offered temporary payments to workers between jobs or provided jobs for the unemployed, which were expected to end with the return to "normalcy." Some were based on a means or poverty test; others were viewed as insurance or entitlements. Nobody thought in simple and direct terms of providing enough money to bring *all* poor people up to some standard of minimum decency. Consequently, the programs overlapped in some ways but failed to integrate in others. Some people who were not poor received payments (especially old age pensions), and others who were living in misery found themselves ineligible for any help.

REDISCOVERY OF POVERTY: KENNEDY AND JOHNSON

In the prosperous era that followed the end of World War II, the scene had changed. The social security system was beginning to pay out large sums to the elderly, and since it was viewed by the public as an insurance program that returned to them in the form of pensions money that they had earlier contributed in the form of taxes on wages, it was not stigmatized as "relief" and was popular among all segments of society. (Actually, it is a program that taxes current workers to pay retirees, but most people do not think of it that way.) The "make-work," or special public jobs, had disappeared. The unemployment compensation system was working smoothly and was taken for granted. When Dwight Eisenhower was elected in 1952, we learned that these programs were no longer connected with a partisan New Deal but were accepted by the Republicans, who had earlier opposed them. The new president's main approach to domestic policy was a desire not to do anything dramatic that might upset the good mood.

Only a few critics seemed worried about the state of domestic economic affairs. Among them was John Kenneth Galbraith, an iconoclast among the economists. He wrote a book called *The Affluent Society* (1958), and some people mistakenly assumed from the title that he was celebrating prosperity. But his main message was that the new affluence was creating new problems that were not being noticed by public policy: a built-in, permanent tendency for inflation; old cities decaying without sufficient investment for proper renewal of public services; and a new type of poverty that was not caused by economic depression and persisted despite economic prosperity. The new poverty, according to Galbraith, affected two types of people: individuals who suffered from one or another personal handicap (bad health, inadequate education, low intelligence) and who should be helped by social security or social welfare but were not being adequately supported; and individuals living in islands of poverty in areas of the country untouched by the new prosperity—for example, the Appalachians, peopled with subsistence farmers and underemployed coal miners, or the inner-city slums to which the rural poor were moving in large numbers. These people and their children needed help from deliberate government action; economic growth by itself was leaving them behind.

Then came a stronger statement: Michael Harrington's *The Other America: Poverty in the United States* (1962). He estimated that over 40 million Americans were poor and that the new prosperity was not helping them but, in many instances, actually making their situation worse.

> One might summarize the newness of contemporary poverty by saying: These are the people who are immune to progress. But then the facts are even more cruel. The other Americans are the victims of the very inventions and machines that have provided a higher living standard for the rest of the society. They are upside-down in the economy, and for them greater productivity often means worse jobs; agricultural advance becomes hunger (1962: 12).

Harrington's book was widely read by members of the new Kennedy administration, including the president and his influential brother, Robert, who had been shocked by the misery he had seen in West Virginia during the electoral campaign. Galbraith became an advisor to the president. The new political leaders found that members of the federal bureaucracy concerned with unemployment and welfare policies knew about a lot of poverty that was not recognized by the public; so did a few members of Congress and their research staffs who had data on malnutrition, disease, alcoholism, drug addiction, and juvenile delinquency. With the arrival of the new administration, poverty and its various forms of pathology came back into public discussion. These leaders were Democrats, heirs to Roosevelt's New Deal, and their political constituencies thought of themselves as liberals devoted to constant social betterment. If the New Deal had not fully solved the problems of poverty, they proposed to move forward with new impetus—but cautiously, since members of this younger generation of Democrats were pragmatists rather than ideologues.

Their approach was to divide the problem into two parts: (1) the situation of the unfortunates with personal handicaps, who might need more help through modest improvements in the welfare system of cash support and social services, including better training for those lacking in job skills; and (2) the able-bodied unemployed, who would be helped most by faster growth in the general economy to reduce the unemployment rate below the 5 to 6 percent level typical of the Eisenhower years. A large tax cut was introduced as a means of stimulating economic expansion, "to get this country moving again." (Note that this approach ignored a central thesis of Harrington: that economic progress itself created as much poverty as it eliminated by introducing new and more efficient technologies in agriculture and industry that increased production but eliminated jobs.)

The macroeconomic policies did have a small effect: In the early 1960s, the unemployment rate went down a little. In the late 1960s, it went down even further as the country became involved in Vietnam and again moved toward a wartime economy. By 1968, only 3.6 percent of the labor force was unemployed, a low rate that has never been matched in later years. The decade as a whole enjoyed the greatest economic prosperity and growth in the nation's history.

The Kennedy and Johnson administrations made important improvements in both the social security and welfare systems. They added three new programs: food stamps (poor families could receive stamps that were redeemable at the grocery store for food); Medicare (insurance to cover hospital and doctor bills for the elderly receiving social security pensions, regardless of their other income); and Medicaid (free medical insurance for some of the poor, especially those receiving AFDC—Aid to Families with Dependent Children, the basic public assistance program), and the preschool program Head Start. Through court challenges, often brought by government legal-aid lawyers, access to welfare benefits became more dependable, open to all who met clearly codified criteria. Civil rights legislation passed in this era opened new opportunities to the minority poor.

Republican Richard Nixon did not share his Democratic predecessors' enthusiasm for expanding social welfare programs, but the momentum established in the Kennedy-Johnson years kept expenditures growing.

FIGURE 10–1 Spending on Social Welfare Programs

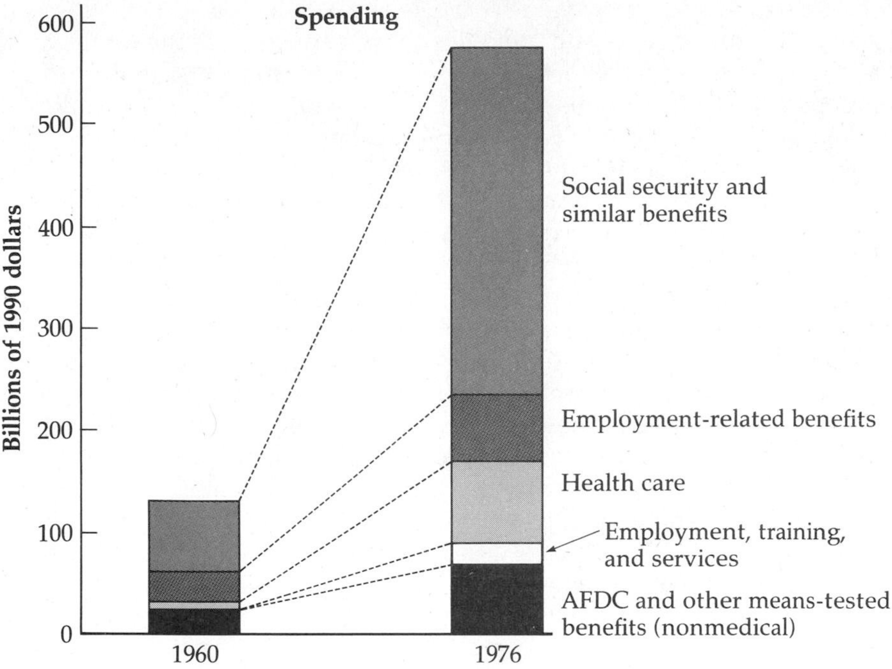

SOURCE: Adapted from Ellwood 1988: 31.

Spending on all social welfare programs, from Social Security to AFDC, almost quadrupled between 1960 and 1976 (see Figure 10–1). This development can only be described as a revolution in the federal budget.

Johnson hopefully called his program for the poor the "War on Poverty." "We stand at the edge of the greatest era in the life of any nation," Johnson told his countrymen in 1965. "For the first time in human history, we have the abundance and the ability to free every man from hopeless want, and to free every person to find fulfillment in the works of his mind or the labor of his hands." (The rest of the passage is reprinted at the beginning of this chapter.)

From the cold perspective of the early 1990s, Johnson's grandiloquent rhetoric sounds quaint, and his optimism appears misplaced. Although the U.S. made steady progress against poverty in the 1960s and early 1970s, we have lost ground since then. More serious, we have lost the hope and commitment that Johnson and many of his contemporaries displayed.

We will take a careful look at trends in poverty and federal policy in the years after Johnson left office. But first, we need to understand how the government measures poverty. This matter was settled at the beginning of the Kennedy administration.

THE OFFICIAL DEFINITION OF POVERTY

When President Kennedy came to office in 1961, he was under pressure from civil rights activists and other liberal constituencies in the Democratic party to take a new look at racial discrimination and persistent poverty. Bureaucratic studies in his administration, supported by academic research, gave substance to Harrington's position that there were still millions of Americans living in misery—indeed, some in hunger. Since the general living standards were rising steadily, it seemed to many that the nation could finally afford to eliminate poverty completely.

Fighting poverty required a coordinated plan that would lift everybody above the threshold of acceptable living standards, both by helping poor people get better jobs and by integrating various parts of the existing system of piecemeal public welfare that left many needy people without assistance. For some, this would mean basic education and job training; for others, improvements in the systems of social insurance and social welfare. In order to measure the size of the problem and to test the effectiveness of remedial measures in the coming years, an official definition of poverty was needed—something the government had never before enunciated. Using the definition, a census could be taken indicating the number of people who were poor; then the specific causes of their poverty could be analyzed, and programs could be designed or redesigned to help them.

The policymakers adopted an old idea from private charities and local public welfare offices, which had always needed guidelines to help them decide who to help and how much to give. There was a tradition going back to the beginning of the century, based on the idea of a "market basket" of cheap but nutritious foods that would keep a family alive, along with other essential items, including enough rent money to keep them from being evicted. Usually, no long-term expenses (such as furniture) were included, as the charity help was considered to be a minimum to tide people over a temporary emergency. Year after year, different groups calculated the items that ought to go into the market basket and then went to the store to find out how much the items would cost. This procedure produced minimum subsistence budgets; people in trouble were helped up to the level of those meager budgets.

The poverty standard finally adopted by the Kennedy administration was designed by Mollie Orshansky (1965), an economist at the Social Security Administration. Orshansky put two pieces of information together from government surveys: the cost of a minimum nutritious diet for a typical family of four and the proportion of income (approximately one-third) that the average family spent on food. Multiplying the price of the food budget by three to allow for nonfood costs, Orshansky calculated an income "poverty line" of approximately $3,000 for a family of four. If you were a member of a four-person family with income below $3,000, you were poor by this standard. On the assumption that family needs vary with size, proportionally higher and lower poverty lines were computed for households that were larger or smaller than the typical four.

Orshansky's poverty line became the official federal standard. Each year, the Census Bureau uses it in conjunction with the annual survey of household income to estimate the number of poor people in the country. (The survey, by the way, measures pretax income, including income from such sources as Social Security and welfare payments.) By the new official standard, 22 percent of Americans were poor in 1960 (U.S. Census 1991b).

Since prices change, the poverty line must be regularly adjusted for inflation; each year, it is increased in proportion to the annual increase in the Consumer Price Index, the government's general measure of inflation. By 1990, the poverty line for a family of four was about $13,000. Note that the poverty line is adjusted for changes in prices but not for changes in the general standard of living. By using this measure, the government is saying something about the meaning of poverty. It is telling us that what makes people poor is their material deprivation in an absolute sense rather than the relative gap between their standard of living and the standard typical of people in the same society. We will come back to this question momentarily.

The Orshansky standard was as reasonable a poverty measure as anyone could come up with at the time. But it was problematic from the beginning and remains a target of political controversy to this day. The trouble started when the proposed measure was being considered by Kennedy's Council of Economic Advisors. Government nutritionists had come up with not one, but two food plans. (Both were based on the dubious assumption that the homemaker was a sophisticated dietitian who never wasted a penny.) The first plan was an emergency diet, suitable to maintain a family for a short period. The second plan, which cost 25 percent more, was designed to provide the nutrition necessary for long-term family health. It was left to the Council of Economic Advisors to decide which food plan to use as the basis for measuring poverty. Adopting the second plan would result in a higher poverty line and, as it turned out, a much higher estimate of the number of impoverished Americans. That was apparently unacceptable to the Kennedy administration. The Council of Economic Advisors chose to base the official poverty line on the emergency food plan. The decision had little to do with nutrition and nothing to do with economics; it was essentially political.

The higher standard, now commonly referred to as "125 percent of the poverty line," is used in some government statistical reports as a broader measure of poverty. The population that falls between the poverty line and 125 percent of the poverty line is sometimes labeled the "near poor," a convention we will occasionally follow in this chapter.

The poverty line was supposed to be based on an objective, scientific standard: minimum nutritional need, plus a proportional allowance for rent and other essentials. But in retrospect, it appears that there is no wholly objective way to specify minimum requirements. Contemporary cultural values always intercede and help define what is appropriate. When the general standard of living goes up, so do ideas about the minimum needed to maintain subsistence. If there was any doubt about this, it was removed by the research of Oscar Ornati and his colleagues (1966), done about the same time the government was defining the poverty standard.

Ornati collected the subsistence budgets that had been used by local social service agencies, going back to 1905. He found that these budgets, which were designed to provide temporary relief for family survival, were typically based on the same notion of a "basic-diet-plus" that was adopted for the federal poverty standard. Ornati discovered that the conception of a subsistence budget tended to expand as national living standards improved. For instance, a 1908 budget for a family of four included 22 pounds of high-protein foods. A 1960 budget assumed that a family of this size needed 55 pounds. The 1908 budget makers supposed that a family needed four rooms but did not require a bath; by 1960, the standard was five rooms and a full bath. And we now assume that electricity, a refrigerator, and a high school education for children are included in a minimum standard of living, but they were not in 1908.

Sociologist Lee Rainwater (1974) further analyzed Ornati's data. He found that, year after year, the price of what the agencies considered subsistence budgets stayed close to 40 percent of the current average disposable income for a four-person family. By no small coincidence, the same was true of the federal standard when it was adopted in the early 1960s. Rainwater went a step further and looked at public opinion data from Gallup polls since 1946. In that year, Gallup started asking people this question: "What is the smallest amount of money a family of four needs to get along in this community?" The average response, from 1946 to 1969, was usually in the range of 50 to 55 percent of the average family disposable income for the year the question was asked.

The conclusion that can be drawn from the research of Ornati and Rainwater is that conceptions of a minimum living standard are inevitably *relative*. They depend on income levels and lifestyles in the communities where they are made. This generalization apparently applies with as much force to the experts as it does to popular opinion. Thus, what we now consider a poverty-line existence would have qualified as middle-class comfort at the beginning of the century and might still be regarded as such in some small town in the Peruvian Andes.[1]

This brings us back to the federal standard. If we could somehow summon the Kennedy administration experts, erase their collective memory of

---

[1] Presumably, there is some nutritional level at which people starve or at least become weakened and highly susceptible to disease. We might call that an absolute standard of poverty, without regard to community standards. But we don't know exactly how to define that standard; in any event, it is not very relevant to a society like our own.

the poverty line they created in the 1960s, and make them do it again in the 1990s, they would almost certainly come up with something closer to our current living standards. But for thirty years, the government has treated the poverty line as an *absolute* standard, one that needs to be adjusted for increasing prices, but not for changing lifestyles.

Most sociologists believe that the government should use a relative standard to measure poverty. Their assumption is that what makes people think of themselves as poor and causes others to regard them as poor is the comparison between their lives and the mainstream lifestyle in their community. The relative standard most frequently proposed is half the median family income. In 1960, the median family income was around $6,000, which meant that the poverty line for a four-person family was, in fact, close to half the median income when it was first minted. By 1990, the basic poverty line had fallen to 38 percent of the median family income.

In recent years, the federal poverty standard has been subject to a different line of criticism—from conservatives. They argue that the official standard, because of a series of technical deficiencies, overstates the amount of poverty (Anderson 1978; Stockman 1984). For example, they claim that people tend to underestimate their income when they respond to government surveys. And they recall that the annual income survey from which poverty is estimated records only *money* income. This was not much of a problem when the poverty standard was established, but since then there has been a vast increase in the amount of *in-kind* aid given to the poor in such forms as food stamps, subsidized housing, and health care. These critics probably have a reasonable argument with regard to food stamps, which are like cash and might well be considered part of income. Unfortunately, they do not mention another change since the 1960s: Increasing taxes, especially payroll and sales taxes, have cut into the incomes of the poor, but taxes are not recorded by the income survey or reflected in the official poverty statistics.

HOW MANY POOR?

According to the various measures of poverty we have been discussing, there are somewhere between 27 and 48 million poor people in the United States (see Table 10–1). By the official standard, there are 33.6 million poor. This figure is based, as we have seen, on Orshansky's minimum subsistence family budget, adjusted for inflation. The basic poverty line for a four-person family is moved up or down to measure poverty in larger or smaller families. The other statistics in Table 10–1 are refinements of the official measure, using the same data from the Census Bureau's income survey but reaching rather different conclusions. All are based on the same poverty-line concept, with adjustments for family size.

The low-end figure of 27.3 million takes the conservatives' criticisms of the official standard to heart. The Census Bureau produced this estimate

TABLE 10–1 Four Measures of Poverty in the U.S., 1990

Poverty Measure	Poverty Line (Family of 4)	Percent Poor	Number Poor (millions)
Official	$13,359	13.5	33.6
Adjusted official[a]	$13,359	11.0	27.3
125% poverty line	$16,699	18.0	44.8
½ median income[b]	$17,677	19.3	48.0

[a]Definition #14, U.S. Census 1991c: 11. Reflects adjustments for in-kind benefits and taxes.

[b]Estimated from U.S. Census 1991b: 30.

SOURCES: U.S. Census 1991a, 1991b, 1991c.

by adding to incomes the imputed value of noncash benefits such as food stamps and Medicare.[2] But the bureau also took into account the income lost to all local and federal taxes.

By the standard of 125 percent of the official poverty line, there are nearly 45 million poor. This number would, of course, have been the official statistic if the Kennedy administration had not chosen to base the poverty line on an *emergency* food budget, but rather on a diet that provides good health over a longer period of time.

A relative standard defines poverty in relation to mainstream standards of living. By the most frequently cited relative standard, there are 48.0 million poor people in the U.S. — a statistic equivalent to the combined population of all the states from the Rocky Mountains to the West Coast. As we noted earlier, when the official poverty line was first established, it was close to one half the median income, the relative standard applied here. By this measure, the U.S. has made no progress against poverty: One fifth of the population was poor in 1960, and one fifth remains poor today.

WHO ARE THE POOR?

The debate about how to measure poverty is endless, and it is not difficult to imagine why. Any serious discussion of poverty in an affluent society, which also regards itself as democratic, inevitably stirs deep political emotions. Just below the surface, the technical dispute about measurement is a debate about the fairness of our political and economic institutions.

---

[2]Assessing the value of Medicare and Medicaid coverage is especially problematic for estimates of this sort. The Census Bureau took into account only what it considered the fungible proportion of Medicare and Medicaid. If a family's income did not exceed the amount needed for food and housing, it was assumed that health coverage did not free up money for other purposes and should not be considered part of their income.

To examine poverty, we will have to settle on some measure. For most purposes in the remainder of this chapter, we will employ the official standard. If we keep its specific defects in mind, the federal standard is not an unreasonable way to measure poverty. It happens also to be the basis of most of the detailed poverty statistics published by government.

Figure 10–2 uses official data to answer a key question: Who are the poor? It starts with the total number of poor people, as measured in

FIGURE 10–2   Distribution of Poverty, 1990

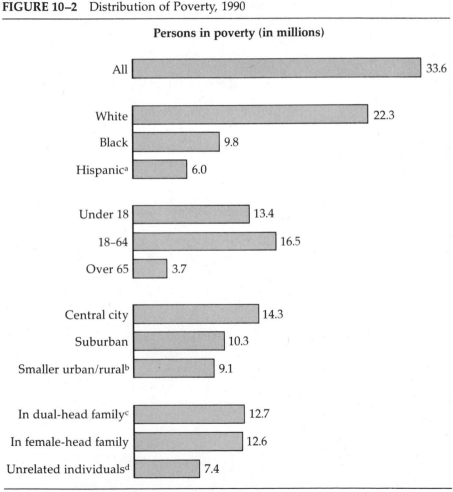

Persons in poverty (in millions)

[a]Hispanic may be any race.
[b]Outside metropolitan areas. Generally under 100,000 population.
[c]Includes small number of members of male-headed families.
[d]Living alone or with nonrelatives.
SOURCE: U.S. Census 1991b.

government surveys (about 33.6 million people) and breaks this figure down in various ways. Here are some of the things we learn:

There are twice as many poor whites as poor blacks.

A high proportion of the poor are children. Few are elderly.

More of the poor live in families headed by married couples than in families headed by women.

Only a minority of the poor live in the central cities of metropolitan areas. The majority are spread out among suburbs, small towns (under 100,000), and rural areas.

There is enough information here to contradict some popular stereotypes. In particular, the typical poor person is obviously *not* a member of a black, female-headed family, living in a big city. As it turns out, only 11 percent of the poor fit that description (U.S. Census 1991b).

Figure 10–3 answers a different kind of question: If you belong to a certain social group, what are your chances of being poor? This life chance is called the *risk* (or *rate* or *incidence*) of poverty. It is simply the percentage of the people in a specific group that fall below the poverty line. Just a quick glance at the graph reveals that there are very large differences in the risk of poverty. Blacks, although they are a minority of the poor, are three times more likely than whites to be poor. Children are at a much greater risk of poverty than adults. Approximately two out of five black and Hispanic children are poor. And whatever their ethnicity, female-headed families have poverty rates far above those of families headed by couples.

As we noted earlier, the measure of poverty employed here counts all money income—including, of course, the transfer payments people receive from the government for welfare, social security, veterans benefits, etc. These payments make a difference. Without them, the poverty rate in 1989 would have been 20 percent instead of 13 percent (U.S. Census 1991b). But the effect of cash transfers is very uneven, as Table 10–2 demonstrates. Note that seniors and female-headed families with children have virtually the same high, pre-transfer poverty rate. But a dramatic gap opens between them after transfers. The poverty rate of children is hardly touched by transfers. It is clear that the only government transfer program that comes close to eliminating poverty is social security for the elderly. The success of that program is not unrelated to the fact that the elderly are well organized in political pressure groups and vote in substantial numbers.

U.S. POVERTY IN INTERNATIONAL PERSPECTIVE

By international standards, poverty rates in the United States are quite high. That is the main conclusion that can be drawn from Table 10–3, which compares the U.S. with other advanced industrial societies. Two measures of poverty are used: the absolute standard employed by the U.S. government and a relative standard, equivalent to half the median income in each

FIGURE 10–3   Risk of Poverty for Selected Groups

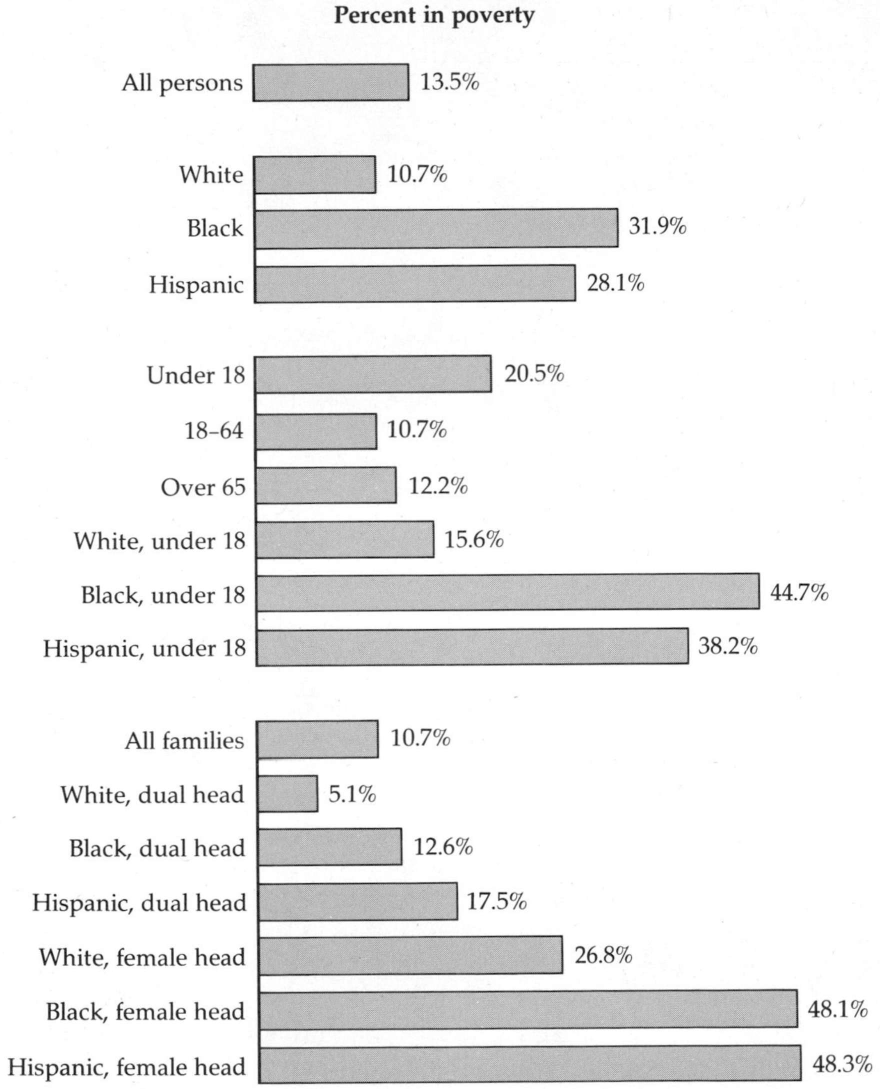

Percent in poverty

Group	Percent
All persons	13.5%
White	10.7%
Black	31.9%
Hispanic	28.1%
Under 18	20.5%
18–64	10.7%
Over 65	12.2%
White, under 18	15.6%
Black, under 18	44.7%
Hispanic, under 18	38.2%
All families	10.7%
White, dual head	5.1%
Black, dual head	12.6%
Hispanic, dual head	17.5%
White, female head	26.8%
Black, female head	48.1%
Hispanic, female head	48.3%

SOURCE: U.S. Census 1991b.

country (adjusted, like the absolute standard, for family size). The first compares countries using a fixed standard of living, converted to the relevant currency. The second varies the standard from one country to the next, assuming that being poor means falling well below the average in your own society.

The United States has the highest absolute and relative poverty rates for children and working-age adults. The comparison with regard to chil-

TABLE 10–2 Poverty Risk Before and After Cash Transfers, 1990

	Before Cash Transfers	After Cash Transfers
All persons	20.5%	13.5%
Under 18	23.5	20.6
Over 65	46.4	12.2
Female-headed families	45.4	37.2
Female head w/children	53.7	47.2

SOURCE: U.S. Census 1991c.

TABLE 10–3 Poverty Rates for Six Countries* (in percent)

	Children		Working-Age Adults		Elderly	
	Absolute	Relative	Absolute	Relative	Absolute	Relative
United States	17.1%	22.4%	10.1%	13.4%	16.1%	23.9%
United Kingdom	10.7	9.3	6.9	5.7	37.0	29.2
Canada	9.6	15.5	7.5	10.7	4.8	17.2
West Germany	8.2	4.9	6.5	4.5	15.4	11.1
Norway	7.6	4.8	7.1	5.4	18.7	5.6
Sweden	5.1	5.0	6.7	6.7	2.1	0.8
Average, excluding U.S.	8.2	7.9	6.9	6.6	15.6	12.7

*Data from 1979 or 1980.
SOURCE: Smeeding et al. 1987, from Burtless 1987: 14.

dren is especially damning. The rates of childhood poverty in the U.S. are between two and three times the average for the other countries. On the other hand, U.S. rates for seniors are close to the middle of the pack—a good deal better than Great Britain, but far behind Sweden and Canada.

TRENDS IN POVERTY

When Lyndon Johnson became president and promised to build a society free from poverty, ignorance, and exploitation, the official poverty rate was falling. It would continue to fall during the presidency of Johnson's successor, Richard Nixon. But the poverty rate more or less flattened out in the mid-1970s and actually climbed in 1980s under Ronald Reagan. Figure 10–4, based on the official poverty standard, traces these shifts. Although the rate has recovered somewhat from the peak it reached during the recession of the early 1980s, the 1990 rate remained at a level well above the low point reached in 1973.

Comparisons with the years before 1970 are problematic, for reasons suggested in our discussion of poverty measures. As we noted earlier, poverty in 1990 was at the 1960 level by a relative measure. On the other hand,

FIGURE 10–4 Poverty Rates, 1960–1990

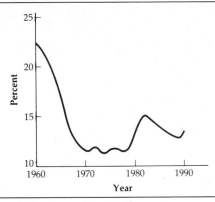

SOURCE: U.S. Census 1991b.

if we had retrospective statistics accounting for noncash transfers and taxes, they might show greater historical progress. But we can be confident of the general picture: declining poverty in the 1960s, stagnation in the 1970s, and growing poverty in the 1980s.

Over the years, as the poverty rate has changed, the composition of the poor population has shifted. Since 1960, the poor population has become somewhat more black, much more Hispanic, and more concentrated in big cities (U.S. Census 1991b; Jencks and Peterson 1991: 7). But the most dramatic changes have come in the age distribution and family structure of the poor. Between 1960 and 1990, the proportion of the poor living in female-headed households doubled (U.S. Census 1991b). Figure 10–5 illustrates the striking reversal that has taken place in the poverty rates of children and seniors, to the advantage of the latter. This shift was, as we will see, the result of federal policies that favored the elderly over families with children.

THE UNDERCLASS AND PERSISTENT POVERTY

Knowing how many people are poor from year to year does not tell us how persistent poverty is on an individual level. Many people fall below the poverty line during a given year as a result of temporary circumstances: loss of a job, physical injury, or divorce. When they get back on their feet, they are no longer poor. Others remain poor for many years. They may be disabled, living on meager retirement incomes, or working at a low-wage job while supporting a large family. (Under the federal standard, large families are more likely to be classified as poor.) Some people are just incapable of holding a job for long, and some have developed a long-term dependence on government benefits that allow them to survive but never approach the mainstream.

FIGURE 10–5 Poverty Rates for Children and Seniors

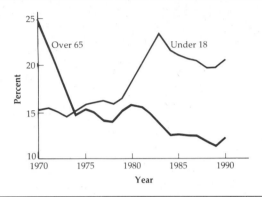

SOURCE: U.S. Census 1991b.

In recent years, interest in the persistence of poverty has turned into concern that the United States may be developing a permanent underclass —a growing class of people who are impoverished and mired in habits and circumstances that prevent them from ever joining the mainstream. This preoccupation has been fed by the increase in poverty in recent years and by the fact that the poor have become more concentrated in large cities, where they are most visible to the national media. We will return to the question of the underclass after we examine the few existing studies of turnover in the poor population.

Accurately measuring the *persistence* of individual poverty is difficult. It requires a "panel study" — finding the same people to ask the same questions, year after year. The Census Bureau has a continuing survey that tracks people for two years. Data from 1984–1985, 1986–1987, and 1987–1988 show that approximately 25 percent of the people who are poor in a given year are not poor the following year. At this rate, there could be a complete turnover in the poverty population in four years. But since the bureau only follows people for two years, we have no way of knowing how many people fell back into poverty shortly after climbing out. One encouraging indication is that most of the people who exited poverty did not remain near the poverty threshold but rose higher than 125 percent of the poverty line (U.S. Census 1989a, 1990e, 1991d).

The University of Michigan's Panel Study of Income Dynamics (PSID) has been tracking a large national sample of families since 1969. Data from the study revealed that during the decade 1969 to 1978, about a quarter of all families in the U.S. were poor for a year or more. Only 3 percent of all families (or 13 percent of poor families) were below the poverty line for eight years or more (Duncan et al. 1984: 41–42).

A more recent PSID analysis looked at long-term poverty among the families of women aged 25–44 who had children (including both female-headed and married-couple families). A woman's family was considered long-term poor if its average income over the six years 1980 to 1985 was below the poverty line. Five percent of the families of such women were long-term poor in the early 1980s, a period of extraordinarily high poverty and unemployment rates. For the more prosperous years 1967–1972, 4 percent of these families were long-term poor.[3] Federal statistics show that families with children have significantly higher poverty rates than the general population. Thus, the long-term poverty rates for all persons must have been lower than the 4 or 5 percent reported for these families.

Studies of the welfare caseload produce a similar impression of continual turnover. AFDC, the main public assistance program, benefits only one third of the poor population, presumably the least self-sufficient of the poor. Beneficiaries are largely single mothers and their families. Ellwood found that most women use welfare for transitional help, but a minority, perhaps a quarter, remain on AFDC for ten years or more. These long-term dependent women are typically never-married mothers who have dropped out of high school and have little work experience (Ellwood 1988: 148–149; see also Duncan et al. 1988).

All these studies demonstrate that there is rapid turnover in the poor population. *Only a minority of the poor, probably not more than 15 percent, are locked into a pattern of long-term poverty.* An even smaller minority of the poor, roughly 8 percent, are long-term dependents on welfare. (This figure represents a quarter of the one third who receive AFDC.)

Can we equate this 15 percent with "the underclass"? That depends on how the term is understood. In the class model we presented at the beginning of this book, *underclass* was used in a broad economic sense to refer to the poor who are loosely connected to or wholly disconnected from the labor market — encompassing most of the people below the federal poverty line. But in the current debate over the underclass, the term is applied to a smaller subset of the poor who are bound to their impoverishment by personal characteristics or structural circumstances. (See Auletta 1982; Murray 1984; Wilson 1987; Jencks and Peterson 1991.)

The underclass label often implies flawed character. The "conventional portrait" of the underclass, according to a *Washington Post* writer, links "extreme poverty, chronic joblessness, welfare dependency, out of wedlock births, female-headed households, [and] high dropout rates. . ." (April 17, 1991). Many would have added crime to this list. University of Chicago sociologist William J. Wilson (1987, 1991) one of the most influential writers on the underclass, more or less accepts this grim portrait, but defines the underclass in terms of marginal economic position coupled with geographic isolation. Wilson has in mind the inner-city poor, living in areas with extremely high concentrations of poverty and caught in a postindustrial economic trap of shrinking job opportunities.

---

[3] We calculated these figures by appropriately weighting and combining separate figures for white and African-American families (Duncan and Rodgers 1989, as reported in Jencks 1991: 35).

How big is the underclass? Wilson, in his contribution to a recent collection of papers on the topic, appears to accept an estimate of 2.4 million people in 1980, of whom 65 percent were black and 22 percent Hispanic (Wilson in Jencks and Peterson 1991: 464). This would mean that the underclass was roughly 8 percent of the poor, 20 percent of the black poor, and 15 percent of the Hispanic poor. Isabel Sawhill (1989) and her colleagues at the Urban Institute estimate the underclass at 1.1 million. They produced this figure by adding up all the poor who live in "underclass areas"—defined as neighborhoods (or census tracts) with very high rates of school dropout, single motherhood, welfare dependency, and joblessness. The underclass, thus defined, accounted for just 4 percent of the poor in 1980. Sawhill reports that 59 percent of the residents of underclass areas were black, and 12 percent were Hispanic. She notes that less than 20 percent of the black poor and fewer than 6 percent of all African-Americans live in such areas.

Is the underclass growing? Both Wilson and Sawhill believe so. The PSID studies on the persistence of poverty would seem to indicate some growth in the underclass. But Jencks (1991) notes that although economically defined poverty and long-term poverty may be on the increase, other indicators of behavior associated with the underclass are declining. He supports this position with long-term data on high school dropout rates, cognitive skills of minority students, and teen birth rates.

One of the difficulties in determining trends in the underclass is distinguishing people's responses to short-range economic fluctuations from their stable accommodations to fixed circumstances. The underclass notion, as it has been employed by most authors, carries the implication of a persistent pattern of behavior among some of the poor. But what does it tell us about trends in the underclass if the numbers of long-term poor expand during periods of extreme unemployment, such as the early 1980s? What does it tell us if big-city minority poverty takes a sudden nose dive when the local labor market approaches full employment? (That is what happened in Boston, as we will see in the next section.) Does this mean that the underclass is disappearing?

THE MYSTERY OF RISING POVERTY RATES

Here is the mystery facing students of poverty: Why did the 1960s progress in reducing poverty falter in the 1970s and 1980s? No one has the definitive answer to this question. But at least three factors are involved: (1) national economic stagnation, (2) a society-wide revolution in family patterns, and (3) shifts in federal policy.

Economic Trends. In early September 1991, the national news media reported a fire in a North Carolina chicken-processing plant that killed twenty-five workers and injured more than fifty others. According to the stories, exits that had been locked to keep employees from stealing chicken had contributed to the death toll. In eleven years of operation, the plant had never been inspected by state or federal safety officials. The workers in the plant were typically single women with children, working eight hours a day for $5 an hour or less—a wage insufficient to raise a family of three above the poverty line (*New York Times* and *Washington Post*, September 4– 11, 1991).

The late 1970s and 1980s saw a proliferation of such low-wage jobs and the loss of many of the better-paying blue-collar jobs. High unemployment and low skills compel many of the poor and near-poor to accept any available job. Relevant trends in employment and earnings are outlined in Figure 10–6. Unemployment rates (decade averages are used, to smooth out yearly gyrations) have increased steadily since the 1950s, but especially since 1970. Unemployment was 60 percent higher in the 1980s than it was in the 1950s.

Average weekly earnings, adjusted for inflation, rose in the 1960s, but by the late 1980s, earnings had slid back to the 1960 level. Of course, the weekly average includes high-paid executives and the workers at chicken-processing plants. The minimum hourly wage, a figure set by federal legislation, gives an indication of what was happening at the bottom. Wages at the lower end of the labor market, even for people who earn a little more than the federal minimum, tend to move up or down with the minimum.

FIGURE 10–6   Trends in Unemployment and Real Earnings, 1960–1990

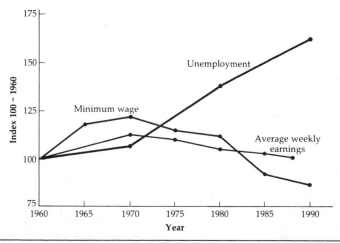

Note: All statistics are indexed to a common scale, with 1960 values equal to 100. Unemployment figures represented are averages for the preceding decade. An index of 150 means an increase of 50 percent over the base period; 80 percent indicates a 20 percent decrease. Weekly earnings are for all nonagriculture wage and salary workers. Earnings and minimum wage are adjusted for inflation. Minimum wage for 1990 reflects a raise to $3.80 on April 1.

SOURCES: U.S. Census 1990c: 409, 412; Council of Economic Advisors 1990: Table C-32.

TABLE 10–4 Work Experience of the Poor, 1990

	Millions	Percent
Total poor over age 16	21.2	100
Worked all year full-time	2.0	9.4
Worked, not full-time all year	6.6	31.1
Did not work in 1990		
Ill or disabled	2.9	13.6
Retired	2.6	12.3
Keeping house	4.1	19.3
Could not find work	0.9	4.2
School or other	2.1	9.9

SOURCE: U.S. Census 1991b.

In the 1960s, the real value of the minimum wage rose faster than average earnings, but by 1990, the minimum wage had sunk to 86 percent of its 1960 value.

The labor market, then, has turned sour. Unemployment is up. Wages are down. These developments affect the poor to the extent that they work or are looking for work.

Just how much do the poor work? Table 10–4 gives an overview of the work experience of poor adults. We find that over 40 percent of poor people over the age of 16 were employed sometime in the course of the year. Remarkably, about 10 percent worked full-time, all year long, without rising above the poverty line. About a third of the poor adults we have been describing are working-age heads of households—husbands, wives, and single heads under 65. Among this group, 20 percent worked forty weeks or more at a full-time job.[4] Many of these people are from households with more than one person in the labor force.

Dual-headed families, which have higher rates of labor force participation, are more decisively affected by economic conditions than are female-headed families. "The poverty of two-parent families," concluded Harvard researcher David Ellwood, "is the poverty of the working poor." Ellwood examined the work experience of dual-headed families with children whose incomes before government assistance placed them below the poverty line. He found that 44 percent of such families had the equivalent of at least one full-time, year-round worker in 1984, a year of very high unemployment (Ellwood 1988: 81, 89).

It is a sad commentary on our economy that many poor people work all year at full-time jobs and do not make enough to lift their families out of poverty. In the 1960s, the number of heads of poor households in this situation was falling fairly rapidly. In the 1980s, the opposite was happening. By 1988, these fully employed heads of poor households and their depen-

---

[4]Schiller 1989: 66 (figures from 1987).

dents numbered over 6 million people, a large chunk of the poor population (Levitan 1990: 20).

About half of poor adults do not work because they are retired, ill or disabled, caring for children, or attending school. In other words, they are not, for the moment, potential workers. The other half are unemployed, underemployed, or working at jobs that pay substandard wages. These people and their dependents might escape poverty under different economic conditions.

In the 1980s, the city of Boston tested the capacity of a rising economy to lift people out of poverty. By 1988, the so-called Massachusetts Miracle had reduced the statewide unemployment rate to 3.3 percent, well below the national average (U.S. Census 1991c: 399). In such a tight labor market, jobs are easier to find, and both wages and working conditions improve as employers compete for workers. What effect did the economic boom have on poverty in Boston? In 1980, the city's poverty rate was higher than the average for U.S. central cities. In the decade that followed, poverty rates climbed in the average central city but fell in Boston. Among Boston families headed by nonstudents under age 60, the rate dropped from 18 percent to 11.5 percent. Among black families in this group, the rate plunged from 29 percent to 13 percent (Osterman 1991: 126). In Boston, it was proved that even tough big-city poverty responds to a healthy labor market.

Using national data for the years 1960–1984, Ellwood (1988: 97, 150) showed that it was possible to predict the annual poverty rate of children with remarkable accuracy using only two indicators: the median earnings of full-time, year-round workers and the unemployment rate. Ellwood's analysis again demonstrates how dependent the poor and the near-poor are on labor market conditions.

Changing Family Patterns. The pressure of economic stagnation on poverty rates has been reinforced by sweeping changes in family patterns. Figure 10–7 summarizes the key trends. Relative to the recent past, Americans are now less likely to marry and more likely to divorce. Children are five times more likely to be born to unwed parents and almost three times more likely to live in female-headed households than they were in 1960. These trends are pervasive in American society. They affect the poor and the nonpoor, blacks and whites, teenagers and their parents and grandparents. Together, they amount to a revolution in American family life.

The key change in family patterns that sums up the others and has contributed to the rise in the poor population is the increase in the number of female-headed families. Over the last thirty years, the proportion of the poor population living in female-headed families has jumped from roughly 20 to 40 percent (U.S. Census 1991b). This development is transforming the lives of children. It is probable, according to one expert, that the majority of the children born today will spend part of their childhood in a female-headed family—and that most of the children who do will experience a period of poverty (Ellwood 1988: 45–47, 67).

FIGURE 10-7 Changing Patterns of Family Life

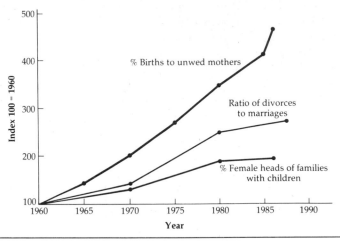

Regarding index, see the notes for Figure 10-6.
SOURCES: U.S. Census 1986a, 1990c.

Why the rapid increase in families headed by women? In part this is because in the course of the last generation, Americans have become much more tolerant of both unwed motherhood and divorce. The popular conception of unwed mothers as irresponsible teenagers is dated and misleading. Girls who become mothers still face tough, poverty-ridden futures; they are disproportionately represented among the long-term welfare-dependent families. But teen birth rates have declined, probably as the result of greater use of contraceptives and access to abortion. There has been some increase in the number of *single* teen mothers, because these girls are less likely to marry when they become pregnant. But the birth rate of single women in their twenties is higher than that for teens and is growing faster. Women in their twenties are responsible for the majority of unwed births (U.S. Census 1990c: 169; Ellwood 1988: 71–72; Jencks 1991: 83–93).

One reason for the shift in behavior and attitudes toward divorce, marriage, and single motherhood is that the choices facing women have changed. Job opportunities for women have expanded, and they are now expected to be capable of supporting themselves. At the same time, the falling wages and rising unemployment rates of men make them less dependable meal tickets. Sociologists William Wilson and Katherine Nickerman (1986) present evidence to suggest that these trends have powerfully affected African Americans, among whom the majority of births are now to unwed women and over 40 percent of families are headed by women (U.S. Census 1990c: 67, 46). They find that the marriageable pool of employed young black males has contracted sharply, especially since the economic slump of the 1970s. Finally, public assistance has become a more dependable source of family support. Although the value of welfare benefits has declined since the 1970s, access to public assistance for families that qualify is more secure than it was in the early 1960s.

If the array of choices facing young women has expanded, the prospects for single mothers remain unenviable. The majority of absent fathers do not pay child support, and those who do contribute pay very little.[5] Employed women, of course, make considerably less on average than men. The conflicts between the nurturer and provider roles, which all working mothers face, are even tougher for employed single mothers, who bear the burdens of childrearing alone. Nonetheless, single mothers are more likely to work full-time than married mothers (Ellwood 1988: 145). Public assistance does not generally provide enough to place a family above the poverty line, and rule changes imposed under the Reagan administration make it impossible for women on AFDC to increase their families' incomes by working — unless they work "off the books" and hide their wages from welfare authorities. The most common way for a single mother to escape AFDC is through marriage; few manage to do so through their own earnings (Ellwood 1988: 148).

Given the difficulties that female heads of households face, it is scarcely surprising that their families are several times more likely to be poor than dual-headed families (see Figure 10–3) or that the growing number of female heads is associated with increasing poverty rates.

Government Policy.   How much can we blame changes in government policy for the increase in poverty since the mid-1970s? Probably not as much as we might be tempted to. As we will see later in this chapter, cuts in social programs under Ronald Reagan in the 1980s certainly made life harder for the poor—especially for the working poor. A study by the Urban Institute estimated that the program cuts and worsening economic conditions were about equally responsible for the jump in poverty rates in the early 1980s (Palmer and Sawhill 1984: 14). But poverty rates began to rise before Reagan was elected. What we have already learned is that long-term changes in our economy and patterns of family life have been pushing poverty rates upward. The government has some power over the economy; it has virtually none over family life.

Liberals typically have great faith in the capacity of the government to change things. Ironically, in recent years, conservative writers have advanced the strongest claims with regard to the power of federal poverty policy. For example, Charles Murray's influential book *Losing Ground* (1984) contended that rising welfare benefits encouraged the formation of dependent female-headed households and were thus responsible for the increase

---

[5] In 1985, only one out of three absent fathers paid any child support. The average payment was $2,220 for the year—under $200 a month (Levitan 1990: 124).

in poverty rates. If Murray were correct, the proportion of female-headed families with children and the poverty rate should both have tumbled in the period from 1972 to 1984, when the value of welfare benefits declined 25 percent, back to 1960 levels.[6] Exactly the opposite happened. The growth in female-headed families accelerated, and the poverty rate climbed steeply. In fact, public assistance is probably the least potent of the forces responsible for the change in family patterns that is affecting the entire society—not just the minority of the poor who receive welfare benefits. Subsequent research has not supported Murray's extravagant claims. (See Garfinkel and McLanahan 1986: 55–63; Ellwood and Bane 1985.)

In recent years, the federal policies that have most influenced poverty rates are not explicitly poverty policies. For example, the dramatic reduction in poverty among the elderly, visible in Figure 10–5, is the direct result of real increases in social security benefits from the 1960s to the mid-1970s (Levitan 1990: 45). The subsequent leveling off of benefits contributed to the reversal of the long-term downward trend in the general poverty rate.

The failure of the government to uphold the value of the minimum wage and maintain the protections of the unemployment insurance system pushed some working poor families below the poverty line, though it is difficult to determine how many. In 1970, the minimum wage was high enough that a steady full-time worker could keep a family of three out of poverty; by 1988, the same worker could just manage to maintain a one-person household above the poverty level.[7] Unemployment insurance can help low-income families from falling below the poverty line. But the proportion of the labor force covered, the value of benefits, and the duration of benefits have all fallen significantly since 1975 (Levitan 1990: 62–65; *Washington Post*, August 29, 1991).

National macroeconomic policies also influence poverty rates. In the 1980s, the Reagan administration focused national policy on reducing inflation rather than maintaining employment and on increasing profits rather than upholding wages. These goals were reflected in tax and spending policies, as well as monetary policy. International trade policy opened the U.S. to increasing foreign competition, reducing both employment and wages in the affected sectors. It is difficult to disentangle the results of these policies from the effects of the very serious problems the U.S. economy was facing at home and abroad, including declining productivity and the erosion of the country's long-held dominant position in international trade. This much is certain: The administration's policies contributed directly to rising unemployment and declining wages. We have already seen how closely poverty rates are tied to these conditions in the labor market.

---

[6] This calculation is based on the combined value of AFDC and food stamps (which did not exist in 1960), adjusted for inflation (Ellwood 1988: 58–59).

[7] U.S. Census 1990c. 1991a. Calculation assumes a forty-hour week and a fifty-week year.

WELFARE REFORM: NIXON AND CARTER

The controversy over welfare policy in the Reagan years had its roots in the late 1960s and 1970s. As we have seen, the expansion of social programs, from AFDC to social security, had made a big dent in the poverty rate by 1970. But as these programs developed, so did opposition to some of them, particularly the means-tested ones that were defined by the public as "welfare." Opponents said that welfare was "growing out of control" and "full of fraud and abuse." By the time Richard Nixon assumed the presidency in 1969, welfare reform had become an salient national issue.

Concern centered on AFDC, partly because it was growing so fast (especially during the decade beginning in 1965) and partly because of widespread suspicion that many families were taking advantage of it who really were not eligible for help—or should not be. In 1960, there were 803,000 families "on welfare." By 1970, the number of families had grown to 2.5 million; by 1975 another million had been enrolled, adding up to more than 11 million individuals, counting children. Three quarters of those families were headed by women; three quarters lived in metropolitan areas; a little less than half were black. In New York City alone, the proportion receiving welfare grew from one in thirty people in 1960 to one in six in 1970. From a program initially designed to help widows, AFDC had grown into a program that helped women and their children who were poor for many reasons; widowhood, divorce, illegitimacy, or the unemployment of husbands.

As suggested above, the migration of many poor farmers *to* the cities and the flight of low-skilled manufacturing jobs *from* the cities were underlying economic causes for the growth in the welfare rolls, despite general prosperity. But there were social and political causes as well. The proportion of black families in the nation that were headed by women jumped from 21 percent in 1960 to 35 percent in 1975 to 43 percent in 1982; there was also an increase in white female-headed families, to 13 percent. Thus, there was some increase in the number of persons who became eligible for welfare. But the largest increase in enrollment came because the administration of the program (not the law) adjusted to political pressure, as indicated by Piven and Cloward:

> The welfare explosion occurred during several years of the greatest domestic disorder since the 1930s—perhaps the greatest in our history. It was concurrent with the turmoil produced by the civil rights struggle, with widespread and destructive rioting in the cities, and with the formation of a militant grass-roots movement of the poor dedicated to combatting welfare restrictions. Not least, the welfare rise was also concurrent with the enactment of a series of ghetto-placating federal programs (such as the anti-poverty program) which, among other things, hired thousands of poor people, social workers, and lawyers who, it subsequently turned out, greatly stimulated people to apply for relief and helped them to obtain it (1971: 198).

In other words, the welfare system served as a mechanism for buying social peace: When those who were left out of prosperity finally began to scream, they were given a modicum of help. By 1975, the process had run its course. A large proportion of the eligibles had been signed up, the rolls had stopped growing, and the resistance movement had grown stronger than the protest movement.

The first suggestion for major reform in the design of the welfare system (beyond mere tinkering with eligibility rules) emerged in the early part of the Nixon administration in 1969. One impetus for change was dismay about the overlap in various programs and the discrepancies in the rules from one area to another, which stemmed from local options in the way the system was administered. In some states, no benefits could be paid to a woman with a husband (or any man) living in the house; in others, she would be eligible if her husband was unemployed. The level of payments under both AFDC and its accompanying Medicaid and food stamps was set very low in the southern states, but in other places, the same programs paid *as much as ten times more*. Although there was little research evidence to support the idea, many officials were convinced that these discrepancies caused people to move north to get higher benefits, and the governors and mayors in northern states demanded changes. Furthermore, the complexity of the total system, which mixed cash benefits, food stamps, and many other noncash benefits and services (family advice, job training, housing and medical subsidies, etc.), created a large bureaucracy of social service workers who were accused of soaking up more and more of the money and interfering more and more in people's lives.

Many experienced administrators in the federal government had been advocating a simplified procedure that would combine various categories of people (mothers, unemployed fathers, elderly not sufficiently supported under social security, permanently disabled) and various types of benefits into a single cash payment based solely on financial need. Furthermore, they wanted it administered according to national standards that would be the same everywhere, probably run by the federal Social Security Administration or possibly even by the Internal Revenue Service. Obviously, they were thinking along the lines of a negative income tax or guaranteed minimum annual income. In principle, it seemed gloriously efficient and appropriate to a computer age that preferred to standardize life according to automatic criteria.

Nixon was elected as a conservative who would stress work rather than welfare, but he hoped to soften his harsh political image by supporting efficient government aid for the poorest families. He liked the idea of simplifying the system by combining several programs into a single national cash grant based on need, if it could be done without too much additional cost to the U.S. Treasury. But Nixon realized that such a concept was politically dangerous, since it implied that people would get something for nothing and that some of those who were able to work might refuse to do so. So he carefully added two stipulations. (1) His program was basically a substitute for AFDC, not a plan for everybody, and thus it would be limited to low-income families with children (with or without a father present). (2) It would contain a work requirement such that able-bodied persons refusing to accept jobs would lose benefits. If cleverly designed, a new scheme could be expected to increase the incentive to work: Instead of a sharp break between those who received welfare and those with jobs who did not, the new program would give grants to everybody who was eligible but would gradually reduce payments as earnings went up. The idea was to guarantee that recipients would always have more cash the harder they worked. The plan was formulated by experts in the bureaucracy coordinated by Daniel Patrick Moynihan, who had served President Kennedy as assistant secretary of labor and was brought back to Washington as Nixon's urban affairs advisor.

The scheme was labeled the Family Assistance Plan and was introduced early in 1969. It would provide a cash grant of $1,600 to a family of four with no income, plus $860 worth of food stamps, bringing the minimum guaranteed income to a level of $2,460 (considerably less than the poverty line that year of $3,800). The following examples show how earned income would reduce government payments until they were phased out completely (for a family of four, excluding food stamps):

Earned Income	Family Assistance	Total Cash Income
0	$1,600	$1,600
$2,000	960	2,960
4,000	0	4,000

When first introduced, the new plan received the support of the National Association of Manufacturers, the AFL-CIO, and every former secretary of health, education, and welfare, both Republican and Democrat. It quickly passed the House of Representatives by a wide margin. But then the troubles began in hearings in the Senate Finance Committee, headed by Senator Russell Long of Louisiana. It was noted that the number of people getting some form of government payment would double, even though most of the newcomers would be getting small grants to supplement low earnings—the working poor. It was also argued that the 50-cent reduction in government benefits for each additional dollar earned was a very high tax rate that might reduce the incentive to work rather than increase it, and furthermore, that such a rate was unfair, since those earning enough to start paying income tax only had to share 14 percent of their earnings with the government. The fact that existing AFDC rules reduced payments 67 cents per dollar earned was not emphasized, probably because it was assumed that most female welfare recipients did not work at all. In fact, many did; as a group, welfare recipients got only half of their total income from government benefits. Furthermore, in order to limit the program, Nixon did not fold into the cash grants most of the other forms of benefits (food stamps, Medicaid, etc.). The result was a series of incongruities, or "notch" problems: When earned income fell a dollar, some people would become eligible for benefits worth hundreds of dollars.

Thus, people with almost identical incomes would be receiving quite disparate forms and amounts of help from the government.

To these arithmetic difficulties were added political issues. The liberals said the plan was much too stingy, offering a level of support far below the poverty line and below that currently paid in many states. They argued that although those states would be required to supplement the plan at first, they would be tempted to phase out the supplements over the years and pay at the new, nationally established minimum level. Furthermore, the liberals were deeply suspicious of any plan introduced by their worst enemy, Richard Nixon. The conservatives answered that any increase in the basic benefit level would very rapidly increase the number of working poor receiving partial payments, that it would be easy to have a third or more of the population getting "handouts," and that the costs to the U.S. Treasury could be enormous. And the southerners were worried that any increase in the existing low levels of AFDC in their region would reduce the supply of poor people available as seasonal workers on farms during the harvest season and would also have the tendency to push up the low local wage rates that were attracting many manufacturing firms to move down from the northern states. The plan was debated to death: It never emerged from the committee and suffered the same fate when reintroduced the following year with some modifications (Moynihan 1973; Anderson 1978).

Nevertheless, the search for a workable form of guaranteed minimum income continued. The government sponsored a series of local experiments to test the impact of various similar schemes on the amount people actually did reduce their efforts to earn money from jobs. The initial results, from an experiment in New Jersey, indicated that work reduction would be minimal and that people liked the new system. But later results from Denver and Seattle were more discouraging: If families were guaranteed enough cash to supplement earnings and bring income up to the official poverty line, but earnings above that line caused a 50-cent-per-dollar reduction in benefits, it was found that husbands worked about 6 percent less than they would under the old system, that wives worked 23 percent less, that female heads of families worked 12 percent less, and that teenagers in the family reduced their part-time work by about one third. Furthermore, there was startling evidence that, instead of encouraging families to stay together by removing the incentive for the father to leave home and make the mother eligible for AFDC, the new system actually increased separations, probably because dissatisfied wives were more sure of support under the experimental rules and felt more free to evict their husbands (*Journal of Human Resources*, Fall 1980).

Before these new research results were fully available, the Carter administration entered office, and yet another team tried to face up to what was then considered to be "the welfare mess." The Carter team came in with a program that was not too different from the Nixon proposals. However, they recognized one significant fact that had been pushed aside earlier: The key problem was not "forcing" people to work but providing jobs when they did not exist. So the Carter plan called for the creation of almost a million and a half jobs, to be added to local governments with federal money, at wages close to the legal minimum. These jobs, along with training programs, would be part of CETA (Comprehensive Employment and Training Act). Benefits to families would follow the Nixon scheme, although at a slightly more generous level. Food stamps would no longer require a cash payment, and the rules would be the same in all parts of the country. Some money would be given to the working poor through a very modest "income tax credit," a form of negative income tax administered by the Internal Revenue Service.

What were the results? The Nixon administration had succeeded in obtaining only one major reform. Categorical aid for the needy elderly, blind, and permanently disabled, previously a hodgepodge of local programs partly subsidized by the federal treasury, was reorganized in 1974 into the federal Supplemental Security Income, or SSI, administered in most states by the Social Security Administration. The Carter administration succeeded in eliminating the cash payment for buying food stamps, expanded the number of people involved in CETA programs, and increased tax credits. But AFDC was not changed; the Congress never seriously debated the issue. The various committees, agencies, and states continued to run their mixture of programs (see Levitan and Taggart 1976; Levitan 1980).

The debates of the Nixon and Carter years made key issues much clearer to the experts:

1. A simple program based mainly on cash grants in the style of a negative income tax runs into certain unavoidable choices that are politically impossible to resolve (at least in the short run):
   a. To keep the costs within bounds, either the initial payment level must be kept low (and liberals protest), or the marginal tax rate on earnings must be set high (which conservatives protest because it reduces work incentive). As soon as the initial payment level begins increasing, the number of working poor eligible for partial payments goes up at a fast rate.
   b. To function with a minimum of "notch" incongruities, all the noncash programs must be folded into a single cash scheme, but that means that special constituencies—in favor of food stamps for nutrition, or medical care for the sick, or housing subsidies to save the cities—all protest.
2. The problem of unemployment among the poor who go on welfare cannot be solved either by a work requirement (which has proven ineffective) or by a relatively low marginal tax rate on earnings to maintain incentive. It is deeply connected with the inadequate supply of jobs at decent wages for low-skilled people, and that problem appears to be getting worse as technology gets more sophisticated.

3. A program that would bring all families in the country up to the
    official (but inadequate) poverty line, and that permitted an ap-
    propriate combination of welfare plus work, is threatening to
    low-wage industries and regions, and they will oppose it. The
    bitter struggle of farmers against those attempting to form labor
    unions to help migrant agricultural workers shows the depth of
    such opposition.

WELFARE REFORM: REAGAN AND BUSH

What was the situation in 1981, when the Reagan administration took control? The two preceding decades had experienced a major growth in the
social programs that had first been organized under Roosevelt's New Deal.
They were expanded to include more people and more problems, particularly medical care and nutrition. Each program had a worthy purpose and
a loyal following. But the total cost had become large, and some of the
programs were getting bad publicity concerning the so-called "welfare
cheats" who made a career of living off the government instead of working
for their keep. Here are some of the figures on costs to the federal government in 1980[8] (Bawden 1984: 75):

Social insurance	
OASDI (old age, survivors, disabled)	$134.8 billion
Medicare (for elderly)	38.3
Unemployment compensation	18.5
Means-tested support	
AFDC (families with children)	8.8
SSI (needy elderly, blind, disabled)	7.0
Food stamps	10.0
Child nutrition	4.7
Housing assistance	4.9
Medicaid (for poor)	15.3
Total	$242.3

If one adds veteran's benefits, job training, school support, scholarships for college, and a few others to the above list of programs, the full
sum comes to $314 billion. That figure is the cost of the "social programs"
of the federal government. They absorbed 54 percent of the budget and
took 10.8 percent of the gross national product (GNP). Back in 1960, before
the major expansion of these programs began, they cost 29 percent of the
federal budget and 4.8 percent of the GNP. Thus, in terms of the productivity of the nation, social expenditures had more than doubled in two
decades.

---

[8]The states contributed another $17.4 bilion to AFDC, Medicaid, and child nutrition.

How many people benefited from these programs in 1980? Taking only the largest of them (Bawden 1984: 78):

Recipients of OASDI	35.6 million
Food stamps	19.4
AFDC: Family units	3.8
Individuals (parents and children)	11.1
Medicaid	21.6
SSI	4.1

Many of these programs, such as AFDC, food stamps, and Medicaid, often served the same people who received multiple benefits. And yet even after cash grants, about 13 percent of the population remained poor; including noncash grants would have reduced that figure to about 8 percent. Without any of the programs, about 22 percent would have been poor (Schwartz 1983).

The Reagan administration began with a new set of goals. They accepted all the difficulties listed above that had defeated attempts to restructure and simplify the system. Indeed, Martin Anderson, who had served briefly in the Nixon administration and then written a persuasive book in 1978 arguing against restructuring, was the man put in charge of domestic planning by President Reagan in the early days of the new administration (see Anderson in Bawden 1984). The new leaders wanted to reduce expenditures, not improve the system. They proposed to tighten all the rules in the existing system to make it cheaper and to eliminate any parts of it that seemed expendable. There were two undefined limits put on this process: (1) Social security was not to be changed much, since it was said to be based on the insurance principle, and people had the right to their benefits (and they represented a huge block of vociferous voters); (2) the "truly needy" should be supported by a "social safety net."

The main Reagan economic thrust was not directed toward helping the poor but rather toward pleasing the middle class, which was restive after years of inflation and was demanding a reduction in government expenditures and taxes. Since it was also part of the Reagan philosophy to increase defense outlays sharply, social programs had to be cut in compensation.

Of course, some programs were supported by organized political groups and proved hard to trim. Outlays for social security and Medicare, which serve the middle class as well as the poor, grew significantly in the 1980s. The major cuts were reserved for programs that only serve the poor, including AFDC, food stamps, and SSI. CETA, which had created many jobs for the poor, was eliminated. Entitlement programs for the poor, such as AFDC and food stamps, were reduced by imposing stricter eligibility requirements and restricting benefits.

By one expert estimate, federal spending on social programs by the middle of the decade was 9 percent less than it would have been under prior policies (Palmer and Sawhill 1984: 13–14). This is not to say that real outlays were reduced all that much. Because both unemployment and the poverty rate shot up in the early 1980s, spending on such programs as AFDC was difficult to control—even when eligibility and benefits were being squeezed. In sum, Reagan did not undo the welfare state that emerged in the 1960s and 1970s, though he did manage to constrain it.

The guiding notion behind Reagan poverty policies is summed up in the image of the safety net. The administration recognized that the public did not want to see hungry children, homeless families, people without warm clothing in winter—did not, in other words, want to see people slip below a *minimal* standard. But the Reaganites did not believe that the government was obligated to help people live at a *decent* standard. Nor did they believe that public policy should aim to reduce inequality or help people climb out of poverty. From this perspective, the administration's apparent indifference to the fate of the working poor, who tend to be the least impoverished of the poor, is not surprising. As it went about reducing spending on social programs, the administration did not hesitate to change the rules governing AFDC and food stamps in ways that sharply reduced the built-in work incentives. After 1981, the full amount of job earnings (less a small allowance for work-related expenses) was deducted from AFDC benefits, reducing the financial reward for labor to zero. The administration made its biggest budget cuts in employment and job-training programs.

The Reagan administration's critics questioned the very existence of a poverty safety net. They pointed to the visible increase in homeless people living on the streets in Washington, D.C. and other big cities and evidence of increasing malnutrition, especially among children. From 1980 to 1984, there was a 30 percent increase in the number of extremely impoverished persons—those living on incomes less than half of poverty level (U.S. Census 1982, 1986b). If there was a safety net, it was set very low.

During Reagan's second term, the administration had to contend with a less malleable Congress and a serious foreign policy scandal, the Iran-Contra affair. As a result, it paid less attention to domestic policy, and there were no major changes in poverty programs.

George Bush, Reagan's vice-president and successor, promised a "kinder, gentler nation" during the 1988 campaign. Whatever Bush's intentions, he was constrained in domestic policy by a dual Reagan legacy: on one side, the strengthened hand of conservatives in the government and the Republican Party, and on the other side, the budget crisis resulting from eight years of unprecedented federal deficits. Nonetheless, the administration undertook an early, high-level review of poverty policy, soliciting ideas from some of the country's best-known experts.

The outcome of the review was inevitable. The President's Domestic Policy Council concluded that all the proposals were either too expensive or too controversial. For example, a program to fight teen pregnancies with birth-control advice was rejected because conservatives would see the proposed in-school clinics as encouraging promiscuity. A proposal for a national child-support system was deemed impractical in a time of tight budget constraints. The system would force absent parents to pay a fixed percentage of their incomes in child support and would commit the government to providing the difference between these payments and a fixed minimum support level.

"We concluded," a White House official told the *New York Times* (July 6, 1991), "that there were no obvious things we should be doing that we weren't doing that would work." The administration was, of course, unwilling to consider transferring funds from its $300 billion defense budget or reversing the tax cuts that had benefited the rich and strained the federal budget since the early 1980s. These moves could have freed up money for fighting poverty and other pressing domestic priorities.

The issue of poverty had clearly come full circle, to a time when it was not on the national agenda. Unlike Kennedy, Johnson, Nixon, Carter, and Reagan before him, Bush planned no new venture in poverty policy. "We have decided," said the Bush official quoted above, "to abjure a glitzy, splashy, high profile announcement of new programs and a grand new strategy." One of his White House colleagues summarized the administration's approach as, "Keep playing with the same toys. But let's paint them shinier" (*New York Times*, July 6, 1991).

AN ANTI-POVERTY PROGRAM

One of the outside participants in the Bush administration's poverty policy review was David Ellwood, whose work we have frequently quoted in this chapter. Ellwood's ideas are based on his own research on poor families with children. He starts from the assumption that we should deal with poverty in a way that strives to be consistent with American values. Americans assume that the poor deserve to be treated with compassion, dignity, and fairness, but they also believe that the welfare system should encourage responsibility and reward work.

Here is the program Ellwood outlines at the end of his book, *Poor Support: Poverty in the American Family:*

1. *"Make work pay so that working families are not poor."* Ellwood suggests that the government raise the minimum wage, expand the earned income tax credit (which gives low-income working families a special tax break on their earnings and provides cash "refunds" for the poorest of the working poor), make the current child-care tax credit similarly refundable, and consider other tax-based benefits for the working poor.

2. *"Ensure that everyone has medical protection."* More than 35 million Americans have no medical insurance. The working poor are the major uncovered population. They do not receive the Medicaid benefits that are extended automatically to families that receive AFDC.

3. *"Adopt a uniform child support assurance system."* Require absent parents to pay a percentage of their income in child support, through a national system of payroll deductions similar to social security. Have the government guarantee a minimum payment, covering any shortfall in collections from the absent parent.

4. *"Convert welfare into a transitional system designed to provide serious but short-term financial, education, and social support for people who are trying to cope with temporary setback."* Families might get eighteen months to three years of assistance designed to help them become independent and self-supporting.

5. *"Provide minimum wage jobs to persons who have exhausted their transitional support."* The government should, to the extent necessary, create jobs for people who have run through their transitional benefits. In effect, Ellwood is calling on the government to guarantee something approaching a full-employment economy (Ellwood 1988: 238).

This is a schematic program, which raises as many questions as it answers. But a national administration with a different set of priorities might begin thinking about poverty by considering proposals such as Ellwood's.

CONCLUSIONS

Much of this chapter has discussed a single issue: How many people in the United States live in misery as a result of inadequate income—how many are poor—and how can they be helped by government action?

But as the discussion unfolded, it became increasingly clear that the very definition of poverty was a crucial and highly debatable issue in itself. For some observers, poverty is a life measured by certain "objective" indicators of malnutrition, poor housing, and lack of medical care. But for others, poverty is a life too far below the mainstream of contemporary America, breeding misery from relative deprivation and social degradation— even though no biological measures of malnutrition and illness may be present. The conservative political view is likely to adopt the first definition and support austere government programs of a "safety-net" variety, so that nobody starves in America. The liberal political view is likely to demand a set of policies that will reduce the distance from the top to the bottom of the social hierarchy; it maintains that excessive inequality is harmful in itself, both to the people who suffer and to the society at large, even if nobody falls below the minimum of the safety net.

Those who measure poverty by minimum "objective" indicators of level of living claim that most poverty has been eliminated from America. Either they use the official statistics that show a reduction of poverty from 20 percent of the population in 1963 to 14 percent in 1990 or they go further and make various adjustments in the statistics to lower the figure to a more acceptable level. The policy consequences of this view were evident in the Reagan-Bush years, when the goal was to provide for "the truly needy" while reducing benefits for others to a bare minimum.

Those who measure poverty in relative terms assert that poverty has not declined, because many people are still far from achieving a mainstream standard of living. By the most common relative definition of poverty (income lower than half the median family income), the percentage of impoverished Americans has been stuck around 20 percent for more than a generation. From the perspective of a relative definition, appropriate policies would have to start with a restructuring of the labor market to reduce the overall unemployment rate and most particularly to increase the demand for workers at lower levels of skill and simultaneously to increase their relative wages. If necessary (and it does seem necessary), they would create government jobs to absorb any workers that the private sector is unable or unwilling to employ. And then they would improve the transfer system to bring all recipients closer to the relative poverty line, whether they work or not.

The study of trends since 1960 in both the incidence of poverty and the effects of government programs helps us to understand some of the causes and effects. Until 1974, poverty was decreasing in the United States, and two major factors were responsible: a booming economy, which gave better jobs to many people—especially younger ones with families to support; and improvements in the system of government transfer payments, which had major impact in helping retired persons. Programs for younger people, especially women with small children, were also helpful, but to a lesser degree. The two factors combined to reduce the overall rate of poverty, as shown above in Figure 10–4. In the early 1970s, progress came to a halt. The poverty rate stagnated in the 1970s and rose in the 1980s.

Why this reversal? At least three factors are involved: (1) economic stagnation, accompanied by rising unemployment and falling real wages, (2) changing family patterns, which produced an increasing number of single-parent families, and (3) government policy, which has become indifferent to the fate of the poor and near-poor.

Federal efforts to respond to poverty began in the Depression of the 1930s under Franklin Roosevelt. In the 1960s, under Presidents Kennedy and Johnson, the government made a substantial new commitment to programs for the poor and the elderly (many of whom were poor). Subsequent administrations considered new approaches to the problem of poverty but did not abandon the basic commitment. Ronald Reagan, however, came to the White House pledged to cut back federal expenditures on social programs. By imposing new limits on poverty programs, he managed to keep poverty spending from skyrocketing in an era of high unemployment and increasing poverty rates. But he was not able to make significant reductions in the size of the budget for social programs. The Bush administration has not shown any inclination to tinker further with poverty policy.

Except among the elderly, U.S. poverty rates are quite high by the standards of the world's advanced societies. Given the character of our economy and the growing number of single-parent families, it is hard to imagine any progress without a renewed national commitment to help the poor.

SUGGESTED READINGS

Auletta, Ken. 1982. *The Underclass*. New York: Random House.
A summary of academic debates about the underclass, combined with a detailed journalistic report on a job training program in New York City and its effects on the young participants.

Ellwood, David T. 1988. *Poor Support: Poverty in the American Family*. New York: Basic Books.
A carefully reasoned, well-researched essay on family poverty and public policy.

Jencks, Christopher, and Paul Peterson. 1991. *The Urban Underclass*. Washington, D.C.: Brookings Institution.
Excellent collection of social science research on poverty generally and the underclass specifically, inspired by (but often critical of) Wilson (see below).

Levitan, Sar. 1990. *Programs in Aid of the Poor*. 6th ed. Baltimore: Johns Hopkins University Press.
Brief, authoritative description of poverty programs. On this topic, also see U.S. House 1991b (and subsequent editions), the so-called Green Book.

McLeod, Jay. 1987. *Ain't No Makin' It: Leveled Aspirations in a Low-Income Neighborhood*. Boulder, Colo.: Westview.
Engaging portrait of black and white teenage boys in a public housing project.

Murray, Charles A. 1984. *Losing Ground: American Social Policy*. New York: Basic Books.
Conservative critique stating that many social programs increase dependency and are self-defeating. Murray's analysis was vigorously criticized by Christopher Jencks, New York Review of Books, May 9, 1985.

Piven, Frances Fox and Richard A. Cloward. 1971. *Regulating the Poor: The Functions of Public Welfare*. New York: Pantheon.
Stimulating analysis of the development of welfare programs in the context of the social and political forces that spawned them.

Rossi, Peter H. 1989. *Down and Out in America: The Origins of Homelessness*. Chicago: University of Chicago Press.
There may be fewer homeless than we imagine and many more extremely destitute people, one step short of being on the streets.

U.S. Bureau of the Census. *Poverty in the United States*. Series P-60.
Basic source of poverty statistics, published annually.

Wilson, William Julius. 1987. *The Truly Disadvantaged: The Inner City, the Underclass, and Public Policy*. Chicago: University of Chicago Press.
A theory of and prescription for ghetto poverty. (See also Jencks and Peterson above.)

11

The American Class Structure: A Synthesis

The 1980s were the triumph of upper America—an ostentatious celebration of wealth, the political ascendancy of the richest third of the population and the glorification of capitalism, free markets and finance. But while money, greed and luxury had become the stuff of popular culture, hardly anyone asked why such great wealth had concentrated at the top and whether this was the result of public policy. . . . That discussion is now unfolding. "[M]oney politics"—be it the avarice of financiers or the outright corruption of politicians—is shaping up as a prime political theme for the 1990s.

Kevin Phillips (1990)

In this chapter, we will pull together some of the key things we have learned about social stratification in the United States. To do that, we will return to our initial emphasis on seeing the stratification system as a hierarchy of discrete social classes. We will also draw some conclusions about the ways the class system has been changing.

In the preface to this book, we suggested that we were going to emphasize several social classes—groups of households at various levels of the socioeconomic hierarchy created by the convergence of key structural patterns. In Chapter 1, we briefly described our own version of that class hierarchy, and in Chapter 2, we reviewed alternative class models growing out of Lloyd Warner's early community prestige research. But for the most part, we have analyzed one or two variables at a time—income, occupation, prestige, class consciousness, etc.—often referring to studies that measure them on continuous scales, from high to low.

Contemporary stratification research typically prefers to use such scales rather than a class hierarchy as its organizing device. This is perhaps so for two reasons. First, computer-based statistical analysis seems to work best with continuous variables. Second, it is hard to find good criteria that permit sharp divisions within each variable that could be used to define clear-cut levels or classes. These two reasons are related to the fact that most empirical studies on American stratification have been descriptive researches seeking intercorrelations among easily observed and measured social variables in the contemporary world. We have only a few studies that seek the underlying structural causes of those variables and the historical trends that change their relative strength.

By contrast, theoretical speculation about historical trends usually finds it more revealing to think about underlying or latent categories rather than superficially observable variables. Often, the categories are abstract or "ideal" types, defined by the clustering of several intertwined characteristics. Thus, Marx alerted us to the basic division between owners and non-owners of the means of production in all societies. But his meaning of the contrast between capitalists and proletarians in modern society depended upon the presence of high technology, an advanced division of labor, and well-organized capital and labor markets. Only in that specific context did he analyze the social and political consequences of the conflict of interest between the two social classes and predict the final outcome. It did not bother him that at any one moment of history, the distinction between the two classes might in fact be blurred, and many individuals might have ambiguous positions (such as land-owning peasants or urban shopkeepers). He was concerned with the trends toward the maturation of capitalism and its eventual overthrow, not with the placement of individuals in the hierarchy. Indeed, his genius consisted in sensing many trends and their implications even before they were fully observable in the real world.

Besides the contrast between history and the present scene, there is another feature that distinguishes between categories and variables: namely, the mental process of symbolization. When they think of the system as a whole, people are likely to truncate a lot of observations into simplifying symbols: the rich, the middle class, and the poor; the bosses and the workers; the people with a college education versus those without; the people with clout and the nobodies; "us" versus "them." These symbols create a limited number of groupings: dichotomies, trichotomies, or occasionally four or five levels. But beyond that, the symbols lose their power as organizing devices for thought and conversation.

In general, respondents tend to talk about the system in terms of these symbols or stereotypes. Yet when they are asked to look more carefully at themselves and their neighbors, the same people can notice many distinctions that are more subtle than the symbolic categories. They describe neighbors who have a little more money than themselves but are not rich; they see some who have lots of money but not much influence; they notice others without much money but with advanced education and good prospects for the future. Thus, both sociologists and citizens can alternate between a broad and categorical style of synthesis and a detailed and minute style of analysis.

The gulf between historical or theoretical material, organized into broad categories, and contemporary or descriptive material, expressed in minute and measurable variables, has plagued the literature on stratification for many years. It makes the theoretical discussions appear abstract and often unprovable and the empirical presentations seem picayune and atheoretical. Yet the two approaches must be integrated in order to provide a satisfactory view of the stratification hierarchy—one that links sophisticated research with the perspectives of the citizens who live in the system and also with the best theorists who have pondered its nature. We make an attempt at synthesis in this chapter. But to do so, we must impose our own views on the data; there is no simple way in which the facts can speak for themselves.

HOW MANY CLASSES ARE THERE?

Readers with good memories will recall our answer to this question from the first chapter: six, but it all depends on your viewpoint. There is no irrefutable answer. The class model we sketched was based on our assumption that the class structure develops out of the economic system. Let's take a closer look at the model and how it was derived.

We start with the recognition that there are three basic sources of income available to households in this country: capitalist property, labor force participation, and government transfers. Labor force participation (which, as we know from the income parade, accounts for most of the income of most of the people) is shaped by the fact that our economy depends on an occupational division of labor organized into bureaucratic units. Occupational placement is linked in turn to educational preparation. Sources of income, along with experiences on the job and in consumption communities, are verbalized as symbols of the system and the niches people occupy within it. One of the key aspects of a person's perception of place in the system is anticipation of change in the near future: Is one stuck, or is there a chance to advance? Another involves the degree of independence in carrying out one's work activities.

Combining the criteria of source of income, occupation, and educational credentials, plus the related processes of symbolization, we can create an "ideal type" picture of the class structure. The several criteria tend to cluster in a pattern that identifies six classes in the contemporary United States:

1. A capitalist class, subdivided into nationals and locals, whose income is derived largely from return on assets.
2. An upper-middle class of university-trained professionals and managers (a few of whom ascend to such heights of bureaucratic dominance that they become part of the capitalist class).
3. A middle class of people who follow orders on the job from those with upper-middle-class credentials, yet have sufficient vocational skills to make good livings and enjoy a comfortable, mainstream style of life. They usually feel secure in their situation and may look forward to some movement up the hierarchy. Most wear white collars, but some wear blue.
4. A working class of people who are less skilled than members of the middle class and work at highly routinized, closely supervised manual and clerical jobs. Their work provides them with a relatively stable income sufficient to maintain a living standard just below the mainstream, but they have little prospect of ad-

vancing in the hierarchy, since they typically lack the necessary educational credentials. Thus, they concentrate on achieving security through seniority rather than promotion.

5. A working-poor class, consisting of people employed in low-skill jobs, often in marginal firms. The members of this class are typically laborers, service workers, or low-paid operators. Their incomes leave them well below mainstream living standards. Moreover, they cannot depend on steady employment, and far from anticipating advancement, they are at risk of dropping into the class below them.

6. An underclass, whose members have little or no participation in the labor force. They may work erratically or at part-time jobs, but their lack of skills, incomplete education, and spotty employment records make it difficult for them to find regular, full-time positions. Some receive income from illegal activities. Many depend on government transfers for their support. Symbolically, their loose relationship to the labor market and dependence on government handouts anchor them at the bottom of the prestige order.

There are two cutting points that are the least obvious: that between the middle and working classes and that between the working poor and the underclass. Let us examine these divisions in some detail.

The distinction between working poor and underclass becomes difficult when we consider the tendency of some individuals to move repeatedly back and forth across this boundary. Yet the distinction seems worth maintaining. As we move up or down in the hierarchy, away from the boundary, the problem of oscillating mobility is less serious. Moreover, the symbolic difference between having a job, even a marginal one, and welfare dependence is clear. Recall, for example, that Coleman and Rainwater's respondents in Boston and Kansas City separated these two groups in their judgments of prestige classes.

The line between middle class and working class has been blurred by trends that have reduced the traditional differences between blue-collar and white-collar employment. A declining income differential, the increasing routinization of clerical tasks, and the corresponding drop in the prestige value of a white collar per se, have all served to close the gap between shop and office. Viewed in terms of major occupational groupings, the problem centers on the sales, clerical, and craft categories. Our way of dividing these between middle and working class is based on a distinction between workers whose jobs are highly routinized, closely supervised, low in prerequisite training or education, and low in pay, and those who are in the opposite situation. On this basis, we had no trouble placing semiprofessional jobs and the lowest-paid managerial jobs in the middle class or operatives in the working class. The assembly-line character of modern office work and the low salaries associated with most clerical jobs led us to place clerical workers in the working class. We split sales workers into two groups: those engaged in retail work and "others." The latter group includes insurance salesmen, real estate agents, manufacturers' representatives, and other people who work quite independently and have much higher incomes than the retail workers. Our decision to place most craft workers and foremen in the middle class is based on similar considerations. They are well paid, skilled, and relatively independent in their work. Moreover, the prestige attached to such occupations places them well above other blue-collar workers.

In summary, we are suggesting a model of the class structure based primarily on a series of qualitative economic distinctions. From top to bottom, they are: ownership of income-producing assets, possession of sophisticated educational credentials, a combination of independence and freedom from routinization at work, entrapment in the marginal sector of the labor market, and limited labor force participation.

Our scheme is illustrated in Table 11–1. If we round off the numbers from the distributions of each variable treated separately and do a little guessing, we can estimate that the capitalist class includes about 1 percent of the population; the upper-middle class, about 14 percent; the middle and working classes, 60 percent; and the working poor and underclass, 25 percent. We can exemplify this model by going into a little more detail about each of the six classes and relate our definitional criteria to material that has been discussed earlier in the book.

CAPITALIST CLASS

The very small class of super-rich capitalists at the top of the hierarchy has an influence on economy and society far beyond their numbers. They make investment decisions that open or close employment opportunities for millions of others. They contribute money to political parties, and they often own newspapers or television companies, thereby gaining impact on the shaping of the consciousness of all classes in the nation. The capitalist class tends to perpetuate itself: It passes on assets and styles of life (including networks of contact with other influential people) to its children. This creation of lineage is of sufficient importance to them that they are active in creating and supporting preparatory schools and universities for their children and for carefully selected newcomers who can be socialized into their world view.

The super-rich operate on the national and international scene. They have less prominent counterparts in local communities—the people who own the local banks, department stores, car dealerships, real estate empires, and newspapers. They too are capitalists and belong in this class, albeit at the margins.

Our definition produces a very small top class: those who own substantial income-producing assets. After a generation or two, those holdings are often distributed among so many heirs that a larger group of less

TABLE 11-1 Model of the American Class Structure: Classes by Typical Situations

Proportion of Households	Class	Education	Occupation of Family Head	Family Income, 1990
1%	Capitalist	Prestige university	Investors, heirs, executives	Over $750,000, mostly from assets
14%	Upper middle	College, often with postgraduate study	Upper managers and professionals; medium business owners	$70,000 or more
60% {	Middle	At least high school; often some college or apprenticeship	Lower managers; semiprofessionals; nonretail sales; craftspeople; foremen	About $40,000
	Working	High school	Operatives; low-paid craftspeople; clerical workers; retail sales workers	About $25,000
25% {	Working poor	Some high school	Service workers; laborers; low-paid operatives and clericals	Below $20,000
	Underclass	Some high school	Unemployed or part-time; many welfare recipients	Below $13,000

rich and less powerful people results. If one studies local communities and counts all those who have a prominent name and live in big houses and belong to the best country club, one will emerge with a larger group (perhaps double or triple our 1 percent). But if one focuses on assets of sufficient size to grant the economic power that we consider crucial, then the groups shrinks in size, which explains the difference between our estimate and those of Warner or Coleman and Rainwater.

UPPER-MIDDLE CLASS

Apart from the very top echelon, the capitalist-proletarian distinction has lost much of its force in modern society. History has proven Marx wrong when he predicted a trend toward simplification into an ever-sharper distinction between the two classes as the driving force of social change. Weber, who lived until 1920, was able to see this more clearly than Marx, who died in 1883. Weber wrote:

> One must therefore distinguish between "propertied classes" and primarily market-determined "income classes." Present-day society is predominantly stratified in classes, and to an especially high degree in income classes. But in the special *status* prestige of the "educated" strata, our society contains a very tangible element of stratification by status. Externally, this status factor is most obviously represented by economic monopolies and the preferential social opportunities of the holders of degrees. . . . Today, the certificate of education becomes what the test for ancestors has been in the past, at least where the nobility has remained powerful: a prerequisite for equality of birth, a qualification for a canonship, and for state office (1946: 301, 241).

Of course, Weber was also somewhat limited by the vision of his epoch. But he noticed that through education, particularly the university degree, one could obtain both the opportunity for an important job in the church or the state and entry into high society, which still had overtones in Weber's Germany from the days of the nobility. In America at the end of the twentieth century, the degree is still the key to high bureaucratic position and to high prestige status in the community, even though the proportion of the population receiving degrees has expanded dramatically. And of course, more people now hold high positions in business than in the church or state.

The more society bureaucratizes, the more it tends to use educational credentials at all levels to sort people out into careers, at least at the beginning. The formality of this process is striking. For example, the current Chinese government civil service, heir to a tradition much older than the current regime, continues to use twenty-four distinct grades or levels of jobs, despite ideological ideas of equality; the United States federal civil service has eighteen grades; and the General Electric Corporation, considered a model of modern management, recognizes twenty-eight levels of managers and fourteen levels of workers. Each of these grades has a different pay scale and different responsibilities. They do not all specify exactly the educational credentials to match the job and the pay, but they usually use educational credentials to sort out beginning applicants into the level that would be most appropriate for them. Afterwards, experience on the job, additional training courses (sometimes at outside schools, sometimes in courses run internally by management), and demonstrated abilities combine to determine who stays put and who moves up.

The upper-middle class is the group in our society most shaped by formal education. A college degree is usually the minimum requirement, and postgraduate study in business management, law, engineering, or medicine is increasingly required. Currently, close to 25 percent of young people get college degrees, and at least half of them pursue some additional training; about 20 percent of all adults have a degree.

If we turn to occupational statistics, we will recall that Table 3–4 showed about 17 percent of the current labor force classified as professionals and technicians and another 13 percent as managers, officials, and proprietors. But we noted that many of the workers in these categories are semiprofessionals, technicians, or low-level managers with modest salaries, limited training, and circumscribed authority. We estimate that only about 14 percent of the total work force has the combination of university degrees, authority and independence on the job, and high income to qualify for the upper-middle class.

The extent to which adolescents in high school (urged on by their parents and teachers) so often strive to prepare themselves for upper-middle class jobs is a clear indication that these positions have become the symbols of success that motivate many Americans. They may not grant prestige equivalent to a title of nobility in the Germany of Max Weber, but they certainly represent the sign of having "made it" in contemporary America. The incomes of households in this group range upwards of $70,000 a year (about twice the mean in 1990) and tend to increase with age. They are sufficient to purchase houses and cars and travel that become public symbols for all to see and for advertisers to portray with words and pictures that connote success, glamour, and high style. Those who have reached this level of success are likely to convince themselves that they deserve what they receive, that they have earned morally just rewards from the diligent use of superior talent. Sometimes they may grow anxious from the strains of competition, but in general they are satisfied that they have achieved a proper share of the American dream.

MIDDLE CLASS

We have remarked before that a stratification hierarchy is clearest (and incidentally, mobility the weakest) at the extremes. When we move toward the center, distinctions become blurred, people move more often during their lives from one slot to another, and symbolizations become ambiguous. This is particularly true at the point where the middle class and the working class intersect—or better, overlap—so the reader should not expect precision of classification.

It takes at least a high school diploma to get most middle-class jobs, but the diploma is a prerequisite more than a guarantee of such employment. About 80 percent of the total adult population has a high school diploma, and many have received some training beyond it short of a four-year college degree. Those with the best schooling have the most chance to become the semi-professionals, technicians, and lower-level managerial people we mentioned above—about 15 percent of the work force. They are joined in the loose grouping we call the middle class with the upper two thirds of those classified as salespeople and craftspeople—another 12 percent of the work force. Typical household incomes for this level would be around $40,000 a year, but there is considerable variation, based particularly on the number of people in the household who are employed. Middle-class positions usually provide greater job security and wider opportunities for advancement than working-class positions, and younger members of the class are likely to be working in situations where some opportunities to advance in the hierarchy are available.

Symbolization of the middle class tends to get confused by an ideological tradition that says that most Americans are middle class. It is a "good," mainstream sort of phrase that a lot of people adopt—including those who are both higher and lower in the hierarchy than the ones we are trying to discuss at this point. Thus, most surveys show 35 to almost 50 percent of our population identifying with the term and only about 2 percent willing to call themselves upper class. If we subtract about 15 percent for the upper and upper-middle classes as we have defined them, then the size of the remaining middle class according to self-identification would be from 20 to 35 percent of the population. Using a composite of various symbols that people use to classify not only themselves but their neighbors, Coleman and Rainwater decided that one third were middle class. Our own estimate would be on this order.

WORKING CLASS

The core of the working class is easy to identify: semiskilled machine operatives, in factories and elsewhere, who make up 15 percent of the work force. But they are joined by lots of others whose work lives and incomes are not markedly different, such as clerks and salespeople whose tasks are routine and mechanized and require little skill beyond literacy and a short period of on-the-job training (some 14 percent of the work force), and the better-paid persons in the service jobs (another 3 percent). Individuals move easily among these classifications. Often, one member of a family wears a blue and another a white collar, and nobody much notices the difference. Households typically earn $25,000 or less.

In opinion surveys, at least half the population usually chooses the label working class for themselves, but evidence indicates that some do so because they particularly dislike certain alternative terms, such as lower class. The detailed procedures of Coleman and Rainwater arrived at a figure of 37 percent for the working class, some of whom we will put among the working poor. Thus, our estimate for the working class comes to around 30 percent of households.

In general, working-class families earn less than middle-class families, and more particularly, they are less secure in their incomes. The working class is more susceptible to layoffs in time of recession, since employers have less invested in their training and experience. Insecurity of work is often combined with a subjective feeling of vulnerability because of lower levels of education. Relatively few members of the group have training beyond a high school diploma, and over a third (especially the older ones) did not graduate. Yet by contrast with those below them in the hierarchy, working-class people generally anticipate that layoffs will be temporary and that they can support their families in a simple but decent manner most of the time.

WORKING POOR

In 1990, the government called 13.5 percent of the population poor, and studies that follow families over time show that in a nine-year period, some 25 percent of them fell below the poverty line at least once. Of the total work force (and many of the poorest and most discouraged people have withdrawn from it), between 6 and 8 percent are likely to be unemployed at any one moment, and about 20 percent are likely to be unemployed at least once during any given year. Thus, it appears that about one fifth of our working families live under duress: They oscillate in income from just above to below the poverty line, they are threatened with periodic unemployment, or they have no chance to work at all. Those among them— probably a little more than half—who are often working but not earning enough money on a steady basis to bring them close to the mainstream style of consumption, we label the working poor.

The working poor include the unskilled laborers, most people in service jobs, and some of the lower-paid factory workers (especially in marginal firms). Many employed single mothers find themselves in this class. Their incomes depend on the number of weeks a year they are employed and on the number of workers in the family. Most families would feel fortunate in a year that brought in $15,000. Some adults have finished high school; a great many have not. They are unable to save money to cover contingencies, so insecurity is a normal part of their lives. The one part of the welfare system that was beneficial to them, the food stamp program, was slashed in the budgets of the Reagan administration, which eliminated participation by most of the working poor. Once retired, members of this class are entirely dependent on their social security pensions, for it is unlikely that they have been enrolled in a private retirement plan that could supplement the government payments.

UNDERCLASS

Those who are seldom employed and are poor most of the time form the underclass in our society. Many suffer long-term deprivation from low education, low employability, low income, and eventually, low self-esteem. Some are kept out of the labor force by age or disability. For a great many, their problems are magnified because they belong to minority groups who are stigmatized and suffer discrimination in the labor market. Or they are single mothers, who must make their way in a job world that pays women less than men. Many who cannot get and keep jobs that pay enough to live on are dependent on the welfare system.

The conditions of life in the underclass are sufficiently difficult and demeaning that it is hard—although not impossible—for children to get enough education and enough hope to climb up to higher levels. The future chances for avoiding a life of poverty for these children are about fifty-fifty.

The descriptions of the six classes just given are summarized in Table 11–1. It is clear that no single variable can be used to delineate these classes, so our synthesis is based on a combination of variables. We believe that they tend to form patterns that are caught by our scheme in a way that is meaningful in two senses. (1) It is congruent with much of the research literature that goes into detail on one or two variables at a time, as well as with the more qualitative community studies that tend to combine many variables into symbolic groupings. (2) It is congruent with the way most Americans tend to see the system and their place within it. Of course, we are thinking here in terms of averages, of typical situations; many individuals and families are hard to place in the scheme, either because they are higher in position on one variable than on another, or because they are mobile, or because two or more members of a family work and they have disparate jobs.[1]

In Table 11–2, we compare our model of the class structure with the class models we examined in Chapter 2, Warner's Yankee City model and the Coleman-Rainwater metropolitan area model from the 1970s. (For further details of these two models, see the capsule summaries on pages 26

---

[1] These difficulties of exact status placement and their consequences for consciousness and behavior have been studied under the phrases "status crystallization" or "status consistency," but the results have been inconclusive (Lenski 1954; Jackson and Burke 1965; Landecker 1981).

TABLE 11-2   Three Class Models Compared

Gilbert & Kahl (National, 1990)		Coleman & Rainwater (Metropolitan, Early 1970s)		Warner (Small City, 1930s)	
Capitalist	1%	Upper upper } Lower upper }	2%	Upper upper } Lower upper }	3%
Upper middle	14	Upper middle	19	Upper middle	10
Middle } Working }	60	Middle Working	31 35	Lower middle Upper lower	28 33
Working poor } Underclass }	25	Semipoor } The bottom }	13	Lower lower	25
Total	100%		100%		100%

and 34.) There are some variations in class labels, but the major differences stem from the fact that their primary emphasis is prestige, whereas ours is economic status. For this reason, they make distinctions between old and new money at the top—in effect splitting our capitalist class in half—and use the traditional blue-collar/white-collar distinction to define the middle and working classes. At the bottom, we adopt Coleman and Rainwater's distinction between those low-income people who depend largely on job earnings and those who do not. We will return to comparisons between these models later in the chapter.

RACE AND CLASS

African Americans have frequently occupied our attention in the preceding chapters. We have examined their incomes, occupations, poverty, social mobility, and political opinions. This focus on the experience of blacks implies that they are not just one more American ethnic group, but somehow a special case.

Our society, black and white, is still haunted by the legacy of slavery. White ethnic groups have come to the United States as immigrants and, over a period of time, merged into the general population. After two or three generations, their class distributions, residential patterns, and life-styles became virtually indistinguishable from the national majority. But by and large, blacks have remained distinct. Those who have moved up the class hierarchy find that race and class contradict each other. They are both African American and successful, and whites have not learned how to take that contradiction in stride. Racism is still a reality in this society, and middle-class blacks probably have the greatest exposure to it. In the lives of blacks who remain stranded at the bottom of the economic pyramid, class and race reinforce each other. These people are both black and poor. Some of them eventually reject the idea of incorporation into the larger society. They yearn for an independent community, with black pride and black power that will give them a new dignity. But the forces of the larger society deny them either full integration or full independence.

In terms of class structure, recent decades have seen extraordinary developments among African Americans. A large, well-educated, increasingly prosperous middle class has emerged. In 1940, 80 percent of black workers were still concentrated in the most menial blue-collar occupational categories. Four decades later, 45 percent held white-collar jobs, including 20 percent in the professional, technical, or managerial categories. In 1990, 30 percent of black families had incomes in excess of $35,000, which is enough to sustain a middle-class lifestyle (U.S. Census 1991a). Young college-educated blacks are now earning salaries very close to those of their white counterparts.

At the same time, about one third of blacks have incomes below the official poverty line (the same proportion who were poor in 1969), and 40 percent have incomes of less than 125 percent of the poverty line (U.S. Census 1991b). In other words, four out of ten black Americans are poor or near-poor. Thus, although many African Americans can now be considered middle or upper-middle class, an even larger number are trapped at the bottom. What we are seeing is the class differentiation of black America—a process that leaves the poorest blacks relatively worse off and more isolated, at the same time as successful members of their race move up the class ladder and out of poor black communities.

Now, more than ever before, black poverty is accompanied by social disintegration. In particular, poor black inner-city neighborhoods are often characterized by distressingly high rates of teen pregnancy, dropping out of school, female-headed families, unemployment (often approaching 50 percent among young people), welfare dependency, substance abuse, and violent crime. Some families manage to maintain stable, working-class lives in these areas, but their circumstances become ever more difficult, both because of the chaos around them and because of an economy in which the supply of decent, dependable working-class jobs is shrinking.

Hopelessness reigns in poor black communities, and young males are especially afflicted. They see no meaningful future for themselves—no jobs, no stable male role models, no apparent path to success. They feel mistreated and exploited by the larger society, and they complain of discriminatory, abusive treatment by police officers. Out of their experiences, they develop an angry group consciousness based on race and class. Some seek identity and security in street gangs. Some become drug dependent. Violence marks their lives.

For black males between the ages of 15 and 25, homicide is the leading cause of death. Incredibly, on any given day, about one of every four young black men is in jail, on probation or on parole—more than are enrolled in college. Allowing for the high proportion of young men from black middle-class families who are in college or the armed forces, the rate of involvement with the corrections system must be closer to one in two for the sons of the black underclass (*Los Angeles Times*, March 7 and July 10, 1990).

Sometimes smoldering anger flares into violence on a wider scale. Rioting breaks out in inner-city ghettoes, accompanied by looting of stores, torching of buildings, and random violence against unlucky outsiders. In their rage and greed, people destroy large parts of their own neighborhoods, losing local retail outlets and jobs in the process. More than a dozen U.S. cities experienced large-scale disturbances in the mid-1960s, including the wave of riots that followed the 1968 assassination of civil rights leader Martin Luther King. Recent explosions of unrest in Miami (1980) and Los Angeles (1992) followed the acquittals of police officers who were accused, on strong evidence, of brutal treatment of black suspects.

Marx, we noted in Chapter 9, identified seven social factors that are likely to produce a class-conscious proletariat. All are present in poor inner-city black neighborhoods save one: organization. The traditional civil rights organizations, such as the NAACP and the Urban League, tend to represent middle-class blacks more than the underclass; labor unions seem largely irrelevant; and social mobility has permitted many of the better-off families who once provided community leadership to flee to middle-class neighborhoods. Without the organizational capacity for structured political protest, the ghetto turns to self-destructive violence.

In 1989, a comprehensive report on black America, produced by a distinguished committee of scholars for the National Research Council, measured the enormous progress of middle-class African Americans. At the same time, the report warned that the conditions existed for new outbreaks of civil disorder in poor black comunities: "The ingredients are there: large populations of jobless youths, an extensive sense of relative deprivation and injustice, distrust of the legal system, frequently abrasive police–community relations, highly visible inequalities, extreme concentrations of poverty, and great racial awareness" (Jaynes and Williams 1989: 31). Their warning (coming twenty years after a similar one from the blue-ribbon Kerner Commission that investigated the disturbnces of the 1960s) presaged the 1992 explosion in Los Angeles, which is likely to be remembered as the worst urban riot of the century in this country.

IS THE AMERICAN CLASS STRUCTURE CHANGING?

The evidence of the preceding chapters leaves little doubt that the answer to this question is yes. Let's take a look at the details. We will use the comparison between our contemporary class model and Warner's Yankee City model from the 1930s as a point of departure for our discussion of change (see Table 11–2). We have already referred to the differences that are rooted in our distinct perspectives on the class order. But the basic shape of the two models is quite similar, suggesting a remarkable stability in the underlying structure over the past half century.

It is a good bet that the class structure of the 1930s would be more familiar to Americans of the 1990s than to their ancestors of the 1880s. But if we examine the classes one by one, we can find evidence of gradual change in the course of recent decades, a process that appears to have accelerated in recent years.

The national capitalist class has been undergoing a steady consolidation since the 1930s. The family-controlled firm is giving way to a system of corporations directed by professional managers. As this happens, wealthy families are diversifying their holdings, and top corporate managers are being drawn into the capitalist class they serve by financial rewards and social ties. Local capitalist classes, like Warner's Yankee City upper classes, have been affected by the tendency of national corporations to move into local markets, often absorbing the banks and department stores through which small-town fortunes were built. The younger generation of this class is moving into professional and managerial careers. Many will enjoy additional income derived from inherited wealth.

These changes are creating the basis for a more cohesive capitalist class, whose members are free from parochial identification with a particular firm, economic sector, or locality. In politics, this tendency is reinforced by the growing influence of business-dominated PACs and policy planning and research organizations that defend the interests of the capitalist class as a whole rather than those of individual capitalists.

Our upper-middle class, like Coleman and Rainwater's in the early 1970s, is larger than Warner's and contains a different mix of occupations. The trends that C. Wright Mills noted a generation ago have continued unabated. The growing organizational and technological complexity of our society increases the demand for specialists with sophisticated training. Salaried managers and professionals replace individual entrepreneurs and independent professionals as the mainstays of the upper-middle class. The gap in income, politics, and lifestyle between this class and the rest of the population has replaced the traditional blue-collar/white-collar division as the most important cleavage in the American class structure.

The middle class is changing in ways that parallel the transformation of the upper-middle class. The petty businessmen in Warner's "lower middle" class are being supplanted by lower-level managers. Teachers, a significant sector of this class as Warner described it, are being joined by large numbers of semiprofessionals from other fields. As the number of routinized, low-skilled jobs has grown, the shop-office distinction, which once divided the middle class from the working class, has been replaced by a more subtle distinction based on preparation for the job and independence at work.

"The disappearing middle class" has become a recurrent theme for political observers, marketing analysts, and pop sociologists. We are convinced, to the contrary, that the middle class is growing. Of course, the middle class can be made to disappear and reappear by manipulating the way it is defined. From our perspective, it is not the middle class but rather the middle-income group that is shrinking. The facts are these. The distribution of income is becoming less egalitarian, with more people finding themselves at the (relative) extremes and fewer in the middle. In the period from 1973 to 1988, the median family income remained stable at about

$32,000 (in 1988 dollars), but the proportion of families near this amount declined. Suppose we define middle income as $25,000 to $50,000 (still using 1988 dollars). The proportion of families in this range declined from 43 to 37 percent, while the proportions both above and below increased. Clearly, the middle-income group has been thinning out (U.S. Census 1990b:14).

Behind this redistribution lie the postindustrial changes in the economy and corresponding shifts in the occupational structure discussed in earlier chapters. High-paying blue-collar jobs in manufacturing and other goods-producing sectors are being replaced by low-wage jobs in the service industries; real wages in manufacturing are shrinking; high unemployment rates have become a permanent feature of the economy. These trends, together with the growing number of female-headed households created by divorce, are sharply reducing the incomes of many working-class families. At the other end of the occupational scale, there are ample opportunities for qualified managers, professionals, and technicians. Many of these new jobs fall into the middle class by our criteria, but they frequently pay enough, especially when combined with a second income, to place a family over the $50,000 benchmark. Thus, the middle class is thriving, while the working class is languishing. Both tendencies push people out of the middle-income range.

Our estimate of the relative size of the bottom classes is much larger than Coleman and Rainwater's but exactly the same as the figure given by Warner for his equivalent lower-lower class. Although none of these numbers should be taken too literally, they reflect the fact that Coleman and Rainwater worked in the early 1970s, a much more prosperous period than the current era or the 1930s, when Warner did his research. Of course, government statistics show that poverty rates have been rising. The economic developments described above, together with persistent high rates of unemployment, are forcing many members of the working class into our working poor category or even the underclass. Some members of the working class, or their offspring, will qualify for middle-class jobs. All these tendencies in the class system are contributing to the erosion of the working class.

As the working class has declined, so has the size and political weight of the labor movement. The growing political efficacy of business and the shrinking influence of labor have destroyed the class balance of power that ruled American politics for a generation after World War II. Like job opportunities and income, political power is shifting upward in the class system.

TOWARD GREATER INEQUALITY

The trends we have been describing point to growing social inequality in the United States. In fact, the evidence we examined in our major stratification variables pointed in the same direction.

Income data give us the clearest picture of change. As we have seen, from the late 1940s, when the government started regular income surveys, until the early 1970s, incomes were usually rising and gradually becoming more equal. Since then, most households have seen their incomes stagnate or even decline, and income has become more concentrated. Many families have been able to stay even only by increasing the number of household members in the labor force. Young families have experienced an absolute decline in the purchasing power of their incomes—an especially bad sign for the future.

The 1980s was a period of stark contrasts in income trends. From 1977 until 1990, while the poverty rate was rising, the average income of the top 1 percent of households jumped from $300,000 to $550,000. In 1979, the average corporate chief executive officer earned twenty-nine times more than the average factory worker; by 1988, CEOs were earning ninety-three times as much. Regressive changes in the federal tax system have exacerbated these growing income disparities.

These developments in income distribution and tax policy have contributed to an abrupt increase in the concentration of wealth. Like the rise in income inequality, this represents a reversal of long-term trends. The greatest gains have been made by the super-rich. We noted that the real net worth of *Forbes'* 400 richest Americans doubled in the course of the 1980s. By the end of the decade, the top 1 percent of households held a greater share of aggregate net worth than the bottom 90 percent.

The shifts in the occupational structure described above are contributing to a lessening of upward mobility in recent years. We found that upward mobility still exceeds downward mobility, and succession of sons to their fathers' status is close to what it was in the 1960s. But especially among younger workers, opportunities appear to be stagnating. The rate of downward mobility has increased, there is less movement up from blue-collar to white-collar positions, and it is becoming harder to move into the better white-collar positions. When we compared the occupations of sons with the occupations of fathers in successive periods, we concluded that the structural sources of upward mobility were shrinking.

Our study of mobility demonstrated that a college degree gives a powerful boost to the probability of career success. We learned that in the 1970s, minorities moved closer to whites, women moved closer to men, and low-income households moved closer to high-income households in their chances of obtaining a college degree. But in the 1980s, progress halted for minorities, and young people from low-income families were slipping backward in their access to higher education.

Our general conclusion, then, is that the American class structure is moving in the direction of greater inequality. The occupational structure is changing in a way that puts the people at the bottom at a distinct disadvantage. Wealth and income are being more concentrated. The poverty rate is rising. The pace of upward mobility is slowing. The balance of political power is shifting toward the privileged.

The floundering U.S. economy is certainly chief among the forces responsible for these regressive trends. But there are others, including the troubles of our families, the nation's unresolved dilemmas of race and gender, and the deterioration of our political institutions. If we do not find a way to address these deep-seated problems, class inequalities will continue to grow.

Bibliography

Aaron, Henry J. 1978. *Politics and the Professors: The Great Society in Perspective.* Washington, D.C.: Brookings Institution.

Abramson, Paul R.; John H. Aldrich; and David W. Rohide. 1991. *Change and Continuity in the 1988 Elections.* Revised ed. Washington, D.C.: Congressional Quarterly Press.

Aiken, Michael, and Paul E. Mott, eds. 1970. *The Structure of Community Power.* New York: Random House.

Alba, R. D., and D. E. Lavin. 1981. "Community Colleges and Teaching in Higher Education." *Sociology of Education* 54:223–247.

Aldrich, Nelson W., Jr. 1988. *Old Money: The Mythology of America's Upper Class.* New York: Vintage.

Alexander, Herbert E. 1980. *Financing Politics: Money, Elections, and Political Reform.* 2nd ed. Washington, D.C.: Congressional Quarterly Press.

Alexander, Herbert E., and Brian A. Haggerty. 1987. *Financing the 1984 Elections.* Lexington, Mass.: Lexington Books.

Allen, Frederick Lewis. 1952. *The Big Change.* New York: Harper & Row.

Anderson, Dewey, and Percy Davidson. 1943. *Ballots and the Democratic Class Struggle.* Stanford, Calif.: Stanford University Press.

Anderson, Martin. 1978. *Welfare: The Political Economy of Welfare Reform in the United States.* Stanford, Calif.: Hoover Institution Press.

Aronowitz, Stanley. 1973. *False Promises: The Shaping of American Working Class Consciousness.* New York: McGraw-Hill.

Auletta, Ken. 1982. *The Underclass.* New York: Random House.

Averch, H. A., et al. 1972. *How Effective Is Schooling? A Critical Review of Research.* Santa Monica, Calif.: Rand Corporation.

Avery, Robert, and Arthur B. Kennickell. 1990. "Measurement of Household Saving Obtained from First Differing Wealth Estimates." Paper presented at the Twenty-First General Conference of the International Association for Research in Income and Wealth.

Avery, Robert, et al. 1984. "Survey of Consumer Finances, 1983" and "Survey of Consumer Finances, 1983: A Second Report." *Federal Reserve Bulletin* 70 (September and December): 679–692, 857–868.

Avery, Robert; Gregory Elliehausen; and Arthur Kennickell. 1988. "Measuring Wealth with Survey Data: An Evaluation of the 1983 Survey of Consumer Finances." *Review of Income and Wealth* 34:339–369.

Bachrach, Peter, and Morton S. Baratz. 1974. *Power and Poverty: Theory and Practice*. New York: Oxford University Press.

Baltzell, E. Digby. 1958. *Philadelphia Gentlemen*. New York: Free Press.

Barnet, Richard J., and Ronald E. Muller. 1974. *Global Reach: The Power of the Multinational Corporations*. New York: Simon & Schuster.

Barnouw, Erik. 1978. *The Sponsor: Notes on a Modern Potentate*. New York: Oxford University Press.

Bawden, D. Lee, ed. 1984. *The Social Contract Revisited*. Washington, D.C.: Urban Institute Press.

Beeghley, Leonard, and John K. Cochran. 1988. "Class Identification and Gender Role Norms among Employed Married Women." *Journal of Marriage and the Family* 50:546–566.

Bell, Daniel. 1976. *The Coming of the Post-Industrial Society*. New York: Basic Books.

Bendix, Reinhard, and Seymour Martin Lipset, eds. 1953. *Class, Status, and Power*. 1st ed. Glencoe, Ill.: Free Press. (2nd ed., New York: Free Press, 1966).

Berg, Ivar. 1970. *Education and Jobs: The Great Training Robbery*. New York: Praeger.

Berle, Adolf A., Jr., and Gardiner C. Means. 1932. *The Modern Corporation and Private Property*. New York: Commerce Clearing House.

Berman, Paul. 1991. "A Union Man from Harvard." *New York Times Book Review*. August 11.

Bielby, William T.; Robert M. Hauser; and David L. Featherman. 1977. "Response Errors of Black and Nonblack Males in Models of Intergenerational Transmission of Socioeconomic Status." *American Journal of Sociology* 82:1242–1288.

Birnbach, Lisa, ed. 1980. *The Official Preppy Handbook*. New York: Workman.

Blau, Francine. 1984. "Women in the Labor Force: An Overview." In *Women: A Feminist Perspective*, edited by Jo Freeman. Palo Alto, Calif.: Mayfield.

Blau, Peter M., and Otis Dudley Duncan. 1967. *The American Occupational Structure*. New York: Wiley.

Bloch, Fred. 1977. "The Ruling Class Does Not Rule: Notes on the Marxist Theory of the State." *Socialist Revolution* 7 (no. 1):6–28.

Bluestone, Barry. 1974. "The Poor Who Have Jobs." In *The Sociology of American Poverty*, edited by Joan Huber and H. Paul Chalfant. Cambridge, Mass.: Schenkman.

Blumberg, Paul M., and P. W. Paul. 1975. "Continuities and Discontinuities in Upper-Class Marriages." *Journal of Marriage and the Family* 37:63–77.

Blume, Marshall E.; Jean Crockett; and Irwin Friend. 1974. "Stock Ownership in the United States: Characteristics and Trends." *Survey of Current Business*, November, 16–44.

Blumenthal, Sydney. 1986. *The Rise of the Counter Establishment*. New York: Times Books.

Blumenthal, Sydney, and Thomas Byrne Edsall, eds. 1988. *The Reagan Legacy*. New York: Pantheon.

Bonjean, Charles M., and Michael D. Grimes. 1974. "Community Power:

Issues and Findings." In *Social Stratification: A Reader,* edited by Joseph Lopreato and Lionel S. Lewis. New York: Harper & Row.

Bott, Elizabeth. 1954. "The Concept of Class as a Reference Group." *Human Relations* 7:259–286.

————. 1964. *Family and Social Network.* London: Tavistock.

Bottomore, Tom. 1966. *Elites in Modern Society.* New York: Pantheon.

Bottomore, Tom, and Robert J. Brym. 1989. *The Capitalist Class: An International Study.* New York: New York University Press.

Boudon, Raymond. 1974. *Education, Opportunity and Social Inequality.* New York: Wiley.

————. 1976. "Comment on Hauser." *American Journal of Sociology* 81:1175–1187.

Bowles, Samuel, and Herbert Gintis. 1976. *Schooling in Capitalist America.* New York: Basic Books.

Boyer, Richard, and Herbert Morais. 1975. *Labor's Untold Story.* 3rd ed. New York: United Electrical Workers.

Bradburn, Norman. 1969. *The Structure of Psychological Well-Being.* Chicago: Aldine.

Braverman, Harry. 1974. *Labor and Monopoly Capital.* New York: Monthly Review Press.

Breiger, Ronald. 1982. "A Structural Analysis of Occupational Mobility." In *Social Structure and Network Analysis,* edited by Peter V. Marsden and Nan Lin. Beverly Hills, Calif.: Sage Publications.

Brody, David. 1980. *Workers in Industrial America: Essays on the 20th Century Struggle.* New York: Oxford University Press.

Bronfenbrenner, Urie. 1966. "Socialization through Time and Space." In *Class, Status, and Power,* edited by Reinhard Bendix and Seymour Martin Lipset. 2nd ed. New York: Free Press.

Brooks, Thomas. 1971. *Toil and Trouble: A History of American Labor.* 2nd ed. New York: Dell.

Burch, Philip H., Jr. 1980. *Elites in American History: The New Deal to the Carter Administration.* New York: Holmes and Meier.

Burnham, James. 1941. *The Managerial Revolution.* New York: John Day.

Burnight, Robert G., and Parker G. Marden. 1967. "Social Correlates of Weight in an Aging Population." *Milbank Memorial Fund Quarterly* 45:75–92.

Burtless, Gary. 1987. "Inequality in America: Where Do We Stand?" *Brookings Review* 5 (Summer):9–16.

Cameron, Juan. 1978. "Small Business Trips Big Labor." *Fortune* 98 (July): 80–82.

Campbell, Angus; Gerald Gurin; and Warren E. Miller. 1954. *The Voter Decides.* Evanston, Ill.: Row, Peterson.

————. 1960. *The American Voter.* New York: Wiley. (Abridged edition, 1964).

Cantril, Hadley. 1951. *Public Opinion.* Princeton, N.J.: Princeton University Press.

Caplow, Theodore. 1980. "Middletown Fifty Years After." *Contemporary Sociology* 9:46–50.

Caplow, Theodore, and Bruce Chadwick. 1979. "Inequality and Life Styles in Middletown, 1920–1978." *Social Science Quarterly* 60:366–368.

Center for Political Studies. "American National Election Study, 1984: Codebook."

Centers, Richard. 1949. *The Psychology of Social Classes: A Study of Class Consciousness.* Princeton, N.J.: Princeton University Press.

————. 1953. "Social Class, Occupation and Imputed Belief." *American Journal of Sociology* 58:546.

Charlot, Monica. 1985. "The Ethnic Minorities Vote." In *Britain at the Polls*, edited by Austin Ranney. Washington, D.C.: American Enterprise Institute.

Clark, Terry. 1971. "Community Structure, Decision-Making, Budget Expenditure, and Urban Renewal in 51 American Communities." In *Community Politics*, edited by Charles Bonjean et al. New York: Free Press.

Cohen, Jere. 1979. "Socio-economic Status and High School Friendship Choice: Elmtown's Youth Revisited." *Social Networks* 2:65–74.

Colclough, Glenna, and E. M. Beck. 1986. "The American Educational Structure and the Reproduction of Social Class." *Sociological Inquiry* 56:456–476.

Coleman, James S., et al. 1966. *Equality of Educational Opportunity*. Washington, D.C.: U.S. Government Printing Office.

Coleman, Richard P., and Lee Rainwater, with Kent A. McClelland. 1978. *Social Standing in America: New Dimensions of Class*. New York: Basic Books.

Collier, Peter, and David Horowitz. 1976. *The Rockefellers: An American Dynasty*. New York: Holt, Rinehart & Winston.

Congressional Quarterly. 1976. *Guide to Congress*. 2nd ed.

_____. 1980. "Democrats May Lose Edge in Contributions from PACs." 38:3405–3409.

Cookson, Peter, and Carolinea Hodges Persell. 1985. *Preparing for Power: America's Elite Boarding Schools*. New York: Basic Books.

Corcoran, Mary. 1978. "Measurement Error in Status Attainment Models." Unpublished paper. Ann Arbor, Mich.: Institute for Social Research.

Corey, Lewis. 1935. *The Crisis of the Middle Class*. New York: Covici-Friede.

_____. 1953. "Problems of the Peace: The Middle Class." In *Class, Status, and Power*, edited by Reinhard Bendix and Seymour Martin Lipset. Glencoe, Ill.: Free Press.

Coser, Lewis. 1978. *Masters of Sociological Thought*. 2nd ed. New York: Harcourt Brace Jovanovich.

Council of Economic Advisors. 1990. *Economic Report of the President*. Washington, D.C.: U.S. Government Printing Office.

Coverman, Shelly. 1988. "Sociological Explanations of the Male-Female Wage Gap." in *Women Working: Theories and Facts in Perspective*, edited by Ann Stromberg and Shirley Harkess. 2nd ed. Mountain View, Calif.: Mayfield.

Coxon, Anthony, and Charles Jones. 1978. *The Images of Occupational Prestige*. New York: St. Martin's Press.

Current Biography. 1982–1988. New York: W. W. Wilson.

Curtis, Richard F., and Elton F. Jackson. 1977. *Inequality in American Communities*. New York: Academic Press.

Daalder, Hans, and Ruud Koole. 1988. "Liberal Parties in the Netherlands." In *Liberal Parties in Western Europe*, edited by Emil Kirchner. Cambridge, England: Cambridge University Press; pp. 151–177.

Dahl, Robert A. 1961. *Who Governs? Democracy and Power in an American City*. New Haven, Conn.: Yale University Press.

_____. 1967. *Pluralist Democracy in the United States*. Chicago: Rand McNally.

Davis, Allison; Burleigh B. Gardner; and Mary R. Gardner. 1941. *Deep South: A Social-Anthropological Study of*

Caste and Class. Chicago: University of Chicago Press.

Davis, James. 1978. "Study of Categorical Data Over Time." *Social Science Research* 7:151–179.

Davis, James Allen, and Tom Smith. 1990. *General Social Surveys, 1972–1990* (machine-readable data file). Chicago: National Opinion Research Center.

Davis, Mike. 1980. "The Barren Marriage of American Labour and the Democratic Party." *New Left Review*, no. 124:45–84.

Demerath, N.J., III. 1965. *Social Class in American Protestantism.* Chicago: Rand McNally.

DeStefano, Linda. 1990. "Pressures of Modern Life Bring Increased Importance to Friendship." *The Gallup Poll Monthly* (March):24–33.

Dinitz, S.; F. Banks; and B. Pasamanick. 1960. "Mate Selection and Social Class: Change in the Past Quarter Century." *Marriage and Family Living* 22:348–351.

Doeringer, Peter B., and Michael J. Piore. 1971. *Internal Labor Markets and Manpower Analysis.* Lexington, Mass.: D. C. Heath.

Domhoff, G. William. 1967. *Who Rules America?* Englewood Cliffs, N.J.: Prentice-Hall.

————. 1970. *The Higher Circles: The Governing Class in America.* Englewood Cliffs, N.J.: Prentice-Hall.

————. 1974. *The Bohemian Grove and Other Retreats.* New York: Harper & Row.

————. 1975. "Social Clubs, Policy Planning Groups, and Corporations." *Insurgent Sociologist* 5 (3):173–195.

————. 1978. *Who Really Rules? New Haven and Community Power Reexamined.* New Brunswick, N.J.: Transaction Books.

Domhoff, G. William, and Hoyt B. Ballard, eds. 1968. *C. Wright Mills and the Power Elite.* Boston: Beacon Press.

Dooley, P. 1969. "The Interlocking Directorate." *American Economic Review* 59:314–323.

Dotson, Floyd. 1950. "The Associations of Urban Workers." Unpublished Ph.D. thesis, Yale University.

————. 1951. "Patterns of Voluntary Association Among Urban Working Class Families." *American Sociological Review* 16:687–693.

Drake, St. Clair, and Horace R. Cayton. 1945. *Black Metropolis.* New York: Harcourt Brace Jovanovich.

Dubofsky, Melvyn. 1980. "The Legacy of the New Deal." *Executive* 6 (Spring):8–10.

Duncan, Gregg J., et al. 1984. *Years of Poverty, Years of Plenty: The Changing Economic Fortunes of American Workers and Families.* Ann Arbor, Mich.: Institute for Social Research, University of Michigan.

Duncan, Gregg, et al. 1988. "Welfare Dependence Within and Across Generations." *Science* 239 (January 29):467–471.

Duncan, Gregg, and Willard Rodgers. 1989. "Has Poverty Become More Persistent?" Institute for Social Research, University of Michigan.

Duncan, Otis Dudley. 1961. "A Socioeconomic Index for All Occupations," and "Properties and Characteristics of the Socioeconomic Index." In *Occupations and Social Status*, edited by Albert Reiss. Glencoe, Ill.: Free Press.

————. 1966. "Methodological Issues in the Analysis of Social Mobility." In *Social Structure and Mobility in Economic Development*, edited by Neil Smelser and Seymour Martin Lipset. Chicago: Aldine.

Duncan, Otis Dudley; Archibald O.

Haller; and Alejandro Portes. 1968. "Peer Influences on Aspirations: A Reinterpretation." *American Journal of Sociology* 74:119–137.

Dye, Thomas R. 1979. *Who's Running America? The Carter Years*. 2nd ed. Englewood Cliffs, N.J.: Prentice-Hall.

_____. 1990. *Who's Running America: The Bush Era*. 4th ed. Englewood Cliffs, N.J.: Prentice-Hall.

Edsall, Thomas B. 1984. *The New Politics of Inequality*. New York: Norton.

Edsall, Thomas, and Mary Edsall. 1991. *Chain Reaction: The Impact of Race, Rights and Taxes in American Politics*. New York: Norton.

Edwards, Alba M., and U.S. Bureau of the Census. 1943. *U.S. Census of Population 1940: Comparative Occupational Statistics, 1870–1940*. Washington, D.C.: U.S. Government Printing Office.

Edwards, Richard. 1979. *Contested Terrain: The Transformation of the Workplace in the Twentieth Century*. New York: Basic Books.

Ehrenreich, Barbara. 1989. *Fear of Falling: The Inner Life of the Middle Class*. New York: Pantheon.

Ellwood, David T. 1988. *Poor Support: Poverty in the American Family*. New York: Basic Books.

Ellwood, David, and Mary J. Bane. 1985. "The Impact of AFDC on Family Structure and Living Arrangements." In *Research in Labor Economics 7*, edited by Ronald Ehrenberg. Greenwich, Conn.: JAI Press.

Erikson, Robert, and John Goldthorpe. 1985. "Are American Rates of Social Mobility Exceptionally High? New Evidence on an Old Issue." *European Sociological Review* 1:1–22.

Farley, Reynolds. 1984. *Blacks and Whites: Narrowing the Gap?* Cambridge, Mass.: Harvard University Press.

Featherman, David L., and Robert M. Hauser. 1978. *Opportunity and Change*. New York: Academic Press.

Forbes. 1990. "The Forbes 400." October 22.

_____. 1991. "What 800 Companies Paid their Bosses." May 27.

Fortune. 1937. "The Industrial War." 14 (November):105–110, 156, 160, 166.

Fraser, Douglas. 1978. "UAW President Fraser Resigns from Labor-Management Group." *Radical History Review* No. 18 (Fall):117–121.

Frears, John. 1988. "Liberalism in France." In *Liberal Parties in Western Europe*, edited by Emil Kirchner. Cambridge, England: Cambridge University Press, pp. 124–150.

Freeman, Richard B. 1976. *The Overeducated American*. New York: Academic Press.

Fussell, Paul. 1983. *Class: A Guide Through the American Status System*. New York: Ballantine.

Galbraith, John Kenneth. 1958. *The Affluent Society*. Boston: Houghton Mifflin.

_____. 1967. *The New Industrial State*. Boston: Houghton Mifflin.

Gallup, George, Jr. 1988. *The Gallup Poll: Public Opinion 1987*. Wilmington, Del.: Scholarly Resources Inc.

_____. 1989. *The Gallup Poll: Public Opinion 1988*. Wilmington, Del.: Scholarly Resources Inc.

Garfinkel, Irwin, and Sara McLanahan. 1986. *Single Mothers and Their Children*. Washington, D.C.: Urban Institute.

Gecas, Viktor. 1979. "The Influence of Social Class on Socialization." In *Contemporary Theories about the Family*, edited by W. R. Burr et al. Vol. I. New York: Free Press.

Geoghegan, Thomas. 1991. *Which Side Are You On: Trying to Be for Labor When It's Flat on Its Back*. New York: Farrar, Strauss & Giroux.

Giddens, Anthony. 1973. *The Class Structure of the Advanced Societies.* New York: Harper & Row.

Gilbert, Dennis. 1981. "Cognatic Descent Groups in Upper-Class Lima (Peru)." *American Ethnologist* 8:739–757.

Ginsberg, Benjamin, and Martin Shefter. 1990. *Politics by Other Means: The Declining Importance of Elections in America.* New York: Basic Books.

Glenn, Norval D. 1975. "The Contribution of White Collars to Occupational Prestige." *Sociological Quarterly* 16:184–189.

Glenn, Norval D., and Jon P. Alston. 1968. "Cultural Distances Among Occupational Categories." *American Sociological Review* 33:365–382.

Goldstein, Marshall. 1962. "Absentee Ownership and Monolithic Power Structures." In *Trends in Comparative Community Studies*, edited by Bert E. Swanson. Kansas City, Mo.: Community Studies.

Goldthorpe, John H., and Keith Hope. 1972. "Occupational Grading and Occupational Prestige." In *The Analysis of Social Mobility: Methods and Approaches*, edited by Keith Hope. Oxford, England: Clarendon.

Green, Mark. 1979. *Who Runs Congress?* 3rd ed. New York: Bantam Books.

Greene, Bert. 1978. *Pity the Poor Rich.* Chicago: Contemporary Books.

Greenstone, J. David. 1977. *Labor in American Politics.* Chicago: University of Chicago Press.

Hacker, Louis. 1970. *The Course of American Economic Growth and Development.* New York: Wiley.

Halberstam, David. 1972. *The Best and the Brightest.* New York: Random House.

Hamilton, Alexander. 1780. Quoted in Arthur J. Schlesinger, Jr., *The Age of Jackson.* Boston: Little Brown (1945), p. 10.

Hamilton, Richard. 1966. "The Marginal Middle Class: A Reconsideration." *American Sociological Review* 31:192–199.

———. 1972. *Class and Politics in the United States.* New York: Wiley.

———. 1975. *Restraining Myths: Critical Studies of United States' Social Structure and Politics.* Beverly Hills, Calif.: Sage.

Harrington, Michael. 1962. *The Other America: Poverty in the United States.* New York: Macmillan.

Harrison, Bennett, and Barry Bluestone. 1988. *The Great U-Turn: Corporate Restructuring and the Polarizing of America.* New York: Basic Books.

Hauser, Robert M. 1976. "On Boudon's Model of Social Mobility." *American Journal of Sociology* 81:911–929.

Herman, Edward S. 1981. *Corporate Control, Corporate Power.* Cambridge, England: Cambridge University Press.

Higham, John. 1963. *Strangers in the Land.* New York: Atheneum.

Hodge, Robert W.; Donald J. Treiman; and Peter H. Rossi. 1966. "A Comparative Study of Occupational Prestige." In *Class, Status and Power*, edited by Reinhard Bendix and Seymour Martin Lipset. 2nd ed. New York: Free Press.

Hodge, Robert W., and Donald Treiman. 1968. "Class Identification in the United States." *American Journal of Sociology* 73:535–547.

Hodges, Harold M. 1964. *Social Stratification: Class in America.* Cambridge, Mass.: Schenkman.

Hodgson, Godfrey. 1975. "Do Schools Make a Difference?" In *The Inequality Controversy: Schooling and Distributive Justice*, edited by D. M. Levine and M. J. Bane. New York: Basic Books.

Hoffman, Saul. 1977. "Marital Instability and the Economic Status of Women." *Demography* 14:67–76.

Hollingshead, August B. 1949. *Elmtown's Youth*. New York: Wiley.

———. 1950. "Cultural Factors in the Selection of Marriage Mates." *American Sociological Review* 15:619–627.

Hollingshead, August B., and Frederick Redlich. 1958. *Social Class and Mental Illness: A Community Study*. New York: Wiley.

Horan, Patrick M. 1978. "Is Status Attainment Research Atheoretical?" *American Sociological Review* 43: 534–541.

Hout, Michael. 1988. "More Universalism, Less Structural Mobility: The American Occupational Structure in the 1980s." *American Journal of Sociology* 93:1358–1400.

Howe, Louise. 1977. *Pink Collar Worker: Inside the World of Woman's Work*. New York: G. P. Putnam's Sons.

Hunter, Floyd. 1953. *Community Power Structure: A Study of Decision Makers*. Chapel Hill, N.C.: University of North Carolina Press.

Hurn, Christopher J. 1978. *The Limits and Possibilities of Schooling*. Boston: Allyn & Bacon.

Inkeles, Alex, and Peter H. Rossi. 1956. "National Comparisons of Occupational Prestige." *American Journal of Sociology* 61:329–339.

Jackman, Mary. 1979. "The Subjective Meaning of Social Class Identification in the United States." *Public Opinion Quarterly* 43:443–462.

Jackson, Elton F., and P. J. Burke. 1965. "Status and Symptoms of Stress: Additive and Interaction Effects." *American Sociological Review* 30:556–564.

Jackson, Elton F., and Richard F. Curtis. 1968. "Conceptualization and Measurement in the Study of Social Stratification." In *Methodology in Social Research*, edited by Herbert M. Blalock, Jr., and Ann B. Blalock. New York: McGraw-Hill.

Jaynes, Gerald David, and Robin M. Williams, Jr., eds. 1989. *A Common Destiny: Blacks and American Society*. Washington, D.C.: National Academy Press.

Jefferson, Thomas. 1821. "Autobiography." In *The Life and Selected Writings of Thomas Jefferson*, edited by Adrienne Koch and William Peden. New York: Modern Library, 1944.

Jencks, Christopher. 1991. "Is the American Underclass Growing?" In *The Urban Underclass*, edited by Christopher Jencks and Paul Peterson. Washington, D.C.: Brookings Institution.

Jencks, Christopher, and Paul Peterson. 1991. *The Urban Underclass*. Washington, D.C.: Brookings Institution.

Jencks, Christopher, et al. 1972. *Inequality: A Reassessment of the Effect of Family and Schooling in America*. New York: Basic Books.

———. 1979. *Who Gets Ahead?* New York: Basic Books.

Johnson, Barry. 1990. "Estate Tax Returns, 1986–1988." *Statistics of Income Bulletin* 9 (Spring):27–62.

Joint Center for Housing Studies of Harvard University. 1991. *The State of the Nation's Housing: 1991*. Cambridge, Mass.: Author.

Journal of Human Resources. Fall, 1980. Special issue on income maintenance programs.

Judis, John B. 1991. "Twilight of the Gods." *Wilson Quarterly* 5 (Autumn):43–57.

Kahl, Joseph A. 1953. "Educational and Occupational Aspirations of Common Man Boys." *Harvard Educational Review* 23:186–203.

———. 1957. *The American Class Structure*. 1st ed. New York: Rinehart.

Kahl, Joseph A., and James A. Davis. 1955. "A Comparison of Indexes of Socio-Economic Status." *American Sociological Review* 20:317–325.

Kalleberg, Arne L., and Larry J. Griffin.

1980. "Class, Occupation and Inequality in Job Rewards." *American Journal of Sociology* 85:731–768.

Kanter, Rosabeth. 1977. *Men and Women of the Corporation*. New York: Basic Books.

Karabel, Jerome, and A. H. Halsey, eds. 1977. *Power and Ideology in Education*. New York: Oxford University Press.

Kassalow, Everett. 1978. "How Some European Nations Avoid U.S. Levels of Industrial Conflict." *Monthly Labor Review* 101 (April):97.

Kaysen, Carl. 1957. "The Social Significance of the Modern Corporation." *American Economic Review* 47:311–319.

Kennickell, Arthur, and Janice Shack-Marquez. 1992. "Changes in Family Finances from 1983 to 1989: Evidence from the Survey of Consumer Finances." *Federal Bulletin* (January):1–18.

Kennichell, Arthur, and R. Louise Woodbum. 1992. "Estimation of Household Net Worth Using Model-Based and Design-Based Weights: Evidence from the 1989 Survey of Consumer Finances." Unpublished paper. Washington, D.C.: Board of Governors of the Federal Reserve System.

Kerr, Clark, and Abraham Seigal. 1954. "The Interindustry Propensity to Strike—An International Comparison." In *Industrial Conflict*, edited by Arthur Kornhauser et al. New York: McGraw-Hill.

Koenig, Thomas. 1980. "Corporate Support for Political Contribution Disclosure." Unpublished paper presented at the American Sociological Association, New York.

Koenig, Thomas; Robert Gogel; and John Sonquist. 1979. "Models of the Significance of Interlocking Directorates." *American Journal of Economics and Sociology* 38:173–186.

Kohn, Melvin L. 1969. *Class and Conformity: A Study in Values*. Homewood, Ill.: Dorsey Press.

———. 1977. *Class and Conformity*. 2nd ed. Chicago: University of Chicago Press.

Kohn, Melvin L., and Carmi Schooler. 1983. *Work and Personality: An Inquiry into the Impact of Social Stratification*. Norwood, N.J.: Ablex.

Kolinsky, Eva. 1984. *Parties, Opposition and Society in West Germany*. New York: St. Martins Press.

Komarovsky, Mirra. 1946. "The Voluntary Associations of Urban Dwellers." *American Sociological Review* 11:689–698.

———. 1962. *Blue Collar Marriage*. New York: Vintage.

Landecker, Werner S. 1981. *Class Crystallization*. New Brunswick, N.J.: Rutgers University Press.

Langerfeld, Steven. 1981. "To Break a Union." *Harpers* 262 (May):16–21.

Laumann, Edward O. 1966. *Prestige and Association in an Urban Community*. Indianapolis: Bobbs-Merrill.

———. 1973. *Bonds of Pluralism: The Form and Substance of Urban Social Networks*. New York: Wiley.

Lavin, David; Richard Alba; and Richard Silberstein. 1981. *Right vs. Privilege: The Open Admissions Experiment at the City University of New York*. New York: Free Press.

Leahy, Robert. 1981. "The Development of the Conception of Economic Inequality." *Child Development* 52:523–532.

———. 1983. *The Child's Construction of Social Inequality*. New York: Academic Press.

Leggett, John C. 1968. *Class, Race and Labor: Working-Class Consciousness in Detroit*. New York: Oxford University Press.

LeMasters, E. E. 1975. *Blue-Collar Aristocrats: Life-Styles at a Working-Class*

Tavern. Madison: University of Wisconsin Press.

Lenski, Gerhard. 1954. "Status Crystallization: A Non-Vertical Dimension of Social Status." *American Sociological Review* 19:405–413.

——. 1966. *Power and Privilege: A Theory of Social Stratification*. New York: McGraw-Hill.

Levine, Donald M., and Mary Jo Bane, eds. 1975. *The "Inequality" Controversy: Schooling and Distributive Justice*. New York: Basic Books.

Levitan, Sar A. 1980. *Programs in Aid of the Poor for the 1980s*. 4th ed. Baltimore: Johns Hopkins University Press.

——. 1990. *Programs in Aid of the Poor*. 6th ed. Baltimore: Johns Hopkins University Press.

Levitan, Sar, and Isaac Shapiro. 1987. *Working But Poor: America's Contradiction*. Baltimore: Johns Hopkins University Press.

Levitan, Sar A., and Robert Taggart. 1976. *The Promise of Greatness*. Cambridge, Mass.: Harvard University Press.

Lewis, Sinclair. 1922. *Babbitt*. New York: Harcourt Brace Jovanovich.

Link, Arthur S., and William Cotton. 1973. *American Epoch*. Vol. I. 4th ed. New York: Knopf.

Lipset, Seymour Martin. 1960. *Political Man*. New York: Doubleday.

——. 1981. *Political Man*. Expanded edition. Baltimore: Johns Hopkins University Press.

Litwack, Leon, ed. 1962. *The American Labor Movement*. Englewood Cliffs, N.J.: Prentice-Hall.

Lopata, Helena Z., et al. 1980. "Spouses' Contributions to Each Other's Roles." In *Dual-Career Couples*, edited by Fran Pepitone-Rockwell. Beverly Hills, Calif.: Sage.

Lord, Walter. 1955. *A Night to Remember*. New York: Henry Holt.

Lorwin, Lewis L. 1933. *The American Federation of Labor*. Washington, D.C.: Brookings Institution.

LTV Corporation. 1990. *A Guide to the 102nd Congress: 1st Session Datebook/Calendar*. Washington, D.C.: Author.

Lubell, Samuel. 1956. *The Future of American Politics*. 2nd ed. Garden City, N.Y.: Doubleday Anchor.

Lundberg, Ferdinand. 1968. *The Rich and the Super-Rich*. New York: Lyle Stuart.

——. 1975. *The Rockefeller Syndrome*. Secaucus, N.J.: Lyle Stuart.

Lynd, Robert S., and Helen Merrell Lynd. 1929. *Middletown*. New York: Harcourt Brace Jovanovich.

——. 1937. *Middletown in Transition*. New York: Harcourt Brace Jovanovich.

Maccoby, Michael. 1976. *The Gamesman*. New York: Simon & Schuster.

MacIntyre, Robert, et al. 1991. *A Far Cry from Fair: CTJ's Guide to State Tax Reform*. Washington, D.C.: Citizens for Tax Justice.

Madison, James (with Alexander Hamilton and John Jay). 1787. *The Federalist Papers*. New York: New American Library (1961).

Makinson, Larry. 1990. *Open Secrets: The Dollar Power of PACs in Congress*. Washington D.C.: Congressional Quarterly.

Malbin, Michael. 1979. "Campaign Financing and the Special Interests." *Public Interest*, no. 56 (Summer): 3–42.

Marx, Karl. 1979. *The Marx-Engels Reader*, edited by Robert C. Tucker. 2nd ed. New York: Norton.

McLeod, Jay. 1987. *Ain't No Makin' It: Leveled Aspirations in a Low-Income Neighborhood*. Boulder, Colo.: Westview.

Miller, Herman P. 1971. *Rich Man, Poor*

Man, rev. ed. New York: Thomas Y. Crowell.

Mills, C. Wright. 1951. *White Collar*. New York: Oxford University Press.

———. 1956. *The Power Elite*. New York: Oxford University Press.

———. 1968. "Comment on Criticism." in *C. Wright Mills and the Power Elite*, edited by G. William Domhoff and Hoyt B. Ballard. Boston: Beacon Press.

Mintz, Beth. 1975. "The President's Cabinet, 1897–1972." *Insurgent Sociologist* 5(3):131–149.

Mortenson, Thomas G. 1991. *Equity of Higher Educational Opportunity for Women, Black, Hispanic and Low Income Students*. Iowa City, Iowa: American College Testing Program.

Mortenson, Thomas G., and Zhijun Wu. 1990. *High School Graduation and College Participation of Young Adults by Family Income Backgrounds: 1970–1989*. Iowa City, Iowa: American College Testing Program.

Moynihan, Daniel P. 1973. *The Politics of a Guaranteed Income: The Nixon Administration and the Family Assistance Plan*. New York: Random House.

Müller, Susan. 1978. "Industrial Structure and Low-Level Earnings: A Reexamination of the Meritocratic Model." Unpublished Ph.D. thesis, Cornell University.

Murray, Charles A. 1984. *Losing Ground: American Social Policy*. New York: Basic Books.

Nagle, John. 1977. *System and Succession: The Social Bases of Political Elite Recruitment*. Austin: University of Texas Press.

Naimark, Hedwin. 1981. *The Development of the Understanding of Social Class*. Doctoral dissertation, New York University.

Nakao, Keiko, and Judith Treas. 1990. "Revised Presige Scores for All Occupations." Chicago: National

Opinion Research Center (unpublished paper).

National Commission on Children. 1991. *Beyond Rhetoric: A New American Agenda for Children and Families*. Washington, D.C.: U.S. Government Printing Office.

Neugarten, Bernice. 1949. "The Democracy of Childhood." In *Democracy in Jonesville*, edited by W. Lloyd Warner. New York: Harper & Row.

Newman, Katherine S. 1988. *Falling from Grace: The Experience of Downward Mobility in the American Middle Class*. New York: Free Press.

Nock, Steven L., and Peter H. Rossi. 1978. "Ascription versus Achievement in the Attribution of Family Social Status." *American Journal of Sociology* 84:565–590.

NORC (National Opinion Research Center). 1953. "Jobs and Occupations: A Popular Evaluation." In *Class, Status, and Power*, edited by Reinhard Bendix and Seymour M. Lipset. Glencoe, Ill.: Free Press.

Oakes, Jeannie. 1985. *Keeping Track: How High Schools Structure Inequality*. New Haven, Conn.: Yale University Press.

Ornati, Oscar. 1966. *Poverty Amidst Affluence*. New York: Twentieth Century Fund.

Orshansky, Mollie. 1965. "Counting the Poor: Another Look at the Poverty Profile." *Social Security Bulletin V* 28, no. 1.

Ossowski, Stanislaw. 1963. *Class Structure in the Social Consciousness*. New York: Free Press.

Osterman, Paul. 1991. "Gains from Growth? The Impact of Full Employment on Poverty in Boston." In *The Urban Underclass*, edited by Christopher Jencks and Paul Peterson. Washington, D.C.: Brookings Institution.

Palmer, John L., and Isabel V. Sawhill,

eds. 1984. *The Reagan Record*. Cambridge, Mass.: Ballinger.

Pechman, Joseph. 1985. *Who Paid the Taxes, 1966–85?* Washington, D.C.: Brookings Institution.

Pen, Jan. 1971. *Income Distribution*. London, England: Allen Lane.

Pennings, Johannes. 1980. *Interlocking Directorates*. San Francisco: Jossey-Bass.

Persell, Caroline Hodges. 1977. *Education and Inequality*. New York: Free Press.

Phillips, Kevin. 1990. *The Politics of Rich and Poor: Wealth and the American Electorate in the Reagan Aftermath*. New York: Random House.

Piore, Michael J. 1977. "The Dual Labor Market and Its Implications." In *Problems in Political Economy: An Urban Perspective*, edited by David M. Gordon. 2nd ed. Lexington, Mass.: D. C. Heath.

Piven, Frances Fox, and Richard A. Cloward. 1971. *Regulating the Poor: The Functions of Public Welfare*. New York: Pantheon Books.

Polsby, Nelson. 1970. "How to Study Community Power: The Pluralist Alternative." In *The Structure of Community Power*, edited by Michael Aiken and Paul E. Mott. New York: Random House.

Pomper, Gerald M. 1989. "The Presidential Power." In *The Election of 1988*, edited by Gerald M. Pomper. Chatham, N.J.: Chatham House Publishers, pp. 153–176.

Rainwater, Lee. 1965. *Family Design: Marital Sexuality, Family Size, and Contraception*. Chicago: Aldine.

————. 1974. *What Money Buys*. New York: Basic Books.

Ramsey, Patricia. 1991. "Young Children's Awareness and Understanding of Social Class Differences." *Journal of Genetic Psychology* 152: 71–82.

Redfield, Robert. 1947. "The Folk Soci-

ety." *American Journal of Sociology* 52:293–308.

Rein, Martin, and Lee Rainwater. 1977. "The Welfare Class and Welfare Reform." Family Policy Notes No. 4. Cambridge, Mass.: Joint Center for Urban Studies, MIT, Harvard.

Reiss, Albert. 1961. *Occupations and Social Status*. Glencoe, Ill.: Free Press.

Reissman, Leonard. 1954. "Class, Leisure and Participation." *American Sociological Review* 19:74–84.

Rieder, Jonathan. 1985. *Canarsie: The Jews and Italians of Brooklyn against Liberalism*. Cambridge, Mass.: Harvard University Press.

Riesman, David. 1953. *The Lonely Crowd: A Study of the Changing American Character*. New Haven, Conn.: Yale University Press.

Ritter, Kathleen, and Lowell Hargens. 1975. "Occupational Positions and Class Identifications of Married Women." *American Journal of Sociology* 80:934–948.

Ross, Howard, 1968. "Economic Growth and Change in the United States under Laissez-Faire: 1870–1929." In *The Age of Industrialization in America*, edited by Frederic Cople Jaher. New York: Free Press.

Rossi, Peter H. 1989. *Down and Out in America: The Origins of Homelessness*. Chicago: University of Chicago Press.

Rubin, Lillian Breslow. 1976. *Worlds of Pain: Life in the Working-Class Family*. New York: Basic Books.

Rubin, Z. 1968. "Do American Women Marry Up?" *American Sociological Review* 5:750–760.

Sawhill, Isabel. 1989. "The Underclass: An Overview." *The Public Interest* 96 (Summer):3–15.

Schiller, Bradley. 1989. *The Economics of Poverty and Discrimination*. 5th ed. Englewood Cliffs, N.J.: Prentice-Hall.

Schreiber, E. M., and G. T. Nygreen.

1970. "Subjective Social Class in America: 1945–1968." *Social Forces* 45:348–356.

Schwartz, John E. 1983. *America's Hidden Success: A Reassessment of Twenty Years of Public Policy*. New York: Norton.

Schwartz, Marvin. 1984–1985. "Preliminary Estimates of Personal Wealth, 1982: Composition of Assets." *Statistics of Income. SOI Bulletin* (Winter):1–7.

—————. 1988. "Estimates of Personal Wealth, 1982: A Second Look." *Statistics of Income Bulletin* 7 (Spring):31–37.

Schwartz, Marvin, and Barry Johnson. 1990. "Estimates of Personal Wealth, 1986." *Statistics of Income Bulletin* 9 (Spring):63–78.

Schwartz, Michael, ed. 1987. *The Structure of Power in America: The Corporate Elite as a Ruling Class*. New York: Holmes & Meier.

Sewell, William H., and Robert M. Hauser. 1975. *Education, Occupation and Earnings*. New York: Academic Press.

Sewell, William H.; Robert M. Hauser; and David L. Featherman, eds. 1976. *Schooling and Achievement in American Society*. New York: Academic Press.

Sewell, William H., and Vimal P. Shah. 1977. "Socioeconomic Status, Intelligence, and the Attainment of Higher Education. In *Power and Ideology in Education*, edited by Jerome Karabel and A. H. Halsey. New York: Oxford University Press.

Simmons, Robert G., and Morris Rosenberg. 1971. "Functions of Children's Perceptions of the Stratification System." *American Sociological Review* 36:235–249.

Smeeding, Timothy, Barbara D. Torrey, and Martin Rein. 1987. "Patterns of Income and Poverty: The Economic Status of the Young and the Old in Six Countries." Paper presented at the Urban Institute, February.

Smith, David H. 1980. *Participation in Social and Political Activities: A Comprehensive Analysis of Political Involvement, Expressive Leisure Time, and Helping Behavior*. San Francisco: Jossey-Bass.

Smith, James D. 1984. "Trends in the Concentration of Personal Wealth in the United States, 1958–1976." *Review of Income and Wealth* 30:419–428.

Spenner, K. I., and D. L. Featherman. 1978. "Achievement Ambitions." *Annual Review of Sociology* 4:373–420.

Stack, Carol B. 1974. *All Our Kin*. New York: Harper & Row.

Statesman's Year-Book. 1979–1990. New York: St. Martin's Press.

Statistics of Income Bulletin. 1988. "Statistics of Income, Domestic Special Studies." Volume 8 (Fall):45–55.

Stendler, Celia Burns. 1949. *Children of Brasstown: Their Awareness of the Symbols of Social Class*. Urbana: University of Illinois Press.

Stockman, David. 1984. "Statement to House of Representatives Sub-Committee on Oversight and on Public Assistance and Unemployment." September 20.

Strole, Leo, et al. 1978. *Mental Health in the Metropolis: The Midtown Manhattan Study*. Revised and enlarged ed. New York: New York University Press.

Stromberg, Ann, and Shirley Harkess. 1988. *Women Working: Theories and Facts in Perspective*. 2nd ed. Mountain View, Calif.: Mayfield.

Strudler, Michael, and Emily Ring. 1990. "Individual Income Tax Returns, Preliminary Data, 1988." *Statistics of Income Bulletin* 9 (Spring): 5–26.

Sweezy, Paul. 1968. "Power Elite or Ruling Class?" In *C. Wright Mills and the*

Power Elite, edited by G. William Domhoff and Hoyt B. Ballard. Boston: Beacon Press.

Szymanski, Albert. 1978. *The Capitalist State and the Politics of Class*. Cambridge, Mass.: Winthrop.

Terkel, Studs. 1974. *Working: People Talk About What They Do All Day and How They Feel About It*. New York: Avon.

Thompson, E. P. 1963. *The Making of the English Working Class*. New York: Vintage.

Thurow, Lester, and Robert Lucas. 1972. *The American Distribution of Income: A Structural Problem*. Washington, D.C.: U.S. Government Printing Office.

Treiman, Donald. 1977. *Occupational Prestige in Comparative Perspective*. New York: Academic Press.

Treiman, Donald, and Patricia Roos. 1983. "Sex and Earnings in Industrial Society: A Nine-Nation Comparison." *American Journal of Sociology* 89:612–650.

Tuchman, Gaye, ed. 1974. *The TV Establishment: Programming for Power and Profit*. Englewood Cliffs, N.J.: Prentice-Hall.

Tutor, Jeannette. 1991. "The Development of Class Awareness in Children." *Social Forces* 49:470–476.

U.S. Bureau of the Census. 1973. *Statistical Abstract of the United States: 1973*.

————. 1975. *Historical Statistics of the United States*. Bicentennial Edition. 2 Parts.

————. 1976. *Bicentennial Statistics, Pocket Data Book, U.S.A.*

————. 1980. *The Social and Economic Status of the Black Population in the United States: An Historical View, 1790–1978*. Current Population Reports, Special Studies Series P-23, no. 80.

————. 1982. *Characteristics of the Population Below the Poverty Level: 1980*. Series P-60, no. 133.

————. 1983. *Statistical Abstract of the United States: 1984*.

————. 1984a. *Statistical Abstract of the United States: 1985*.

————. 1984b. *Money Income and Poverty Status of Families and Persons in the United States: 1983—Advance Data*. Current Population Reports. Series P-60, no. 145.

————. 1984c. *1980 Census of Population, United States Summary, Section A: United States*. Characteristics of the Population. Series PC80–1–D1-A.

————. 1986a. *Statistical Abstract of the United States, 1987*.

————. 1986b. *Characteristics of the Population Below the Poverty Level: 1984*. Series P-60, no. 152.

————. 1988. *Educational Attainment in the United States: March 1987 and 1986*. Current Population Reports. Series P-20, no. 428.

————. 1989a. *Money Income of Households, Families, and Persons in the United States: 1987*. Current Population Reports. Series P-60, no. 162.

————. 1989b. *Transition in Income and Poverty Status: 1984–85*. Series P-70, no. 15-RD-1.

————. 1989c. *Voting and Registration in the Election of November 1988*. Series P-20, no. 440.

————. 1990a. *Money Income and Poverty Status in the United States, 1989*. Current Population Reports. Series P-60, no. 168.

————. 1990b. *Trends in Income, by Selected Characteristics: 1947 to 1988*. Current Population Reports. Series P-60, no. 167.

————. 1990c. *Statistical Abstract of the United States: 1990*.

————. 1990d. *Measuring the Effect of Benefits and Taxes on Income and Poverty: 1989*. Current Population Reports. Series P-60, no. 169-RD.

————. 1990e. *Transition in Income and*

Poverty Status: 1985–86. Series P-70, no. 18.

————. 1990f. *Household Wealth and Asset Ownership: 1988.* Current Population Reports. Series P-70, no. 22.

————. 1991a. *Money Income of Households, Families, and Persons in the United States: 1990.* Series P-60, no. 174.

————. 1991b. *Poverty in the United States: 1990.* Series P-60, no. 175.

————. 1991c. *Measuring the Effect of Benefits and Taxes on Income and Poverty: 1990.* Series P-60, no. 176-RD.

————. 1991d. *Transition in Income and Poverty Status: 1987–88.* Series P-70, no. 24.

————. 1991e. *Voting and Registration in the Election of November 1990.* Series P-20, no. 453.

U.S. Department of Justice. 1990. *Sourcebook of Criminal Justice Statistics: 1989.*

U.S. Department of Labor. 1978. *Employment and Training Report to the President.*

————. 1980. *Handbook of Labor Statistics.*

————. 1981a. *Employment and Earnings.* 28 (January).

————. 1981b. *Educational Attainment of Workers, March 1979.* Special Labor Force Report 240.

————. 1984. *Employment and Earnings.* January.

————. 1990a. *Outlook 2000.*

————. 1990b. *Employment and Earnings.* January.

————. 1991. *Employment and Earnings.* January.

U.S. General Accounting Office. 1990. *Voting: Some Procedural Changes and Informational Activities Could Increase Turnout.*

U.S. House of Representatives, Committee on Ways and Means. 1990. *Tax Progressivity and Income Distribution.* March 26.

————. 1991a. *Overview of the Federal Tax System.* WMCP-102–7.

————. 1991b. *Background Material and Data on Programs within the Jurisdiction of the Committee of Ways and Means.*

U.S. Internal Revenue Service. 1988. *Statistics of Income.* Fall.

————. 1990. *Statistics of Income.* Spring.

U.S. Public Health Service. 1990. *Prevention Profile, Health, United States, 1989.*

USA Today. 1990. "Financial Disclosure of the U.S. Congress." May 31.

Vanfossen, Beth Ensminger. 1977. "Sexual Stratification and Sex-Role Socialization." *Journal of Marriage and the Family* 39:563–574.

Veblen, Thornstein. 1934. *The Theory of the Leisure Class.* New York: Modern Library. (First published 1899).

Velez, William. 1985. "Finishing College: The Effects of College Type." *Sociology of Education* 58:191–200.

Wachtel, Howard M., and Charles Betsey. 1972. "Employment at Low Wages." *Review of Economics and Statistics* 44:121–129.

Wallace, Phyllis A., ed. 1982. *Women in the Workplace.* Boston: Auburn House.

Walton, John. 1970. "A Systematic Survey of Community Power Research." In *The Structure of Community Power,* edited by Michael Aiken and Paul E. Mott. New York: Random House.

Warner, W. Lloyd, and Paul S. Lunt. 1941. *The Social Life of a Modern Community.* New Haven, Conn.: Yale University Press.

Warner, W. Lloyd, et al. 1949a. *Democracy in Jonesville.* New York: Harper & Row.

————. 1949b. *Social Class in America.* Chicago: Science Research Associates.

————. 1973. *Yankee City*. Abridged Edition. New Haven, Conn.: Yale University Press.

Wattenberg, Ben J. 1974. *The Real America*. Garden City, N.Y.: Doubleday.

Weber, Max. 1946. *From Max Weber: Essays in Sociology*, edited by H. H. Gerth and C. Wright Mills. New York: Oxford University Press.

Who's Who in America. 1980. 41st ed. Chicago: Marquis.

Who's Who in American Politics. 1979. 7th ed. New York: Bowker.

Whyte, Martin King. 1990. *Dating, Mating and Marriage*. New York: Aldine de Gruyter.

Whyte, William H. 1952. *Is Anybody Listening?* New York: Simon & Schuster.

Wilson, William J. 1980. *The Declining Significance of Race*. 2nd ed. Chicago: University of Chicago Press.

————. 1987. *The Truly Disadvantaged: The Inner City, the Underclass, and Public Policy*. Chicago: University of Chicago Press.

————. 1991. "Public Policy Research and *The Truly Disadvantaged*." In *The Urban Underclass*, edited by Christopher Jencks and Paul Peterson. Washington, D.C.: Brookings Institution.

Wilson, William Julius, and Katheryn Nickerman. 1986. "Poverty and Family Structure: The Widening Gap Between Evidence and Public Policy Issues." In *Fighting Poverty: What Works and What Doesn't*, edited by Sheldon Danziger and Daniel H. Weinberg. Cambridge, Mass.: Harvard University Press.

Wolfinger, Raymond E. 1973. *The Politics of Progress*. Englewood Cliffs, N.J.: Prentice-Hall.

Wool, Harold. 1976. *The Labor Supply for Lower-Level Occupations*. New York: Praeger.

Wright, James. 1989. *Address Unknown: The Homeless in America*. New York: Aldine de Gruyter.

Zeitlin, Maurice. 1974. "Corporate Ownership and Control: The Large Corporation and the Capitalist Class." *American Journal of Sociology* 79:1073–1119.

————. 1980. *Classes, Class Conflict and the State*. Cambridge, Mass.: Winthrop.

Zweigenhaft, Richard. 1975. "Who Represents America?" *Insurgent Sociologist* 5 (3):119–130.

Name Index

Subject Index

Credits

This constitutes an extension of the copyright page. The numbers in parentheses after each entry are the page numbers on which the reprinted material appears.

Blau, Peter M., and Otis Dudley Duncan. 1967. *The American Occupational Structure.* New York: The Free Press, a Division of Macmillan, Inc. Copyright © 1967 by Peter M. Blau and Otis Dudley Duncan. Reprinted by permission of The Free Press. (165–170)

Bott, Elizabeth. 1954. "The Concept of Class as a Reference Group." *Human Relations* 7:259–286. Reprinted by permission of Plenum Publishing Corporation. (241–242)

Centers, Richard. 1949. *The Psychology of Social Classes: A Study of Class Consciousness.* Princeton, N.J.: Copyright © 1949, © renewed by Princeton University Press. Reprinted by permission of Princeton University Press. (234, 235, 239)

Coleman, Richard P., and Lee Rainwater. 1978. *Social Standing in American Society.* Copyright © 1978 by Basic Books, Inc. Reprinted by permission of Basic Books, a division of HarperCollins Publishers Inc. (33–38)

Dahl, Robert A. 1961. *Who Governs? Democracy and Power in an American City.* New Haven, Conn.: Yale University Press. Reprinted by permission of Yale University Press. (193–195)

Davis, Allison; Burleigh B. Gardner; and Mary R. Gardner. 1941. *Deep South: A Social-Anthropological Study of Caste and Class.* Chicago: University of Chicago Press. Copyright 1941 by The University of Chicago. Reprinted by permission of The University of Chicago Press. (29–30)

Featherman, David L., and Robert M. Hauser. 1978. *Opportunity and Change.* New York: Academic Press. Reprinted by permission of Academic Press and David L. Featherman. (154, 155, 158)

Halberstam, David. 1972. *The Best and the Brightest.* New York: Random House. Reprinted by permission of Random House. (213–214)

Harrington, Michael. 1962. *The Other America: Poverty in the United States.* New York: Macmillan. Reprinted by permission of Macmillan Publishing Co., Inc. (271)

Hunter, Floyd. 1953. *Community Power Structure: A Study of Decision Makers.* Copyright 1953 The University of North Carolina Press. Reprinted by permission of The University of North Carolina Press. (191)

Jackman, Mary. 1979. "The Subjective Meaning of Social Class Identification in the United States." *Public Opinion Quarterly* 43:443–462. Reprinted by permission of Elsevier Science Publishing Co., Inc. (44)

Laumann, Edward O. 1966. *Prestige and Association in an Urban Community.* Indianapolis: Bobbs-Merrill. Reprinted by permission of Bobbs-Merrill Co., Inc. (119, 126, 127)

LeMasters, E. E. 1975. *Blue-Collar Aristocrats: Life Styles at a Working-Class Tavern.* Madison: University of Wisconsin Press. Reprinted by permission of University of Wisconsin Press. (123)

Lewis, Sinclair. 1922. *Babbitt.* New York: Harcourt Brace Jovanovich. Reprinted by permission of Harcourt Brace Jovanovich. (21)

Lynd, Robert S., and Helen Merrell Lynd. 1929. *Middletown.* New York: Harcourt Brace Jovanovich. Copyright 1929 Harcourt Brace Jovanovich, Inc.; copyright 1957 Robert S. and Helen M. Lynd. Reprinted by permission of Harcourt Brace Jovanovich. (53–55)

———. 1937. *Middletown in Transition.* New York: Harcourt Brace Jovanovich. Copyright 1937 Harcourt Brace Jovanovich, Inc.; copyright 1965 Robert S. and Helen M. Lynd. Reprinted by permission of Harcourt Brace Jovanovich. (57–60)

Marx, Karl. 1979. *The Marx-Engels Reader,* edited by Robert C. Tucker. 2nd ed. New York: W. W. Norton. Reprinted by permission of W. W. Norton & Company, Inc. (6)

Mills, C. Wright. 1956. *The Power Elite.* New York: Oxford University Press. Reprinted by permission of Oxford University Press. (61, 199–201)

Mortenson, Thomas G. 1991. *Equity of Higher Educational Opportunity for Women, Black, Hispanic and Low Income Students.* Iowa City, Iowa: American College Testing Service. Reprinted by permission of American College Testing Service. (180, 184)

Piven, Frances Fox, and Richard A. Cloward. 1971. *Regulating the Poor: The Functions of Public Welfare.* New York: Pantheon Books, a division of Random House, Inc. (293)

Rainwater, Lee. 1965. *Family Design: Marital Sexuality, Family Size, and Contraception.* Chicago: Aldine. Reprinted by permission of Lee Rainwater. (120)

Ritter, Kathleen, and Lowell Hargens. 1975. "Occupational Positions and Class Identifications of Married Women." *American Journal of Sociology* 80:934–948. Reprinted by permission of The University of Chicago Press. (238)

Rubin, Lillian Breslow. 1976. *Worlds of Pain: Life in the Working-Class Family.* New York: Basic Books. Copyright © 1976 by Lillian Breslow Rubin. Reprinted by permission of Basic Books, a division of HarperCollins Publishers Inc. (122–124, 136).

Warner, W. Lloyd, and Paul S. Lunt. 1941. *The Social Life of a Modern Community.* New Haven, Conn.: Yale University Press. Reprinted by permission of Yale University Press. (24–26)

Warner, W. Lloyd, et al. 1949. *Social Class in America.* New York: Harper & Row. Reprinted by permission of Harper & Row Publishers, Inc. (27, 129, 130)